Deutsche Forschungsgemeinschaft

Transient Phenomena in Multiphase and Multicomponent Systems

WILEY-VCH

Deutsche
Forschungsgemeinschaft

Transient Phenomena in Multiphase and Multicomponent Systems

Edited by
Franz Mayinger
and Bernd Giernoth

Research Report

Deutsche Forschungsgemeinschaft
Kennedyallee 40, D-53175 Bonn, Federal Republic of Germany
Postal address: D-53175 Bonn
Phone: ++49/228/885-1
Telefax: ++49/228/885-2777
E-Mail: (X.400): S=postmaster; P=dfg; A=d400; C=de
E-Mail: (Internet RFC 822): postmaster@dfg.de
Internet: http://www.dfg.de

Library of Congress Card No.: applied for

A catalogue record for this book is available from the British Library.

Die Deutsche Bibliothek – CIP Cataloguing-in-Publication-Data
A catalogue record for this publication is available from Die Deutsche Bibliothek.

ISBN 3-527-27149-X

Printed on acid-free and chlorine-free paper.

Cover Design and Typography: Dieter Hüsken
Composition: K+V Fotosatz GmbH, D-64743 Beerfelden
Printing: betz-druck gmbh, D-64291 Darmstadt
Bookbinding: Wilhelm Osswald & Co., D-67433 Neustadt

Printed in the Federal Republic of Germany

Contents

Numerical Methods

Preface

In spite of the fact that thermo- and fluid-dynamics of multiphase systems have a long tradition in the international research, there is few known about transients in multiphase flow. This is especially true for multicomponent systems due to the lack of experimental and also theoretical information. Thermodynamic equilibrium is frequently assumed, and differences in temperature and concentration between the phases are neglected in the literature. On the other side, newly developed inertialess measuring techniques – i.e. on an optical or capacitive basis – became available, which allow to gain detailed insights even in highly transient flows. These experimental possibilities and the findings gained with these techniques, together with a tremendous development in computer-hardware and -software, made it possible to develop theoretical models for a better and more reliable description of transient multiphase flows.

This situation gave rise to the Deutsche Forschungsgemeinschaft to initiate a „Schwerpunktprogramm" to study transient phenomena in multiphase systems with one or several components. Input for this research program came also from the increasing industrial demand in more knowledge about thermo- and fluid-dynamic effects of non-equilibria during strong transients. This demand resulted from increasing requirements for risk- and safety-analysis of industrial plants in chemical- and energy-engineering. The results of the scientific efforts performed within this concentrated research initiatives – lasting from 1992 to 1998 – are presented in this book.

Consideration was given to a well balanced treatment of experimental and theoretical approaches to study multiphase phenomena under transient conditions. Single component two-phase mixtures were investigated as well as multicomponent ones. The results are presented here.

Besides thermo- and fluid-dynamic transients, also instabilities and oscillatory behaviour were investigated. The development of non-invasive and inertialess measuring techniques was also reinforced. Special emphasis was given to the mathematical modelling of thermo- and fluid-dynamic phenomena during transients.

The book deals with macroscopic as well as microscopic transient situations. Flow pattern and flashing were studied developing from stagnant liquids and in pipe flow. Cavitation, drop impact on hot walls and flashing in slits are examples of micro-studies presented in the book.

A large part of the book deals with numerical methods for describing transients in two-phase mixtures. These modelling activities deal with instabilities, flow in injector nozzles, cavitation in accelerated liquids, and reach up to transient phenomena in double front detonations. Last but not least, new developments in measuring techniques are presented, such as particle image velocimetry with two synchronised video cameras and capacitive tomography.

Engineers and physicists being confronted with two-phase flow problems in designing or operating technical plants find new and valuable information for their work. Researchers get the state of the art in the field of transient two-phase flow and post-graduate students will benefit when studying this book.

The financial support from the Deutsche Forschungsgemeinschaft for performing the research is highly appreciated.

Munich, August 1999 F. Mayinger

Depressurisation and Cavitation

1 Transient Pressure Release of Supercritical Gases

Rudolf Eggers, Bernhard Gebbeken and Andreas Fredenhagen *

1.1 Background and Introduction

Separation processes with supercritical gases as solvents have been introduced mainly in food industries but also increasingly in chemical and pharmaceutical industries during the last 25 years [1]. Preferably supercritical carbon dioxide is used due to its selectivity to non-polar substances, non-inflammability, economic and environmental aspects. Further, gas mixtures have been investigated and applied in order to change the solvent power. Finally, in these processes, the supercritical CO_2 is often loaded with water in phase equilibria because many of the materials to be separated are of some natural moisture content. Supercritical fluid (SCF) processes are carried out under high pressure conditions, normally within a range of 5 MPa and 50 MPa. At the end of those processes and in case of emergency, pressure vessels have to be released as quick as possible but without any unsafe operating conditions like excessive vessel temperatures or partially blocking and plugging the safety valves and venting lines. Hence, blowdown predictions of pressurised and supercritical gases had to be provided both for controlled operation of the high pressure vessels and for emergency considerations.

In consequence, a research programme was performed at the Technical University of Hamburg-Harburg in order to get detailed knowledge of the transient behaviour of the fluid and the vessel during blowdown processes of initially supercritical gases. Basic research on the phenomena of vessel release processes of initially saturated conditions has been done by Mayinger [2]. So far, few experimental data were available for pressure release of typical supercritical gases. Preliminary studies are carried out by Eggers and Green [3]. Haque et al. [4] accomplished some experiments to measure the temperature distribution within large pressure vessels during release of mixtures of carbon dioxide and nitrogen. Further investigation of vessel release of reacting and

* Technische Universität Hamburg-Harburg, Arbeitsbereich Verfahrenstechnik II, Wärme- und Stofftransport, Eißendorfer Str. 38, D-21073 Hamburg, Germany

non-reacting two phase systems have been done by Friedel et al. [5] using low viscous refrigerants, and by Thies [6] who used a higher viscous two phase system consisting of water/carbon dioxide and small amounts of polymers (Luviskol).

1.2 Pressure Release Conditions and Supercritical State

According to Figure 1.1, the vapour pressure line of CO_2 ends at the critical point at p=7.38 MPa and T=31.05 °C. The region above these values represents the supercritical state. Starting a release process out of this area depending on initial pressure and temperature, the vapour pressure line may be reached through the liquid region leading to release vaporisation or through the gaseous region leading to release condensation. This dependence is illustrated by the T,s-diagram in Figure 1.2 with three initial conditions to start a release process:

- State A: The vessel contains liquid and vapour in phase equilibrium. After a top vented pressure release is started, in the first period only vapour is blown off. In consequence of the decreasing pressure inside the vessel, boiling of the liquid phase starts to develop. The liquid-vapour interface

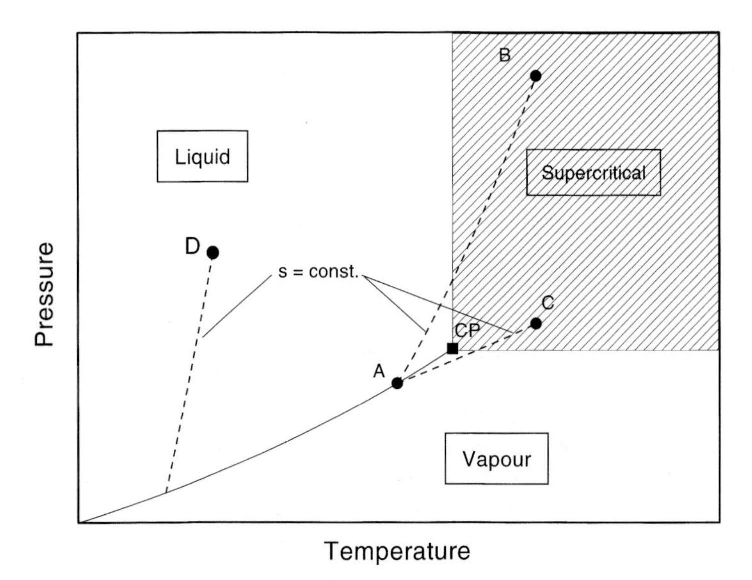

Figure 1.1: Initial states in p,T-diagram.

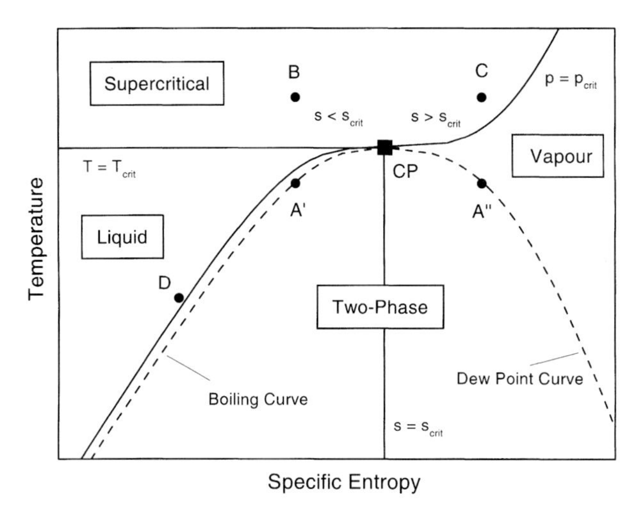

Figure 1.2: Initial states in *T,s*-diagram.

rises due to the growth of vapour bubbles. Depending on the initial liquid level and the velocity of the release process, two phase flow venting may occur.

- State B: The vessel contains dense gas in supercritical state with an entropy data of $s_B < s_{crit}$. If a rapid release process of a one-phase fluid is assumed to be nearly isentropic, the boiling curve will be crossed and subsequently a vapour release process of the liquid phase with two phase venting is to be expected.
- State C: The vessel contains dense gas in supercritical state of $s_C > s_{crit}$. Hence, the dew point curve will be crossed and a release condensation of the gaseous phase will be initialised.

During the pressure release of supercritical gases, two periods can be observed: the "supercritical/subcooled blowdown" and the "saturated blowdown". The supercritical/subcooled blowdown period terminates when the system pressure drops below the saturation pressure and flashing occurs. The void fraction depends on time and location within the vessel. The saturated blowdown phase lasts until ambient pressure is reached. Using carbon dioxide as release fluid, dry ice will be formed if the pressure drops below the triple point pressure of 0.518 MPa within the two phase region. The research project to be presented here deals with transient release processes of pure carbon dioxide [7] and mixtures of carbon dioxide and nitrogen. Additionally, those conditions have been investigated, when the carbon dioxide is saturated with water. In any case, the phase equilibria diagrams have to be known.

1.3 Experimental Facility

The main component of the pressure release installation is a vertical cylindrical high pressure vessel with the following technical data:

- inner volume: $V = 50$ l,
- inner diameter: $d = 0.242$ m,
- height to diameter ratio: $H/d = 4.5$,
- wall thickness: $b = 0.036$ m,
- maximum temperature: $T = 120\,^{\circ}\mathrm{C}$.

Figure 1.3 shows a flow sheet of the pressure release test plant. The temperature in the vessel is maintained by a heating device. Two high pressure diaphragm pumps are installed to perform the filling process of the vessel. The

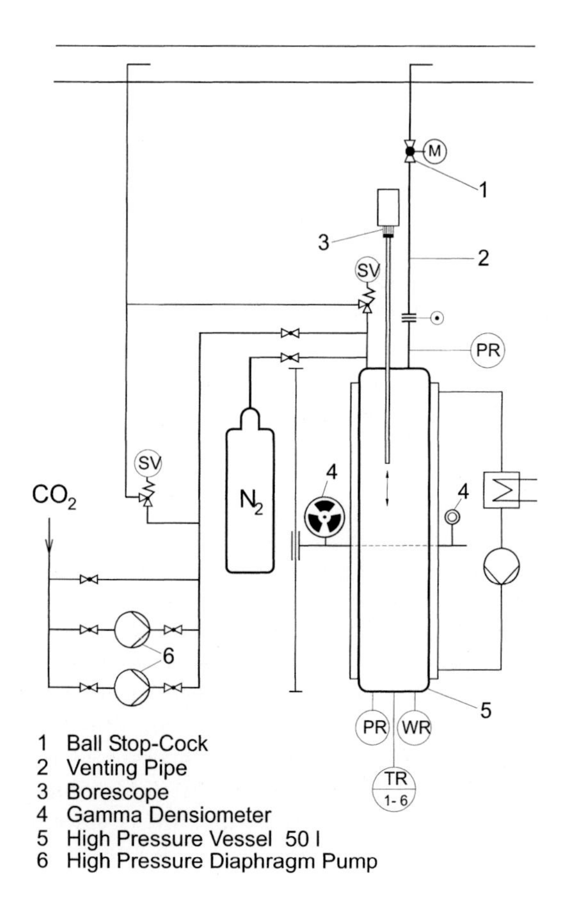

Figure 1.3: Experimental apparatus.

venting pipe of the variable cross sectional area ($A_{max}=50\ mm^2$) is connected to the top. To assure a controlled initiation of the release process, a quick-opening ball valve is used. For visual observation of the in-vessel phenomena, a pressure capsulated moveable borescope has been developed. It is vertically conducted through the top quick closure of the vessel [7]. A CCD camera is connected to the borescope outside the vessel. The obtained video pictures and the measured signals are synchronised. Thus, the height of the flashing liquid layer can be detected directly.

During the experiments, the transient data of pressure, mass flow rate and the void distribution along the axis of the vessel have been measured. The temperature profiles of the vessel walls have been measured in a previous program [3]. Following a brief description of the measurements are given, for details see [8, 9].

Pressure (PR): A piezoresistive pressure transducer is attached to the bottom of the vessel. Axial pressure gradients have been proved to be very low due to the large ratio of the volume of the vessel to the cross sectional area of the venting pipe. The maximum experimental error due to thermal drift and non-linearity of the sensor was $\Delta p=0.08$ MPa.

Temperature (TR 1–6): Six thermocouples (NiCr-Ni) of 0.5 mm thickness have been installed at axial positions from 0.9 m down to 0.05 m above the bottom of the vessel. The maximum experimental error was 0.3 K.

Mass flow (WR): The released mass flow rate, either single or two phase flow, changes the weight of the entire facility that has been measured by means of load cells. To avoid measurement errors that are caused by momentum transfer, a glide bearing section has been mounted into the venting pipe. The mass flow rate in the venting pipe is to be obtained by calculating the derivative of the mass signal with respect to time. The maximum experimental error of the mass signal itself was 0.038 kg.

Void fraction: The measurements of the average fluid density in a cross sectional area enables the calculation of the void fraction as long as thermal equilibrium between the flashing liquid (l) and gaseous phase (g) exists:

$$\bar{\varepsilon}(h) = \frac{\bar{\rho}(h) - \rho_l(p)}{\rho_g(p) - \rho_l(p)} . \qquad (1)$$

The measurements of the average density has been done by a gamma densitometer. The gamma source (Cs-137) and the scintillation counter have been installed on a support system moveable along the vertical axis. The axial density distribution has been obtained by repeating the experiments and measuring the average fluid density at various positions. The exactness of the saturation densities decreases as the thermodynamic state approximates the near critical region. At a temperature of $T=293$ K ($p\approx 6$ MPa), the maximum error of the void fraction is 0.06.

1.4 Results

According to Figure 1.1, transients of pressure, temperature, void fraction and mass flow are reported from experiments with pure carbon dioxide starting the pressure release at four different thermodynamic states. The relevant data are listed in Table 1.1.

All following results have been attained from release experiments that were run with the maximum cross sectional area of 50 mm^2.

Pressure: The pressure transients for experiment A–D are shown in Figure 1.4. At time $t=0$ s, the release process is started. When saturated fluid is

Table 1.1: Initial conditions for the experiments shown.

Experiments	p_0 MPa	T_0 K	ρ_0 kg/m^3	Thermodynamic conditions
A	6.3	297	$\rho_l=724$ $\rho_g=232$	Saturation, liquid
B1	25	313	880	Supercritical $s-s_{crit}<0$
B2	15	313	780	Supercritical $s-s_{crit}<0$
B3	10	313	623	Supercritical $s-s_{crit}<0$
C	7.5	313	234	Supercritical $s-s_{crit}>0$
D	10	278	949	Subcooled

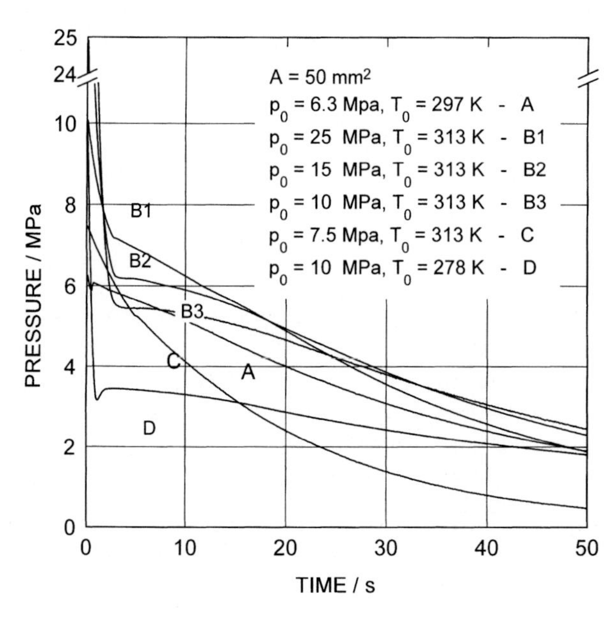

Figure 1.4: Pressure transients for release of carbon dioxide from various initial states.

used (A), at first gaseous phase flows out and the vessel pressure drops. Simultaneously, release condensation of the liquid phase appears. The rate of vaporisation leads to a slightly increased pressure. The bubbling rises the liquid level combined with a strong ebullition. Thus, a two phase outflow follows. After 15 s, a liquid level is formed again and the fluid flow in the venting pipe is of single phase.

In Experiment B1–B3, the saturation conditions are reached at system pressures of $p/p_{crit}=0.75$, 0.85, 0.97. This first period of supercritical pressure loss is very rapid. The experimental saturation data belong to constant entropy and in consequence the assumption of nearly isentropic change of state during the first release period is justified. As soon as the saturation conditions are reached, the flashing is initiated.

This provokes a sudden decrease of the pressure change rate. Considering the blowdown experiment from initially subcooled conditions (experiment D), the onset of flashing happens at much lower vessel pressure. Saturation conditions are reached after the vessel pressure has dropped below $p/p_{crit}=0.49$. Shortly before the minimum pressure is reached, the bubble nucleation is initiated. In contrast to what is measured in experiment B1–B3, a recovering of the vessel pressure (~ 0.3 MPa) can be observed after the bubble nucleation has started. The pressure remains almost constant for a few seconds. Subsequently, a slow decrease can be observed.

Temperature: The measured temperature transients for experiment A, B2 and D are shown in the Figures 1.5–1.7. Furthermore, the saturation temperatures calculated by the measured vessel pressure are shown. During the supercritical/subcooled blowdown period, a fast decrease of the fluid temperature can be observed. For experiment D, a considerable thermodynamic non-equilibrium appears. A liquid superheat of up to 4 K is measured. After approxi-

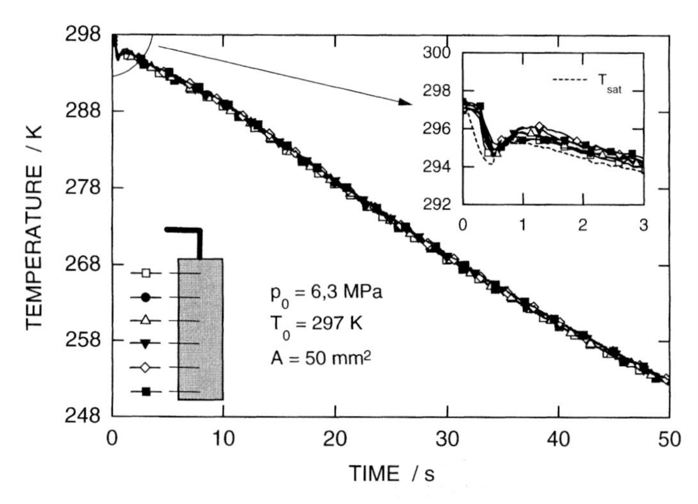

Figure 1.5: Temperature transients for release of carbon dioxide from experiment A.

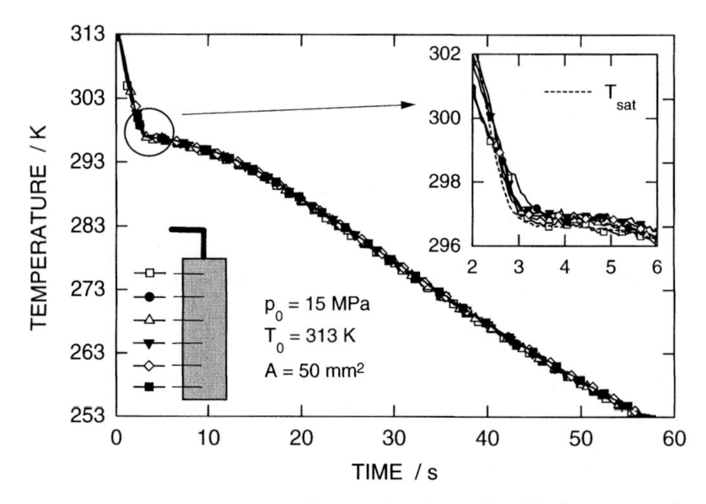

Figure 1.6: Temperature transients for release of carbon dioxide from experiment B2.

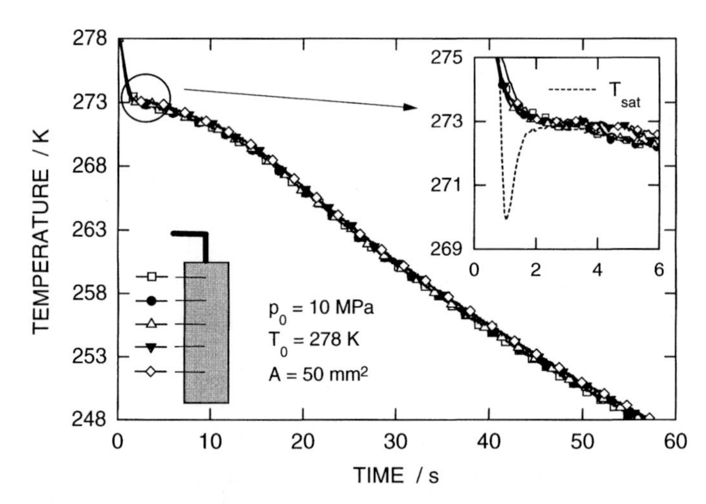

Figure 1.7: Temperature transients for release of carbon dioxide from experiment D.

mately 3 s, saturation conditions are almost reached. Subsequently, no axial temperature gradients are measured within the fluid. During the experiments B1–B3, the liquid superheat is within the range of the measuring error. This is due to the fact that a sufficient number of small bubbles is generated by nucleation and the amount of latent heat is low such that it can be transported by conduction and convection without considerable delay.

Void distribution: The void distribution is measured along the axis of the pressure vessel. Different profiles are obtained by varying the initial fluid conditions as well as the rate of depressurisation [8]. The void distributions ob-

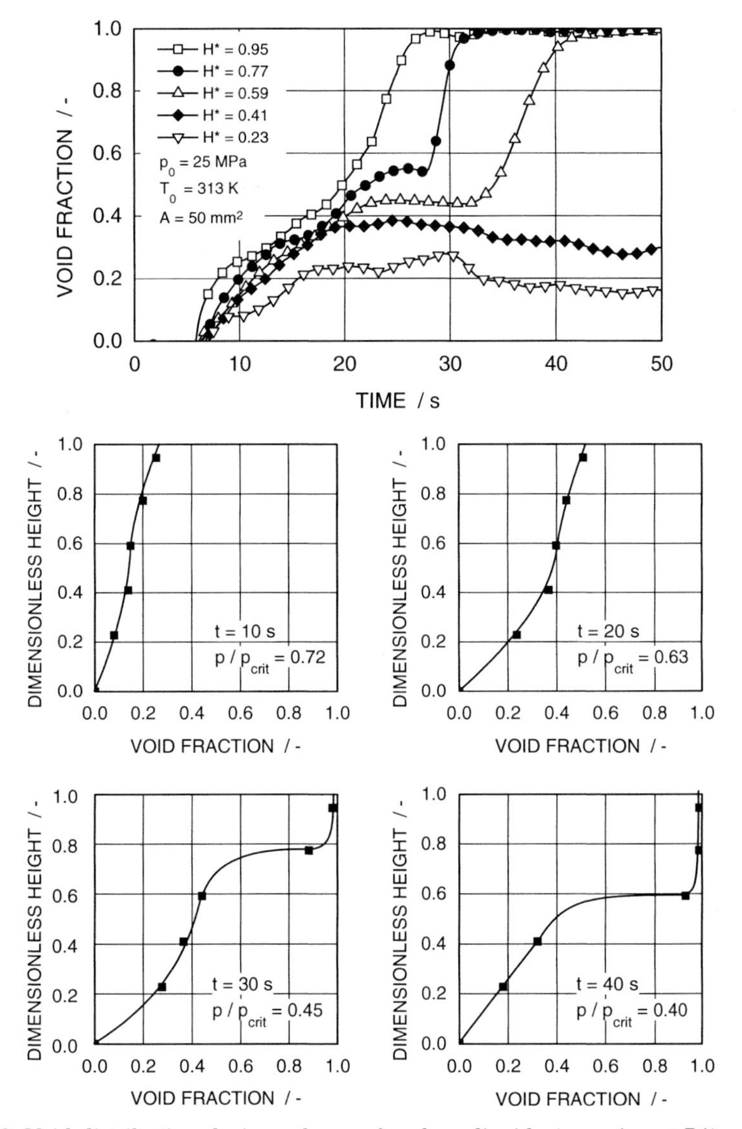

Figure 1.8: Void distribution during release of carbon dioxide (experiment B1).

tained by experiment B1–B3 are shown in the Figures 1.8–1.10. As long as the fluid condition is single-phase, the void fraction is defined to be $\varepsilon=0$. During the vessel release, the void fraction at the bottom of the vessel is defined to be $\varepsilon=0$ as well.

For experiment B1 (Figure 1.8), an almost linear increase of the void fraction can be observed after $t=10$ s. After $t=20$ s, there is a strong increase in the bottom region of the vessel and a small increase in the middle and in the top region. After $t=23$ s, the two-phase flow in the top of the vessel becomes un-

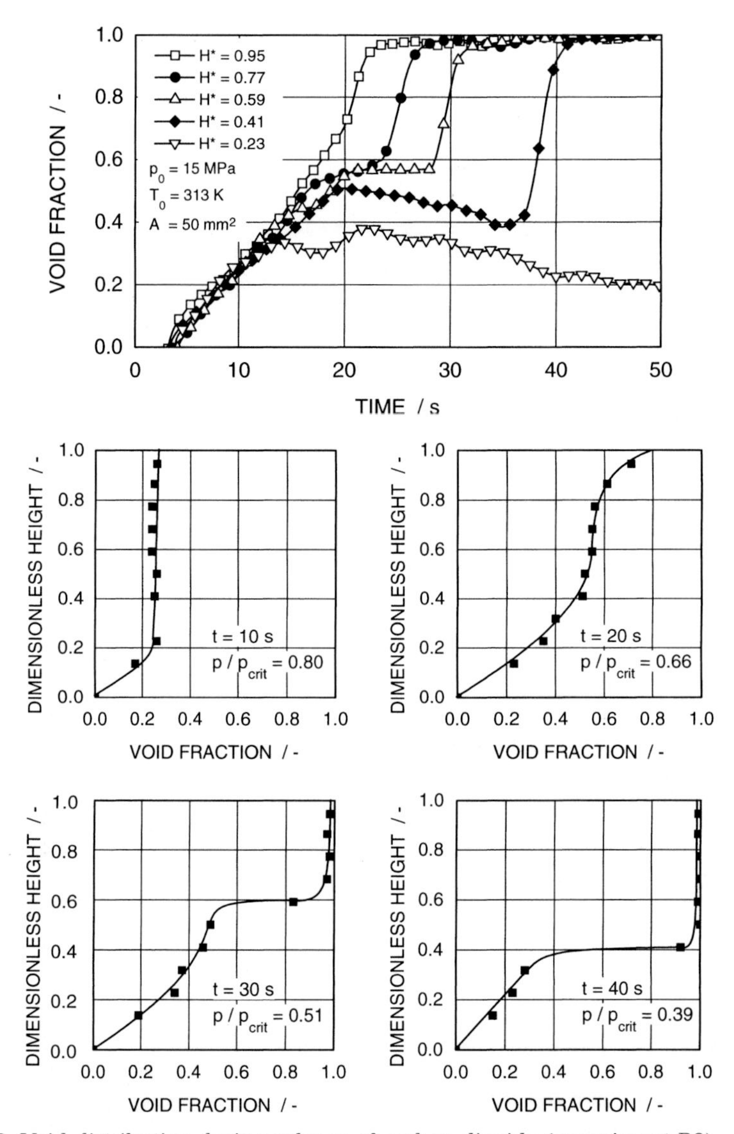

Figure 1.9: Void distribution during release of carbon dioxide (experiment B2).

stable, and a mixture level develops such that the void fraction at this location increases up to $\varepsilon = 1$. Subsequently, an almost linear increase of the void fraction can be observed from the bottom to the mixture level.

For experiment B2 (Figure 1.9), the void fraction profile is constant after $t = 10$ s. This is due to the low relative motion of the two phases. Up to this time, no considerable amount of vapour phase has been transported to upper regions which would result in a local increase of void fraction. The measured void fractions are higher than in experiment B1. Subsequently, the relative ve-

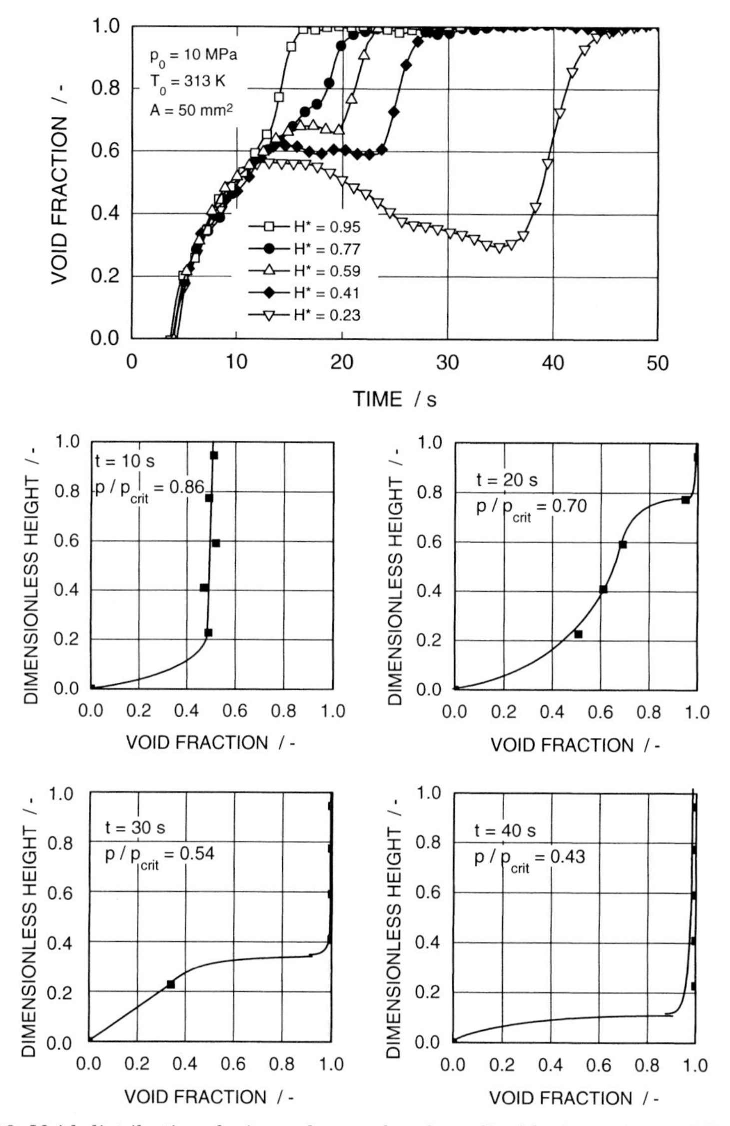

Figure 1.10: Void distribution during release of carbon dioxide (experiment B3).

locity between the liquid and the vapour phase increases. Void fraction gradients along the axis of the vessel can be observed. After $t=20$ s, a mixture level starts to develop. A linear increase of the void fraction is obtained for the bottom and for the top region. In the middle region, no increase with respect to height is observed. Subsequently, for the region above the mixture level, a void fraction close to one and for the region below the mixture level an almost linear increase of the void fraction is obtained.

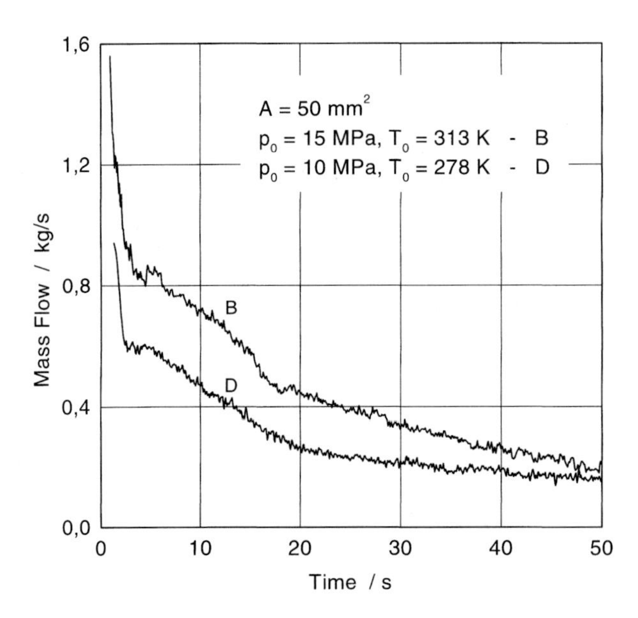

Figure 1.11: Mass flow for different initial states.

For experiment B3 (Figure 1.10), the void fraction profile is constant after $t = 10$ s as it is in experiment B2 although the measured void fractions are larger. At $t = 14$ s, a mixture level develops. The void fraction in the top of the vessel increases towards $\varepsilon = 1$.

For larger initial pressure, saturation conditions are reached at lower process pressure. The lower the saturation pressure the larger is the relative velocity between the coexistent phases. Therefore, after the onset of flashing, void gradients are evident for the case $p_0 = 25$ MPa, whereas for the case $p_0 = 10$ MPa the void profile is constant. Furthermore, the overall void level is lower in case of higher initial pressure.

Mass flow rate: In Figure 1.11, the outgoing mass flow rate in the venting pipe is shown for experiment B2 and D. The measured mass flow transient is similar to the pressure transient. However, besides the vessel pressure, the outgoing mass flow rate is influenced by the void fraction in the top of the vessel. For both experiments a decrease of the change rate of the mass flow can be observed when the flashing starts inside the vessel. A further decrease is obtained at the time when the mixture level is developing and the flow pattern at the inlet of the venting pipe changes from two-phase to single-phase. Larger mass flow rates are measured during experiment B2 due to the larger pressure inside the vessel.

Finally, the influence of a mixing component on the pressure release is of importance. Figure 1.12 shows the pressure profile of a carbon dioxide/nitrogen mixture with $x_0 = 0.1$ as the initial mole proportion of nitrogen. At the beginning of the process, a steep drop of pressure and temperatures inside the vessel oc-

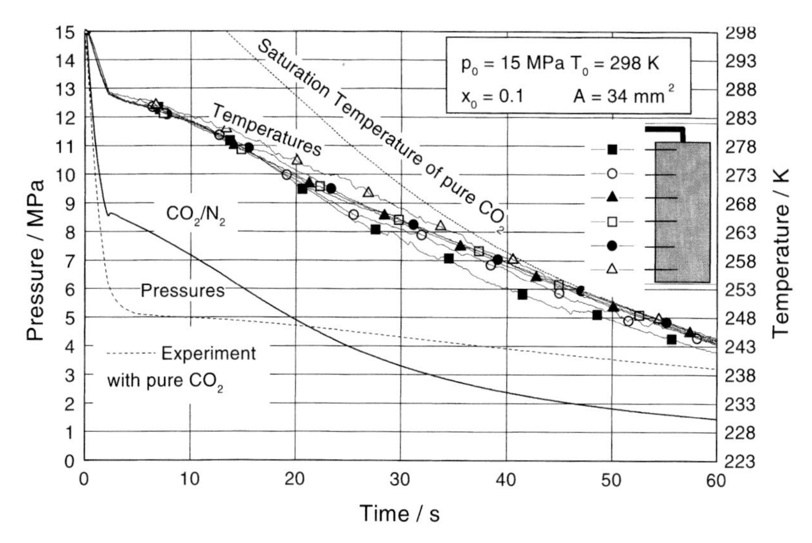

Figure 1.12: Pressure and temperature transients during blowdown of a CO_2/N_2 mixture with 10 Mole-% N_2.

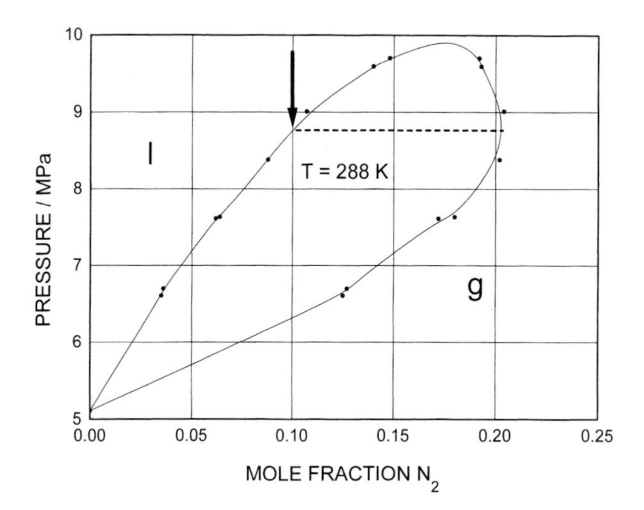

Figure 1.13: Phase equilibrium of CO_2/N_2 mixture.

curs. After about 2 s, the thermodynamic conditions (pressure, temperature) are such that for the initial N_2-content of the liquid, a liquid-vapour equilibrium state is reached (Figure 1.13) and flashing occurs. After the onset of flashing, an axial temperature profile forms. This is caused by the separation of the two components during the process. The nitrogen, being more volatile, is enriched in the vapour phase and blown out faster. The N_2 fraction diminishes with increasing expansion time. This can be seen by comparing the measured tem-

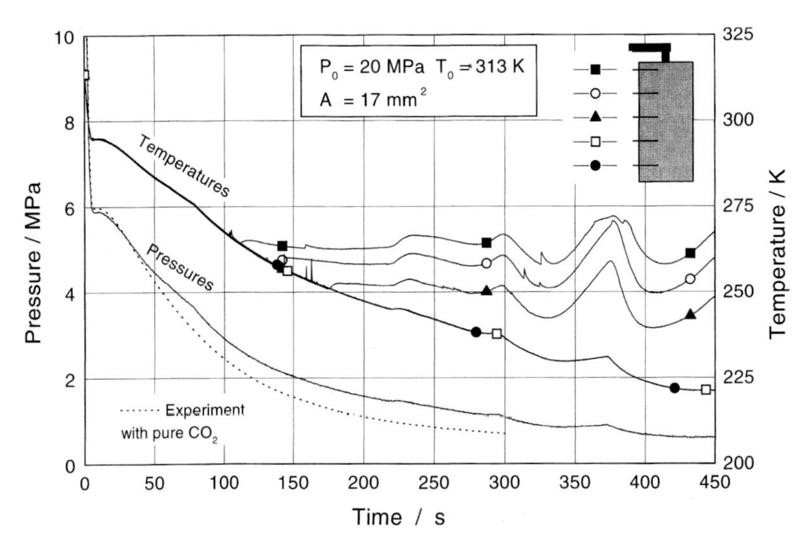

Figure 1.14: Pressure and temperature transients during blowdown of CO_2 saturated with H_2O.

peratures with the saturation temperature of the pure carbon dioxide calculated from the measured pressure. After about 40 s, the temperature in the bottom region of the vessel reaches the saturation line. At this moment no nitrogen is left at this axial level in the vessel.

In Figure 1.14, the temperature and pressure transients for a blowdown process with carbon dioxide saturated with water are shown. Compared to the blowdown of pure carbon dioxide, this process is slower and less even. However, similar to the previous experiments without water, the measured temperature is always the saturation temperature of pure carbon dioxide in the region of the vessel where two phases coexist. For the temperature in the bottom region of the vessel, this is the case during the whole period shown. Therefore, the water has a negligible influence on the saturation temperature. However, the peaks in the pressure transient, which can also be seen in the saturation temperature, point out that the flow through the venting pipe is inhibited significantly by the water. This blocking indicates a formation of a solid phase (ice or hydrate) in the pipe.

1.5 Modelling of the Transient Pressure Release

In this section, we present a model to calculate the devolution of the pressure release [9]. This model is based on an ansatz from Prasser [10]. The assumptions are the following:

- homogeneous pressure inside the vessel (calculation 0-dimensional),
- void fraction depends on vertical position (calculation 1-dimensional),
- description of momentum transfer by a drift flux model,
- thermodynamic equilibrium of the two phases in the vessel,
- heat transfer from the wall in the fluid (heat conduction in the wall 1-dimensional),
- critical flow in the venting pipe,
- isentropic flow with slip $S=1$ in the venting pipe,
- thermodynamic equilibrium in the venting pipe.

The mass and energy balances for the vessel are:

$$\frac{dM}{dt} = -\dot{M}_{out}, \tag{2}$$

$$\frac{d(hM - pV)}{dt} = -\left(\dot{M}_{out}\, h_{out}\right) + \dot{Q}_W. \tag{3}$$

The mass flow $\dot{M}_{out} = \dot{M}_{crit}$ is calculated with the Homogeneous Equilibrium Model (HEM) [11] (see below). The heat flow $\dot{Q}_W$ is calculated with the assumptions:

- inner wall temperature is equal to fluid temperature,
- no heat transfer from outside the vessel,
- no heat transfer in the pure gas phase.

The pressure and the total void fraction are calculated by solving the differential Equations (2) and (3) assuming thermodynamic equilibrium.

The local mass and energy balances of the liquid and the gas phase are:

$$\frac{\partial\left((1-\varepsilon)\rho_l + \varepsilon\rho_g\right)}{\partial t} + \frac{\partial\left(w_l(1-\varepsilon)\rho_l + w_g\,\varepsilon\,\rho_g\right)}{\partial z} = 0, \tag{4}$$

$$\frac{\partial\left((1-\varepsilon)\rho_l u_l + \varepsilon\rho_g u_g\right)}{\partial t} + \frac{\partial\left(w_l(1-\varepsilon)\rho_l h_l + w_g\,\varepsilon\,\rho_g h_g\right)}{\partial z} = q_W. \tag{5}$$

The velocity of the gas w_g in these equations can be eliminated using a drift flux model:

$$w_g = \frac{(1-\varepsilon)C_0 w_l + w_{gj}}{1-\varepsilon C_0}. \tag{6}$$

To calculate the drift velocity w_{gj}, the models of Kataoka and Ishii (with consideration of the viscosity) [12], Labuntsov [13] and Sonnenburg [14] are used and investigated.

The level of the continuous liquid phase is calculated by integration of the liquid content over the vertical vessel axis until the total liquid mass in the vessel is reached:

$$M_l(z) = \int_0^z A_V \rho_l (1-\varepsilon(\bar{z}))d\bar{z}. \tag{7}$$

The critical mass flow is calculated by the Homogeneous Equilibrium Model. The main assumptions have been mentioned before. Momentum and energy balance result for a isentropic and frictionless flow in:

$$\dot{m}(p) = [\varepsilon(p)\rho_g(p) + (1-\varepsilon(p))\rho_l(p)]\sqrt{2[h_0 - x(p)h_g(p) - (1-x(p))h_l(p)]}. \tag{8}$$

In the critical section, the mass flow must be maximal. Therefore, the critical mass flow is found by solving Equation (9):

$$\left(\frac{d\dot{m}}{dp}\right)_{crit} = 0. \tag{9}$$

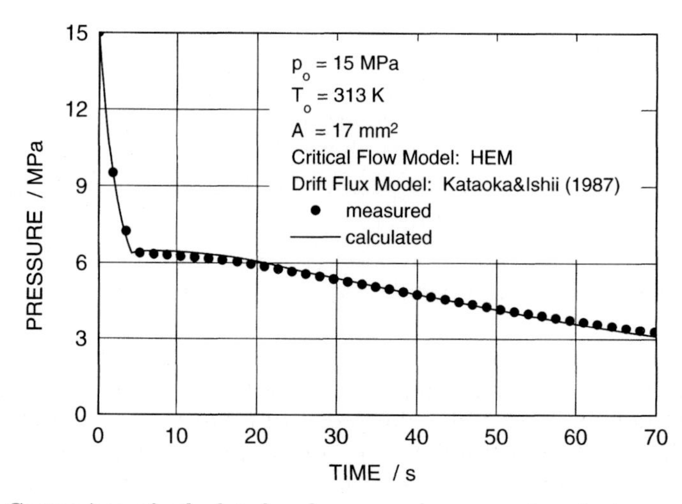

Figure 1.15: Comparison of calculated and measured pressure transients.

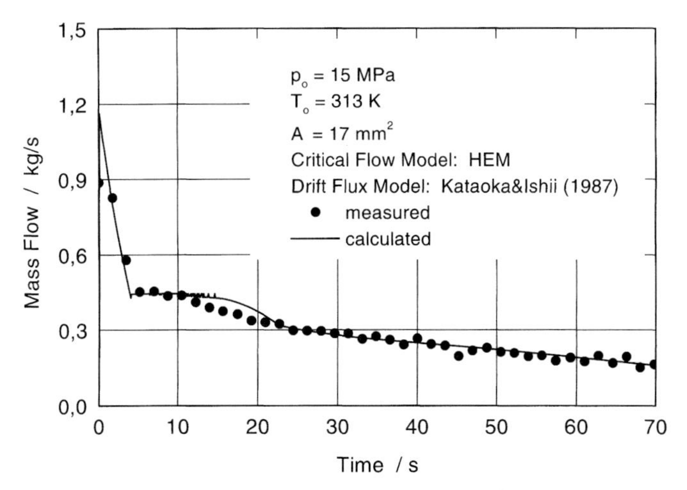

Figure 1.16: Comparison of calculated and measured mass flow.

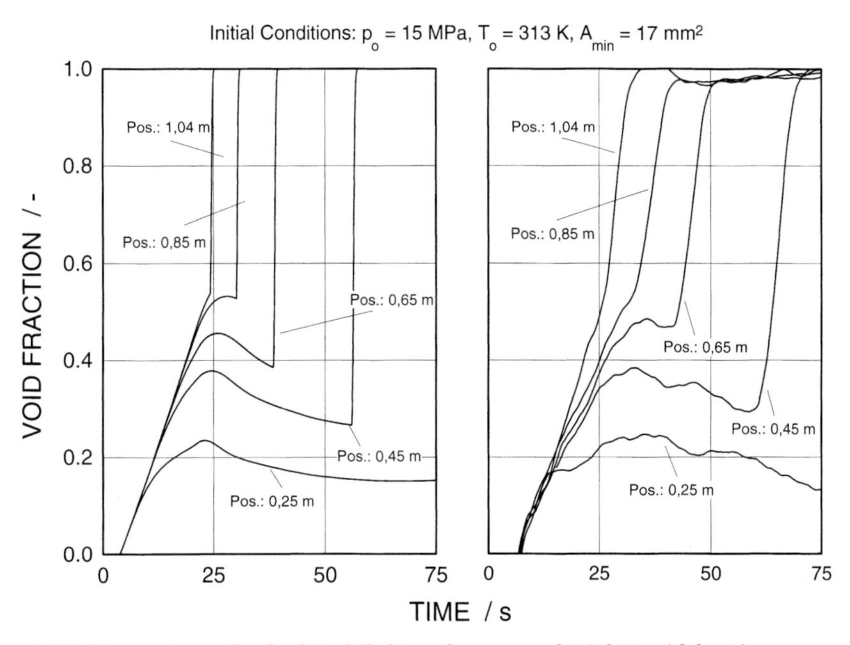

Figure 1.17: Comparison of calculated (left) and measured (right) void fractions.

A comparison between modelling and experiment is shown in Figures 1.15–1.17. The agreement is good for the pressure (Figure 1.15) and the mass flow (Figure 1.16). The void distribution (Figure 1.17) represents well the qualitative form of the measured devolution, but the rise of the void fraction to $\varepsilon = 1$ is too fast.

It shows that the application of the drift flux models from Kataoka and Ishii [12] and from Labuntsov [13] results in a good representation of the experiments, whereas the model of Sonnenburg [14] results in a too high drift velocity.

1.6 Summary and Outlook

The pressure release of pressurised carbon dioxide from a top vented vessel has been investigated. During the experiments, pressure, mass flow, axial temperature profile, and axial void profile were measured. The investigations show that the process inside the vessel can be divided in two parts: In the first part, the fluid is single phase. This expansion can be treated as isentropic. In the second part, the fluid is two phase. In this phase, the gas and liquid are approximately in thermodynamic equilibrium.

A model for the simulation of the pressure release is presented. Herein, the drift velocity between the phases is calculated with the drift flux models of Kataoka and Ishii or Labuntsov and the mass flow through the venting pipe with the Homogeneous Equilibrium Model.

Further the influence of mixing components has been examined. During the expansion of nitrogen/carbon dioxide mixtures, a separation between the two components occurs. This process can be calculated from the measured temperatures. Dissolved water can inhibit the release process significantly. This could be caused by formation of carbon dioxide hydrate in the venting pipe. Therefore, further investigations are necessary.

Other topics of interest, which have not been investigated yet are:

- dry ice formation in the venting line and
- transient release process with an additional solid phase.

List of Symbols

A minimal cross sectional area in the venting line
A_V cross sectional area of the vessel
C_0 radial distribution parameter
h enthalpy
M fluid mass in the vessel
$\dot{M}$ mass flow
$\dot{M}_{crit}$ critical mass flow
$\dot{m}$ mass flow density
p pressure

$\dot{Q}_w$ heat flow from the wall into the vessel
s entropy
T temperature
t time
u internal energy
V volume
w velocity
z vertical coordinate
ε void fraction
ρ density

Subscripts

l liquid
g gaseous
0 initial state
crit critical point

References

[1] M.A. McHugh, V.J. Krukonis: *Supercritical Fluid Extraction* (2nd Ed.). Butterworth-Heinemann, Boston-USA, 1994.
[2] F. Mayinger: *Two-phase flow phenomena with depressurization; consequences for the design and layout of safety and pressure relief valves.* Chem. Eng. Process **23** (1988) 1–11.
[3] R. Eggers, V. Green: *Instationary Heat Transfer in a Pressure Vessel during Expansion Process.* Proc. of the 9th Int. Heat Transfer Conf. (Jerusalem, Israel) **4** (1991) 401–405.
[4] A. Haque, S. Richardson, G. Saville, G. Chamberlain: *Rapid depressurization of pressure vessels.* J. Loss Prev. Proc. Ind. **3** (1990) 3–7.
[5] L. Friedel, N.J. Kranz, G. Wehmeier, F. Westphal: *Theoretical and experimental analysis of venting-induced processes in reacting and non-reacting two-phase systems.* Chem. Eng. And Proc. **34** (1995) 71–78.
[6] A. Thies: *Die thermo- und fluiddynamischen Vorgänge infolge der Druckentlastung chemischer Reaktoren.* Fortschrittsberichte VDI, Reihe 3 Nr. 416, VDI-Verlag Düsseldorf, 1995.
[7] B. Gebbeken, R. Eggers: *Thermohydraulic phenomena during vessel release of initially supercritical carbon dioxide.* Proc. of the first Int. Symp. Two-Phase Flow Modelling and Experimentation, Rome. G.P. Celata, R.K. Shah (Eds.), Edizioni ETS Pisa **2** (1995) 1139–1146.
[8] B. Gebbeken, R. Eggers: *Blowdown of carbon dioxide from initially supercritical conditions.* J. Loss Prev. Proc. Ind. **9** (1996) 285–293.

[9] B. Gebbeken: *Thermohydraulische Vorgänge in Behältern während der Druckentlastung von CO_2 aus dem überkritischen Zustand.* Ph.D. thesis, Technische Universität Hamburg-Harburg, 1997.

[10] H.-M. Prasser: *Ein mathematisches Modell zur Zweiphasenströmung in einem zylindrischen Druckgefäß bei kleinen Lecks.* Kernenergie **25** (1982) 294–298.

[11] H. Isbin, J. E. Moy, A.J.R. DaCruz: *Two-phase, steam-water critical flow.* AICHE Journal **3** (1957) 361–365.

[12] I. Kataoka, M. Ishii: *Drift flux model for large diameter pipe and new correlation for pool void fraction.* Int. J. Heat Mass Transfer **30** (1987) 1927–1939.

[13] D.A. Labuntsov, I.P. Kornyukhin, E.A. Zakharova: *Vapour concentration of a two-phase adiabatic flow in vertical ducts (engl.).* Therm. Eng. (Tepleoenergetika) **15** (1968) 78–84.

[14] H.G. Sonnenburg: *Entwicklung eines umfassenden Drift-Flux-Modells zur Bestimmung der Relativgeschwindigkeit zwischen Wasser und Dampf.* Gesellschaft für Reaktorsicherheit Bericht Nr. GRS-A-1752, 1991.

2 An Experimental Study of Evaporation Waves in Adiabatic Flashing

Gerrit Barthau and Erich Hahne *

Abstract

An experimental study on the inception and propagation of evaporation waves in adiabatic flashing of Refrigerant 11 (CCl_3F) has been performed in glass cylinders with inner diameters 32 mm $\leq d \leq$ 252 mm. The formation of evaporation waves was observed at depressurization rates as low as $\Delta p/\Delta t \approx 1$ bar/s. Additional experiments with metallic inserts in the test tubes indicate that the presence of metal/liquid contact lines promotes the formation of evaporation waves and stabilizes their propagation into the superheated liquid.

2.1 Introduction

When a pressurized liquid in equilibrium with its vapour in a closed container is suddenly depressurized, the whole bulk of the liquid becomes superheated (metastable) with respect to the decreasing pressure and adiabatic flashing will occur. Such flashing processes can occur either intentionally, e.g. in steam accumulators, or undesiredly, e.g. during blowdown in chemical reactors, or due to pipe fracture in liquefied gas containers. The governing parameters for the resulting thermo-hydraulic processes are the amount of pressure decay and the depressurization rate. In steam accumulators, the depressurization rate is in the range of some bar/h, whereas in a guillotine break of larger pipes, depressurization rates of up to Mbar/s can occur.

* Universität Stuttgart, Institut für Thermodynamik und Wärmetechnik, Pfaffenwaldring 6, D-70569 Stuttgart, Germany

The unique characteristic of such adiabatic flashing processes is the withdrawal of the latent heat of evaporation from the *internal energy* of the superheated liquid. By its very definition, a phase change process is bound to an interface and, consequently, the intensity of the process is limited either by transport of energy to this interface, or by transport of interface to the energy, or by the formation of new interface near the energy. Therefore, extent and distribution of liquid/vapour interface in the system are of paramount importance for the flashing intensity.

In many models for the flashing process, it is postulated that heterogeneous or homogeneous nucleation, followed by classical bubble growth occurs [1–7]. There is, however, clear evidence that a totally different thermo-hydraulic phenomenon – namely an evaporation wave – can develop also in a flashing process [8–15].

When the depressurization occurs from the vapour side and is not strong enough to initiate homogeneous nucleation, and when the container walls do not offer nucleation sites for heterogeneous nucleation anywhere, the phase change process is necessarily restricted to the liquid level, as "still evaporation" in the first instance. At a given threshold superheat (depending on the depressurization rate), the initially smooth liquid level seems to undergo a hydrodynamic instability triggered by a nucleation process in the liquid/wall contact region. This instability spreads rapidly over the entire surface, generating a jet of aerosol and resulting in a dramatic increase of the vapour production. Subsequently, a liquid/vapour interface (the evaporation wave front) penetrates into the stagnant superheated liquid with a velocity being some orders of magnitude smaller than the velocity of sound in the liquid.

The first systematic study of such evaporation waves was performed by Grolmes and Fauske [8] with water, methyl alcohol and refrigerants R 11 and R 113 as test fluids in glass tubes with 2 to 50 mm diameter. It was found that in smaller diameter tubes a higher superheat was necessary to initiate evaporation waves.

In a thorough study, Hill and Sturtevant [13] succeeded in documenting photographically the mode of inception of evaporation waves from the liquid surface. They also showed that there is actually a sharp, rough interface between the superheated liquid below and the aerosol-like two-phase flow above. Their experiments were performed with R 12 and R 114 in a glass tube of 25.4 mm diameter with choked and unchoked outlet conditions for the two-phase flow.

Reinke and Yadigaroglu [15] conducted experiments on evaporation waves with propane, butane, water and R 134a in glass tubes of 25–80 mm diameter, in a funnel-shaped channel and a channel with rectangular cross section, also made of glass. They found the propagation velocity of the evaporation wave to be not related to the channel diameter and channel shape.

In all these experiments, the depressurization was initiated by cutting a diaphragm, which results in nearly unrestricted (large) discharge areas and therefore in high initial depressurization rates of approximately 4000 bar/s to 12 000 bar/s.

In the present study, it should be explored, whether the occurrence of evaporation waves must be considered also in "technical systems", especially at lower depressurization rates (restricted discharge areas), in tubes of larger diameter, and in the presence of metallic elements simulating the effect of metallic container walls.

2.2 Experimental

The experimental facility is shown schematically in Figure 2.1. The test tubes are thickwalled glass cylinders with inner diameters d_i=32, 52, 102, 152 and 252 mm, arranged concentrically in a glass container with 305 mm inner diameter and 1000 mm height. The temperature of the test section is controlled by thermostated water circulating in this container. A length of around 8 cm from the bottom of the test tubes remains unheated in order to avoid parasitic wall nucleation from the bottom stainless steel flange during depressurization. The top of the test tubes is connected to the low pressure reservoir by a stainless steel hose with 50 mm nominal width. A fast opening solenoid valve (50 mm nominal width) and a special ball valve are used to control the flow. The low pressure reservoir is a 0.2 m^3 stainless steel container equipped with an internal heat exchanger. The hose between experimental container and ball valve is

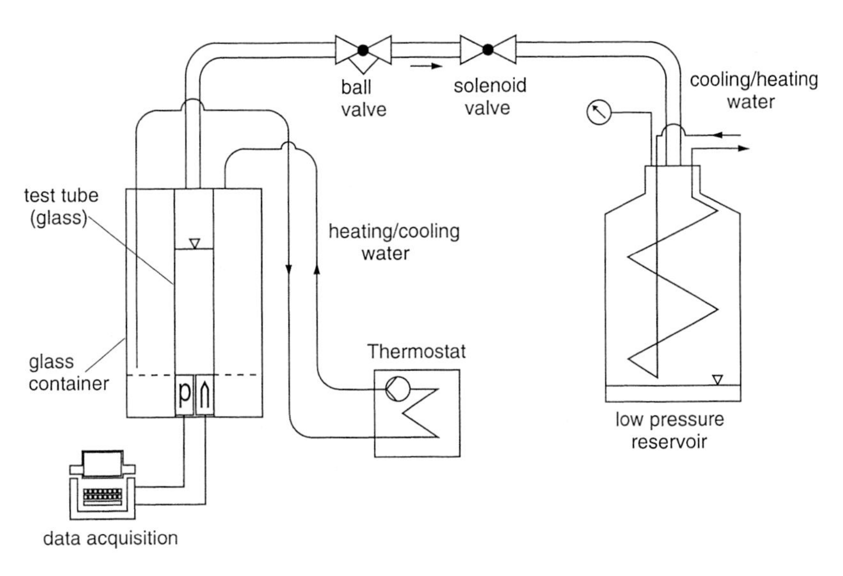

Figure 2.1: Schematic drawing of the experimental set-up.

heated electrically by a strip heater, the ball valve and the solenoid valve are heated by hot air, using temperature-controlled fan-forced heaters.

The axial temperature distribution in the test liquid is measured with a calibrated thermocouple soldered into the top of a stainless steel tube (diameter 5 mm). This tube is arranged in the bottom flange and can be moved up and down.

The pressure in the test tube is measured with a fast miniature pressure transducer (Kulite XTC-2-190M-VG; measuring range –1 to 7 bar; natural frequency $\approx$80 kHz), arranged in the bottom flange and is recorded by a data acquisition system (Gould Windograf) with 800 samples/s. Calibration of the pressure measurement equipment showed an accuracy better than 0.03 bar. The switch-on signal of the solenoid valve and its signal for 80%-open position are recorded also by "event markers" on the Windograf.

Each experimental run is recorded with a standard video-system (50 half-frames/s). In order to synchronize the video recordings and the pressure data, an illuminated diode, connected to the activation circuit of the solenoid valve, is placed in the view field of the video-camera.

The fully open ball valve provides a discharge area of A=334 mm^2 with circular cross section and can be equipped with additional laser cut seat apertures allowing for precise control of small discharge areas. The maximum discharge area of A>804 mm^2 is obtained when the ball valve is dismounted from the connection hose.

All experiments discussed here were performed at an initial pressure (before depressurization) of p_i=4.3 bar; this corresponds to an initial equilibrium temperature of ϑ_i=72 °C. An exception is presented in Figure 2.5, where the results of experiments with p_i=6.7 bar (ϑ_i=90 °C) are also shown.

2.3 Experimental Results

2.3.1 Glass tubes without metallic inserts

In Figure 2.2 a, the experimental results for the smallest diameter test tube (d_i=32 mm) and different discharge areas A are shown. The actuation of the solenoid valve is taken as zero point for time t; p is the measured base pressure. The initial pressure in the low pressure reservoir is p_R.

For all tests, the initial liquid level height in the test tube is 58±3 cm, which guarantees an isothermal liquid zone in the test liquid of at least 40 cm height. During the first 0.15 s after opening the solenoid valve, the base pressure remains nearly constant in all runs. This is partly due to the finite opening time of the solenoid valve, but also due to the fact that the outflowing vapour mass has to be accelerated.

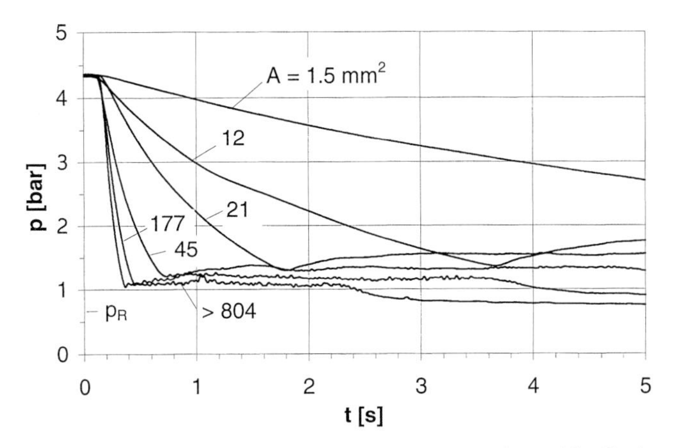

Figure 2.2 a: Pressure traces for the 32 mm diameter test tube with discharge area A as parameter (reservoir pressure $p_R = 0.7$ bar). No metallic insert.

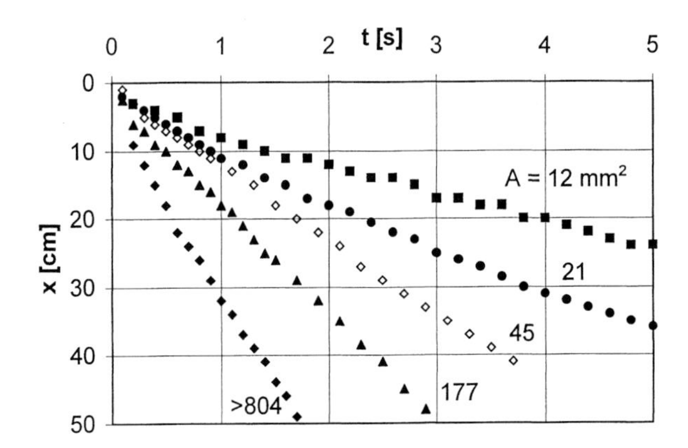

Figure 2.2 b: Evaporation wave front position vs. time for the tests of Figure 2.2 a.

In the test-run with the very small discharge area $A \approx 1.5\ \text{mm}^2$, the liquid level remains smooth and quiescent, emitting vapour in "still evaporation" only, and without the inception of an evaporation wave. Schlieren patterns below the liquid level indicate that heat is fed to the liquid-vapour interface by transient natural convection.

In test $A = 12\ \text{mm}^2$, initially, also "still evaporation" occurs on the liquid level at a very small average depressurization rate $\Delta p/\Delta t \approx 1$ bar/s. After $t \approx 3.6$ s, an evaporation wave originates at $p_{\text{min}} = 1.35$ bar and continues to penetrate downwards into the superheated liquid with the pressure increasing again.

For the larger discharge area $A = 21\ \text{mm}^2$, an evaporation wave originates at $p_{\text{min}} = 1.3$ bar; a quasi-steady equilibrium between vapour production and two-phase outflow is established at a pressure of $p_w \approx 1.5$ bar. We define this pressure as "wave pressure p_w" as the average, nearly constant, pressure existing with an evaporation wave.

For $A > 804\ \mathrm{mm}^2$ (when the ball valve was removed), the evaporation wave originates after $t \approx 0.4$ s at $p_{min} \approx 1.0$ bar, the depressurization rate being $\Delta p/\Delta t \approx 8$ bar/s. The superheat for the inception of the wave is $\Delta\vartheta \approx 50$ K. Wave pressure and consequently superheat of the bulk liquid remain nearly constant until $t \approx 2.3$ s when the evaporation wave enters the unheated bottom liquid of the test tube and dies off.

The superheat $\Delta\vartheta = \vartheta_i - \vartheta_t$ is defined as the temperature difference between the initial saturation temperature in the pressurized container $\vartheta_i = \vartheta_{sat}(p_i)$, and the instantaneous saturation temperature $\vartheta_t = \vartheta_{sat}(p)$ at the instantaneous pressure p.

In Figure 2.2 b, the position of the evaporation wave front in the test tube is shown for the tests of Figure 2.2 a. Position $x = 0$ cm corresponds to the initial liquid level, the zero point for time t corresponds to that moment when the evaporation wave starts at the liquid level (e.g. at 3.6 s for $A = 12\ \mathrm{mm}^2$). From this figure, the average velocity of the wave front is found to be in the range of 5 cm/s ($A = 12\ \mathrm{mm}^2$) to 30 cm/s ($A > 804\ \mathrm{mm}^2$).

As can be seen in Figure 2.2 a, the "ripples" in the pressure traces during the life-span of the evaporation waves become more pronounced for larger discharge areas A (i.e. lower wave pressures p_w; higher superheats $\Delta\vartheta$; higher propagation velocities). In order to find out, whether these higher frequency pressure fluctuations result from the evaporation front itself or from downflow two-phase phenomena, an additional series of experiments has been performed. Here, the ball valve was kept fully open ($A = 334\ \mathrm{mm}^2$), but the reservoir pressure was varied in the range $0.7 \le p_R \le 1.8$ bar.

Figures 2.3 a and 2.3 b show the results of these experiments (5 tests). For the highest reservoir pressure $p_R = 1.8$ bar and a corresponding superheat of $\Delta\vartheta \approx 31$ K of the test liquid, no evaporation wave appears; the liquid level remains in "still evaporation" for approximately 70 s (not shown in Figure 2.2 a), then a parasitic nucleation site is activated at the wall of the test tube.

For $p_R \le 1.7$ bar ($\Delta\vartheta \ge 33$ K), evaporation waves develop, which is in good agreement with the findings of Grolmes and Fauske [8] for higher depressurization rates (cf. Figure 2.5). Again, the ripples in the pressure traces become more pronounced for lower wave pressures p_w and higher propagation velocities.

With respect to wave pressure $p_w \approx 1.3$ bar and propagation velocity of the wave front $w \approx 11$ cm/s, run $A = 45\ \mathrm{mm}^2$ (Figure 2.2) lies just in between the wave pressures for run $p_R = 1.1$ bar and run $p_R = 1.3$ bar (Figure 2.3), both with about a 7 times larger discharge area $A = 334\ \mathrm{mm}^2$. As can be concluded from Figure 2.3 b, the flow in run $p_R = 1.1$ bar and run $p_R = 1.3$ bar is not choked in the discharge area because, otherwise, a further reduction of the reservoir pressure to $p_R = 0.7$ bar could not result in a further increase of the propagation velocity of the evaporation wave and therefore in a higher mass flux in the discharge area. For run $A = 45\ \mathrm{mm}^2$, however, the pressure ratio in the discharge area is approximately $p_R/p_w \approx 0.7/1.3 < 0.55$, indicating that the two-phase flow

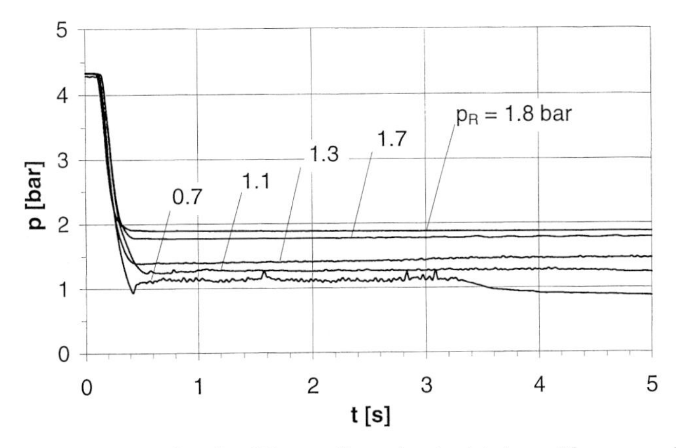

Figure 2.3 a: Pressure traces for the 32 mm diameter test tube with reservoir pressures p_R as parameter ($A = 334\ \text{mm}^2$). No metallic insert.

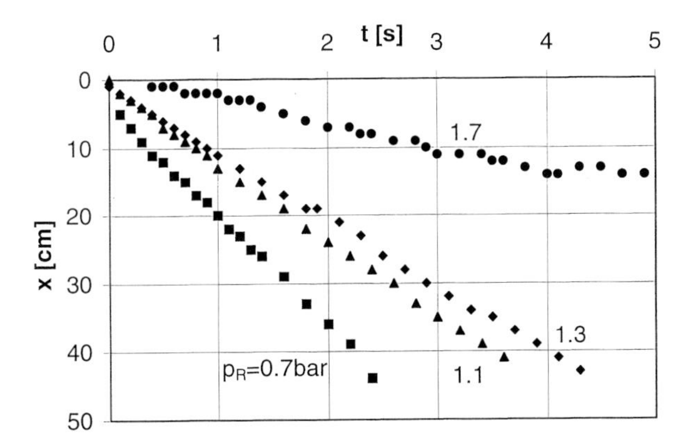

Figure 2.3 b: Evaporation wave front position vs. time for the tests of Figure 2.3 a.

in the discharge area might be choked [16]. All 3 tests show similar higher frequency pressure fluctuation patterns ("ripples"). This seems to indicate that these higher frequency pressure fluctuations (which also have been observed in the experiments by Grolmes and Fauske [8], Hill and Sturtevant [13], Reinke and Yadigaroglu [15]) result rather from the evaporation front itself than from downflow two-phase phenomena.

In Figure 2.4, some pressure traces are shown for the largest diameter test tube ($d_i = 252$ mm) and for different initial liquid level heights h. The discharge area ($A = 334\ \text{mm}^2$) and the reservoir pressure ($p_R = 0.7$ bar) were kept constant.

The dotted parts of the pressure traces of tests $h = 500$, 581 and 656 mm indicate the appearance of a parasitic wall nucleation, which occurred in the majority of the experiments with the largest test tube. Close inspection of the video recordings of these runs shows that nearly always only a single bubble

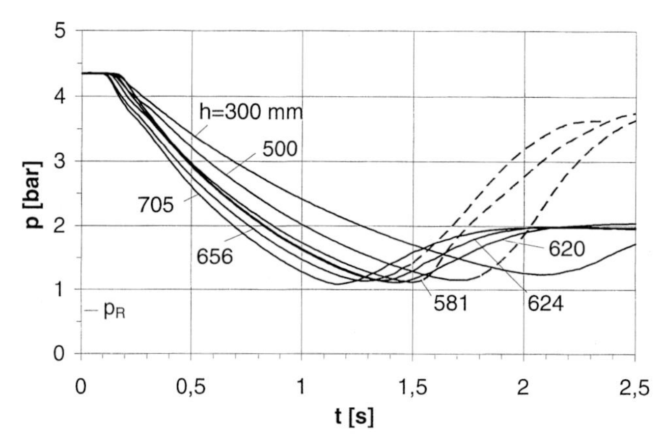

Figure 2.4: Pressure traces for the 252 mm diameter test tube with initial liquid level height h as parameter (p_R=0.7 bar; A=334 mm^2). No metallic insert.

(not a bubble chain) nucleated somewhere at the wall. This may reveal that the nucleation site is an accidental agglomeration of inert gas precipitated from the test liquid. It should be mentioned that for the high superheat $\Delta\vartheta=46$ K, even sites with critical radii as low as $r_c \approx 75$ nm (according to $r_c=2\sigma/\Delta p$) can be activated. As can be observed from the pressure traces in Figure 2.4 (dotted curves), the nucleation of a single bubble from the container wall results in a steeper and higher pressure recovery than the occurrence of an evaporation wave from the liquid level. The video recordings show that such a single bubble grows within approximately 0.5 s to diameters of ≈20 cm (cf. [17]), occupying nearly the entire cross section of the test tube. The initially smooth liquid/vapour interface of the bubble becomes disturbed and rough for bubble diameters >5–10 cm, indicating for the inception of evaporation wave-like transport phenomena now at the *bubble's* interface. The growth and rise of this vapour mass results in a strong internal circulation of the superheated liquid in the container and therefore in a strongly increased vapour production.

The good reproducibility of the minima of the pressure traces in Figure 2.4, characterizing the inception of evaporation waves, is attributed to the fact that the interior of the test facility was not exposed to ambient air during this experimental series. Therefore, the adsorption kinetics at the test tube's surface and also the amount of inert gases, dissolved in the test liquid, remained nearly constant during these series of runs, resulting in reproducible nucleation on the glass/liquid contact line. For the sake of clarity, only 6 of the 13 runs for level heights $500 \le h \le 705$ mm are shown, the pressure minimum being for all 13 runs $p_{min}=1.12\pm0.04$ bar, corresponding to a superheat $\Delta\vartheta=46\pm1$ K.

In Figure 2.5, the superheats $\Delta\vartheta = \vartheta_i - \vartheta_{sat}(p_{min})$ for the inception of evaporation waves are summarized for different test tube diameters with p_i=4.3 bar and p_i=6.7 bar being the initial pressures of the saturated liquid. As discussed for Figure 2.2 a, $\Delta\vartheta$ also depends on the depressurization rate $\Delta p/\Delta t$;

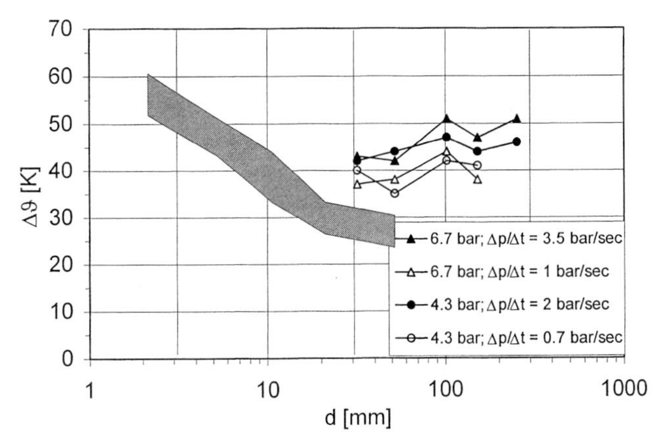

Figure 2.5: Superheat $\Delta\vartheta$ for the inception of evaporation waves vs. test tube diameter d with initial pressure p_i and depressurization rate $\Delta p/\Delta t$ as parameter. No metallic insert.

therefore, the data are shown for the highest and the lowest $\Delta p/\Delta t$ common to all runs with different diameter test tubes. The experiments with the different diameter tubes have been conducted in the sequence d=52, 152, 32, 252, 102, 32 mm over a period of approximately 2 years, and the test fluid has been evaporated and condensed more than 100 times during this period. Despite of some earlier results for the 32 mm test tube showing higher superheats than the later ones, it is felt that the whole data set is superimposed by a small "time-and/or purity-effect", which caused higher $\Delta\vartheta$ at later experiments. Taking this into account, it is concluded from the results of the present experiments with R 11 that a consistent influence of the test tube diameter on the superheat $\Delta\vartheta$ for the inception of evaporation waves is *non-existent or rather weak* in the range $30 \leq d \leq 250$ mm.

The dashed zone for $2 \leq d \leq 50$ mm in Figure 2.5 represents the experimental results of Grolmes and Fauske [8] for R 11. A comparison with these data is difficult, due to different definitions of the superheat $\Delta\vartheta$. It might be possible that the higher superheats for the smaller test tube diameters in the Grolmes/ Fauske data set [8] can be explained on the basis of the Borda-Carnot relations (pressure drop in sudden cross section area enlargements).

2.3.2 Glass tubes with metallic inserts

In order to simulate the effect of metallic container walls on the inception and propagation of evaporation waves, metallic inserts were arranged in the larger diameter test tubes. In one series of experiments, a 12 mm diameter stainless steel tube was arranged in the 152 mm and in the 252 mm test tube to penetrate through the liquid level. In another series, two rods, one made of stainless steel and one of copper (diameter = 10 mm), were arranged in the 152 mm diameter test tube also penetrating through the liquid surface.

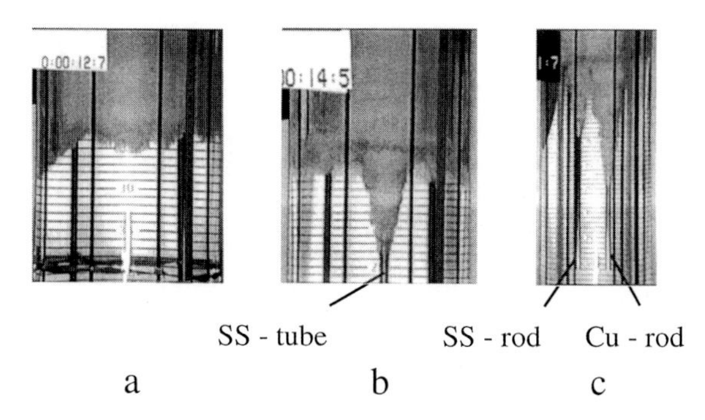

Figure 2.6: Evaporation waves without and with metallic inserts in the test tube. a) 252 mm diameter test tube without metallic inserts. b) 252 mm diameter test tube with stainless steel insert. c) 152 mm diameter test tube with stainless steel insert (left) and copper insert (right).

In most depressurization runs, boiling fronts originated from the contact line of the liquid level with the metal rods (solid/liquid/vapour contact) and penetrated into the superheated liquid. The term "boiling front" here denotes the front part of an evaporation zone, growing along the metal surface into the superheated liquid. The boiling front is followed by a cone-shaped evaporation wave front in the liquid.

Figure 2.6 a shows an evaporation wave in the 252 mm test tube without metallic inserts [1], and Figure 2.6 b shows a boiling front on the stainless steel (SS) insert in the same test tube. Since the propagation velocity of the boiling front on the metallic surface is higher than the propagation velocity of the evaporation wave in the liquid, the cone-shaped liquid/two-phase interface develops.

Figure 2.6 c shows the stainless steel rod (left) and the copper rod (right) in the 152 mm test tube. Here, the front first originated on the copper rod, the propagation velocity is higher on copper than on steel. Very seldom heterogeneous nucleation occurred in the lower part of the bulk liquid on the metallic surfaces. This was a very unexpected result.

Figure 2.7 shows the pressure traces for the 152 mm test tube with the stainless steel and the copper rod for different discharge areas A. For comparison, the positions of the pressure minima for the corresponding experimental runs without metallic inserts (dash-dotted line) and with the single stainless steel tube (dashed line) are also shown. As can be seen from the position of the pressure minima (dotted line, dashed line), boiling fronts on metallic inserts originate at higher pressures (i.e. at lower superheats) than evaporation waves in glass tubes (dash-dotted line).

When the boiling front spreads along the metal rods into the superheated liquid, the cone-shaped liquid/two-phase interface is larger than the interface

[1] The six vertical connecting rods and the rule, visible in Figure 2.6, are arranged outside the test tube.

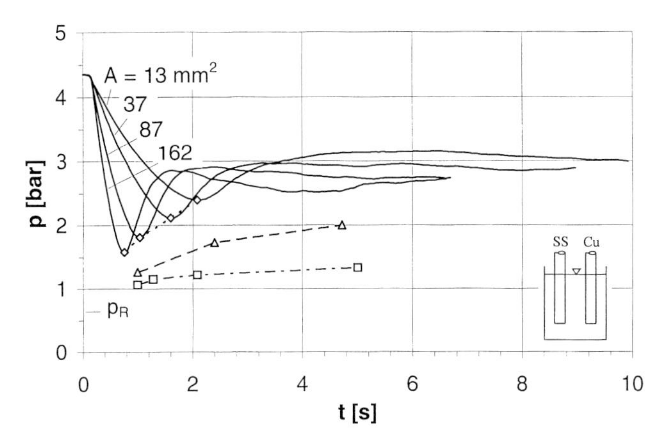

Figure 2.7: Pressure traces for the 152 mm diameter test tube with 2 metal rods (stainless steel, copper) with discharge area A as parameter. Position of pressure minima: △ single stainless steel tube; □ without metallic inserts.

in the runs without metallic inserts (Figure 2.6), resulting in larger vapour production and therefore higher pressure recovery. Higher pressure recovery (i.e. higher wave pressure p_w) on the other hand results in lower superheats of the metastable liquid and therefore in a reduction of parasitic wall nucleation probability – in other words, in a self-protection of the wave process.

2.4 Discussion

Concerning the *inception* of evaporation waves, Hill and Sturtevant [13] note: "At the highest superheats, individual nucleation sites rapidly initiate at random spots on the liquid free-surface and at the glass/liquid contact line... . At intermediate superheats, nucleation begins only at many sites on the glass/liquids contact line... . At low superheats, nucleation begins at one or more sites on the glass/liquid contact line and propagates across the surface."

It is exactly this "low superheat mode", which we have observed in the present study at much smaller depressurization rates than Hill and Sturtevant [13].

As can be concluded from the experimental results presented in Figure 2.5, the inception of the liquid level instability at the glass/liquid contact line, as a statistical nucleation process, seems – if at all – only weakly related to the length of this contact line ($d_i \cdot \pi$). One might expect that a larger length provides a higher probability for the occurrence of a "most favourable" nucleation site than a smaller length. Nucleation therefore should start at a lower superheat. Comparing the 252 mm diameter tube with the 32 mm, the difference in length of the contact line is about an order of magnitude, but the superheat for the great length is even somewhat higher than for the small length. Since the data presented in Figure 2.5 have been evaluated with the same depressurization rates $\Delta p/\Delta t$ (corresponding to $\Delta\vartheta/\Delta t$) for the different diameter tubes, the superheat-time relation (i.e. the "age") for the different contact lines ($d_i \cdot \pi$) is also the same.

Since nucleation appears to be not influenced by contact length and depressurization time, we conclude that the nucleation process on the glass/liquid contact line is not dominated alone by stochastic molecular effects (i.e. by "events"/(length·time)), at least in the range of nominal superheats $35\,K \leq \Delta\vartheta \leq 50\,K$.

The fact that in the experiments with metallic inserts, the boiling front always originates at lower superheats (compared to glass) from the (shorter) metal/liquid contact line, indicates that metal/liquid contact lines (compared to glass/liquid) either offer a better deterministic nucleation behaviour or/and a higher *local* superheat in the meniscus region. The term "deterministic nucleation" here refers to nucleation phenomena in a macroscopic framework depending on surface roughness, contact angles, gas pockets... (cf. [18]).

As can be seen in Figure 2.7, the inception superheat is lowest for stainless steel/copper, higher for SS alone and highest for glass; the SS/Cu arrangement in its inception behaviour being governed by the copper rod, as discussed for Figure 2.6c. While these differences in the inception superheat can be explained with different deterministic nucleation behaviour alone, the fact, that, in each of the 3 series, the inception superheat decreases with increasing inception time, requires additional explanation:

The superheat $\Delta\vartheta$, defined and discussed in the preceding sections, always referred to the *bulk* liquid state. In reality, the inception of a boiling front or of an evaporation wave in the solid/liquid contact region is governed by the *local* superheat in this meniscus region. Here, the liquid film, adjacent to the solid wall, becomes continuously thinner, and its superheat is governed by transient conduction in the *solid*. Therefore, the thermal properties of the wall material (thermal conductivity, specific heat capacity, density) are supposed to play a dominant role for the inception of evaporation waves and boiling fronts.

It is interesting to note the statement of Grolmes and Fauske [8] in the discussion of their paper: "...Upon depressurization, the tube wall provides a line heat source as the liquid cools by evaporation. In this regard, we would not be surprised to find that thermal properties of the tube wall, as well as diameter, also influence the transition from evaporation to flashing...".

The quasi-steady *propagation* of evaporation waves (i.e. at constant wave pressure) in small diameter glass tubes ($d_i \leq 52$ mm) has been found to be governed by two different modes:

At high depressurization rates, the macroscopic shape of the liquid/two-phase interface is concave, indicating the propagation process to be governed by "far-wall phenomena", depending only on the thermodynamic and transport properties of the *fluid*. This propagation mode has been observed by Hill and Sturtevant [13], Reinke and Yadigaroglu [15] and also in the present study.

At low depressurization rates, the macroscopic shape of the liquid/two-phase interface is convex, indicating the propagation being governed by "near-wall phenomena". Taking into account the observations with metallic inserts, discussed in section 2.3.2, these "near-wall phenomena" can be identified as a *boiling front* on glass.

It is somewhat puzzling to note for glass that, at high depressurization rates, the evaporation wave velocity is higher than the boiling front velocity, whereas at low $\Delta p/\Delta t$ the opposite holds.

The propagation of evaporation waves in the larger test tubes ($d_i = 102$, 152, 252 mm) has been studied only at very low depressurization rates. Here, the macroscopic shape of the wave front is highly irregular, neither concave nor convex, the propagation proceeds sometimes along the glass wall but mostly "fingers" into the superheated liquid. Schlieren patterns in the superheated liquid below the wave front indicate that the two-phase flow cannot totally remove the cooled liquid from the front of the wave.

As can be seen in Figure 2.6 c, the boiling front velocity is higher on copper than on stainless steel and, as can be seen in Figure 2.6 b, the velocity is higher on SS than on the glass wall of the test tube. It might be that the propagation velocity is influenced somewhat by the surface roughness (the metal rods have been finished with identical emery paper), but it appears to us that the boiling front velocity strongly depends on the thermal properties of the *wall* material (especially its thermal conductivity).

It is interesting to note that the boiling fronts on the metallic inserts observed in the present study of adiabatic flashing processes strongly resemble boiling fronts observed in diabatic experiments, dealing with boiling inception at high superheats [19, 20].

2.5 Summary

The results of the present experimental study can be summarized as follows:

- Adiabatic flashing of a highly wetting liquid (Refrigerant 11) in glass containers is apt to result in evaporation waves, generating a homogeneous downstream two-phase flow with vapour as the continuous phase and liquid droplets as the dispersed phase.
- Evaporation waves did occur also in larger diameter systems (at least up to $d = 250$ mm), the superheat for the inception being (nearly) independent on the diameter.
- Evaporation waves can originate also at low depressurization rates (down to $\Delta p/\Delta t \approx 1$ bar/s).

- Metallic inserts, penetrating the liquid level, promote the inception of boiling fronts, resulting in evaporation waves with large liquid/two-phase interface. It is likely that such boiling fronts originate also from the liquid level contact line in metallic containers.
- Therefore, front-like phenomena (evaporation waves and boiling fronts) should be taken into account in modelling of flashing processes, supplementing the assumptions of homogeneous or heterogeneous nucleation as the main source of liquid/vapour interface.

Acknowledgements

This experimental investigation was supported by the Deutsche Forschungsgemeinschaft, in Schwerpunktprogramm "Transiente Vorgänge in mehrphasigen Systemen mit einer oder mehreren Komponenten". The test liquid was supplied by Hoechst Company.

The authors gratefully acknowledge this support.

References

[1] A.A. Kendoush: *The delay time during depressurization of saturated water.* Int. J. Heat Mass Transfer **32** (1989) 2149–2154.

[2] J. Bartak: *A study of the rapid depressurization of hot water and the dynamics of vapour bubble generation in superheated water.* Int. J. Multiphase Flow **16** (1990) 789–798.

[3] P. Deligiannis, J.W. Cleaver: *Determination of the heterogeneous nucleation factor during a transient liquid expansion.* Int. J. Multiphase Flow **18** (1992) 273–278.

[4] E. Elias, P.L. Chambre: *Flashing inception in water during rapid decompression.* J. Heat Transfer **115** (1993) 231–238.

[5] K.M. Akyuzlu: *A parametric study of boiloff initiation and continuous boiling inside a cryogenic tank using a one-dimensional multi-node nonequilibrium model.* HTD-Vol. 262, Phase Change Heat Transfer, ASME, 1993, pp. 9–14.

[6] B. Boesmans, J. Berghmans: *Non-equilibrium vapour generation rate and transient level swell during pressure relief of liquefied gases.* 10th Int. Heat Transfer Conf., Brighton, Vol. **3** (1994) 281–286.

[7] P. Deligiannis, J.W. Cleaver: *Blowdown from a vented partially full vessel.* Int. J. Multiphase Flow **22** (1996) 55–68.

[8] M.A. Grolmes, H.K. Fauske: *Axial propagation of free surface boiling into superheated liquids in vertical tubes.* 5th Int. Heat Transfer Conf., Tokyo, Vol. **IV** (1974) 30–34.

[9] H.J. Viecenz: *Blasenaufstieg und Blasenseparation in Behältern bei Dampfeinleitung und Druckentlastung.* Diss. Univ. Hannover (Germany), 1980.

[10] H. Chaves: *Phasenübergänge und Wellen bei der Entspannung von Fluiden hoher spezifischer Wärme.* Mitt. Max-Planck-Institut f. Strömungsforschung Nr. 77, 1984.

[11] K. Lund: *Druckentlastung von mit Flüssigkeit befüllten Apparaten.* Diss. Univ. Hannover (Germany), 1986.

[12] P.A. Thompson, H. Chaves, G.E.A. Meier, Y.G. Kim, H.D. Speckmann: *Wave splitting in a fluid of large heat capacity.* J. Fluid Mech. **185** (1987) 385–414.

[13] L.G. Hill, B. Sturtevant: *An experimental study of evaporation waves in a superheated liquid.* IUTAM Symp. on Adiabatic Waves in Liquid-Vapor Systems, Göttingen (Germany), 1989, pp. 25–37.

[14] J.E. Shepherd, S. McCahan, J. Cho: *Evaporation wave model for superheated liquids.* IUTAM Symp. on Adiabatic Waves in Liquid-Vapor Systems, Göttingen (Germany), 1989, pp. 3–12.

[15] P. Reinke, G. Yadigaroglu: *Surface boiling of superheated liquids.* Two-Phase Flow Modelling and Experimentation, Rome, Vol. **II** (1995) 1155–1162.

[16] C.J. Leung: *A generalized correlation for one-component homogeneous equilibrium flashing choked flow.* AIChE J. **32** (1986) 1743–1746.

[17] G. Barthau, E. Hahne: *Fragmentation of a vapour bubble growing in a superheated liquid.* Proc. EUROTHERM Seminar No. 48: Pool Boiling 2, Paderborn (Germany), 1996, pp. 105–110.

[18] Md. Alamgir, J.H. Lienhard: *Correlation of pressure undershoot during hot-water depressurization.* J. Heat Transfer **103** (1981) 52–55.

[19] B.P. Avksentyuk, V.V. Ovchinnikov: *A study of evaporation structure at high superheatings.* Russian J. Engineering Thermophysics **3** (1993) 21–39.

[20] J. Fauser, J. Mitrovic: *Heat transfer during propagation of boiling fronts in superheated liquids.* Proc. EUROTHERM Seminar No. 48: Pool Boiling 2, Paderborn (Germany), 1996, pp. 283–290.

3 Flashing of Binary and Ternary Mixtures in Highly Transient Pipe Flow

Franz Mayinger and Johann Wallner *

3.1 Introduction

The discharge of subcooled or saturated liquid from a high pressure condition into a low pressure environment is an important aspect in safety assessment studies related to plants in power and process industries. In flashing flows, a limitation of the discharge rate may occur, and the analysis of these critical flow phenomena becomes difficult due to both thermal and mechanical non-equilibrium effects. Additional to that, the preferential evaporation of the more volatile component has to be considered in the case of multicomponent mixtures, which are often participating in reaction processes of the chemical industries. Up to now, there is only limited information on flashing phenomena in multi-component mixtures because most of the experimental investigations related the peculiar behaviour of mixtures were concerned with the heat transfer in boiling systems. Studies showed that boiling of binary and multicomponent mixtures is markedly different from single-component boiling in that the driving force for the heat transfer is in turn governed by mass transfer because of the different compositions of the gas and liquid phase. Thus, the evaporation rate can be severely retarded in the mixtures because the rate of mass diffusion is usually much lower in comparison to the heat diffusion in the liquid phase. This fact is not considered by most of the basic two-phase models for multicomponent mixtures, which assume thermal equilibrium flow conditions offering no heat and mass transfer resistance [1, 2].

The object of this study is the investigation of the behaviour of hydrocarbons and their mixtures in flashing pipe flow and in particular mechanical and thermal non-equilibrium effects.

* Technische Universität München, Thermodynamik A, Boltzmannstr. 15, D-85748 Garching, Germany

3.2 Flashing and Boiling of Multicomponent Mixtures

The transition from liquid to two-phase mixture initiated by flashing is a continuous process, which occurs in several stages. The superheat of the liquid entering the pipe is increased by the pressure drop caused by friction and acceleration. It is supposed that in technical systems heterogeneous wall nucleation dominates the inception of flashing. The degree of superheat, which is required for the onset of nucleation, depends on the depressurisation rate and the surface conditions of the pipe wall. The nucleation process is considered to result in a bubbly two-phase mixture. With increasing the void fraction, a transition to slug flow or to the churn-turbulent flow regime for higher velocities and finally annular flow occurs.

Studies on flashing flows of water through nozzles or short pipes showed that the length of the nucleation zone is a strong function of the flow velocity and the depressurisation rate [3]. The contribution of bubbles generated downstream is much smaller due to the difference in residence time. Therefore, the development of the void fraction depends on the wall dominated nucleation zone. Because of the lack of basic information on the nucleation phenomena under flashing conditions, the bubble number density as a parameter of major importance was either empirically correlated [4] or considered to be a constant [5, 6]. Riznic and Ishii [7] proposed a vapour generation model in flashing flow based on Kocamustafaogullari's correlation [8] for the nucleation site density in pool and convective boiling. The model considers that in flashing flows the bulk liquid is superheated, while in case of convective boiling, only the boundary layer at the wall is superheated, and the nucleation sites and the growing bubbles are exposed to fluctuations between (T_W-T_{sat}) and 0 due to nucleation, evaporation and liquid convection. The rapid growth of the generated bubbles is assumed to be dominated by diffusion [9], or is described by conduction and convection [5]. The interfacial heat transfer in the bubbly flow regime is often derived from bubble growth laws proposed for pool boiling [10]. Since the mechanism of bubble generation and bubble growth in nucleate and convective boiling was successfully transfered to the inception of flashing, the peculiar behaviour of binary mixtures in boiling, which is considered to result from the different compositions of the vapour and the liquid phase, is briefly discussed:

A study on the heat transfer in nucleate boiling for binary mixtures showed for a constant heat flux besides a reduced heat transfer coefficient a lower nucleation site density in comparison to the pure fluid [11]. Although an increasing superheat is connected with a rapidly increasing nucleation rate, the higher superheat for the binary mixture due to the lower heat transfer coefficient reduces the nucleation site density.

The growth of a vapour bubble in a binary mixture is more complicated in comparison to pure fluids due to the different compositions of the vapour and the liquid phase [12]. When assuming a phase equilibrium, the concentration of the more volatile component in the vapour bubble y is greater than that in the liquid phase x near the bubble interface. For a pure fluid, the growth of a single bubble in a uniformly superheated liquid is controlled by the pressure difference between the bubble and the surrounding liquid and the heat conduction towards the phase interface. Because of the different composition of the vapour and the liquid phase and the preferential evaporation of the more volatile component, the bubble growth in a binary mixture is additionally influenced by the mass transport of the more volatile component through the liquid. In consequence, the concentration of the less volatile component is increasing in the liquid close to the surface of the bubble. This causes a rise of the local boiling temperature, which is higher than the saturation temperature of the initial liquid.

Scriven [13] was the first to develop an analytical model for bubble growth in a binary liquid mixture. His model comprises a spherical bubble in an initially superheated liquid. Looking at the asymptotic stage of bubble growth, he considered heat and mass transfer by one-dimensional radial conduction and convection to the bubble interface. The expression derived for large superheats

$$R(t) = \left(\frac{12}{\pi}\right)^{1/2} \cdot Ja \cdot (at)^{1/2} \cdot \left[1 - (y - x) Le^{1/2} \frac{c_{pf}}{h_v - h_l} \frac{dT}{dx}\right]^{-1} \tag{1}$$

reverts the Plesset and Zwick solution [14] for pure fluids. The additional bracketed term represents the reduced bubble growth of a binary mixture when compared to a pure fluid. The Lewis-Number $Le = \frac{a}{D}$ – the ratio of thermal diffusivity and the mass diffusion coefficient – compares heat conduction and mass diffusion. Since the thermal diffusivity is an order of magnitude higher than the mass diffusion coefficient, the thermal boundary layer is much thicker than the diffusion shell.

3.3 Experimental Installation

The experimental facility (Figure 3.1), in which the blowdown tests were performed, consists of three main components:

- a pressure vessel containing the fluid – propane, butane or a mixture of them – under saturated conditions,
- a depressurisation pipe, through which the flashing fluid flows, equipped with a replaceable orifice of various diameters at its downstream end, and
- a low pressure containment system, in which the depressurised two-phase mixture is captured and the vapour is condensed by a refrigerative cooling equipment.

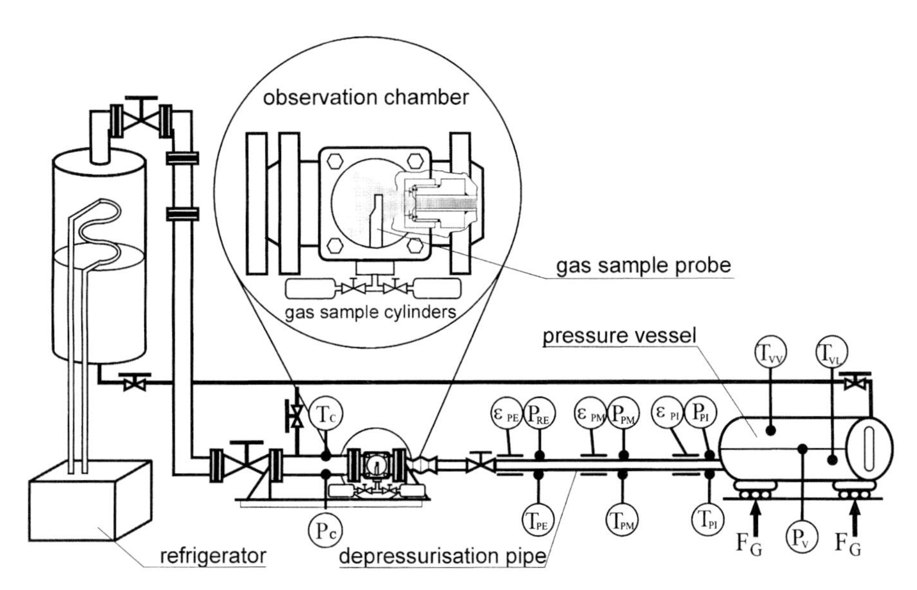

Figure 3.1: Experimental installation.

Three different pipe lengths were used having an L/D-ratio of 70, 170 and 240, while the pipe diameter was kept constant with 10 mm. Three different sharp edged orifices were mounted at the downstream end of the pipe, having diameters of 3, 5 or 8 mm, respectively. In addition to that, experiments with open pipe without orifice were performed.

Before initiating the depressurisation, the high pressure part and the low pressure system of the facility are separated by a burst disc mounted behind the orifice at the end of the pipe. The burst disc is designed in such a way that it can withstand a pressure difference, which corresponds to approximately the half of the initial pressure in the vessel. Therefore, a small room behind the burst disc has to be pressurised to about 5 bar. This is done by pressurising the small volume between the large ball valve and the burst disc up to a pressure of half the vessel pressure. After suddenly opening the large ball valve, the pressure in the observation chamber decreases rapidly and finally, the disc, which cannot withstand the increasing pressure difference, bursts and starts the blowdown.

The fluid- and thermodynamic conditions in the pipe during the depressurisation are the main objects of interest for the investigation. During the blowdown

- pressure
- temperature
- void fraction
- inventory of fluid in vessel and
- composition of the vapour phase (for mixtures)

are observed.

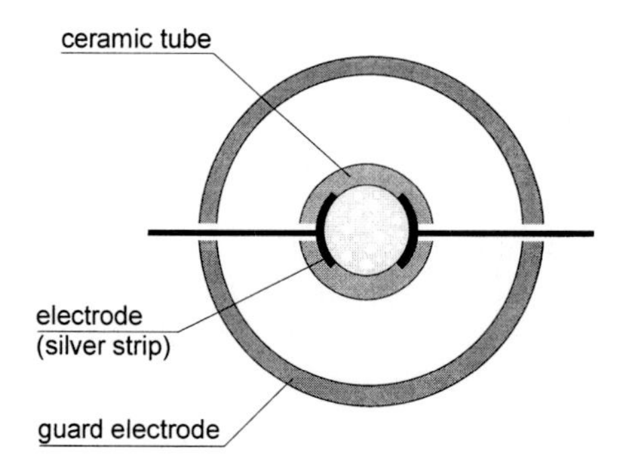

Figure 3.2: Non-intrusive impedance void fraction sensor.

Vessel and pipe are mounted on a moveable carriage so that the momentum of the flashing flow leaving the pipe through the orifice could press the vessel against a pressure gauge acting as a dynamometer. Four other dynamometers supporting the pressure vessel from below continuously indicate the mass inventory of the fluid in vessel. A pressure sensor and up to 6 thermocouples at different levels are installed in the vessel.

The sensor units in the pipe consisting of a pressure gauge, a thermocouple and a void fraction sensor were fitted up to 3× depending on the length of the pipe. Additionally, the back pressure in the observation chamber and the temperature in the containment system are recorded.

The local and temporal value of void fraction is a key parameter of the experiment. Therefore, a special impedance void fraction sensor (Figure 3.2) acting on a capacitive basis and using the different dielectric constants of vapour and liquid was developed. Two silver coatings on the inner side of a ceramic tube form the non-intrusive electrodes of the sensor. The capacitance signal of the impedance sensor was measured by an oscillator-bridge (500 kHz) with a

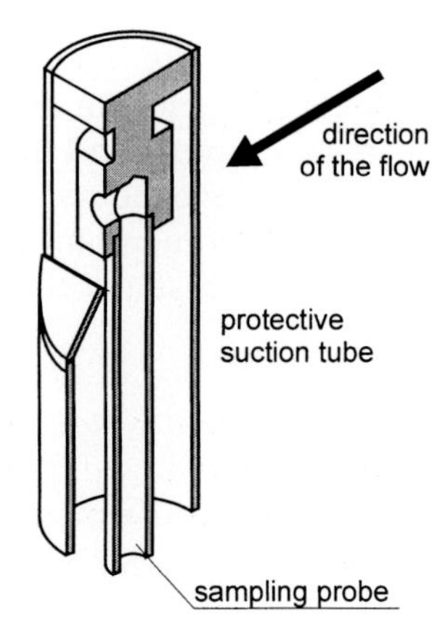

Figure 3.3: Gas sample probe.

resolution of 1.25 fF and a sampling rate of 500 Hz and correlated with the void fraction using the well-known Maxwell equations.

For binary and ternary mixtures, a sampling probe (Figure 3.3) was installed in the observation chamber to get information about the composition of the gas phase. Evacuated sampling cylinders controlled by solenoid valves drew off the gas phase exiting the pipe through the orifice. To avoid a falsification of the sample by liquid, a protective suction pipe drew off the liquid film streaming round the sampling probe. The composition of the samples was determined by a gas chromatograph.

3.4 Experimental Procedure

Experiments are performed to investigate the influence of

- the molar composition of the binary mixture,
- the difference of the boiling point temperatures of the components, and
- a third component in a ternary mixture.

For reference, blowdowns with pure substances were carried out. The test fluids and the compositions of the mixture are shown in Table 3.1:

Table 3.1: Composition of test fluids.

Mixture	Substances	Mole fraction
One-component	Propane	1
	n-butane	1
Two-component	Propane/n-butane	1:3, 1:1, 3:1
	Propane/n-pentane	1:1
	Propane/n-hexane	1:1
Three-component	Propane/n-butane/n-pentane	1:1:1

The temporal course of the experiment is demonstrated for a pure fluid by means of the time history of the pressure and the void fraction (Figure 3.4) since the pressure in the vessel and along the pipe as well as the void fraction are major quantities for the understanding of the blowdown and the evaporation processes. Before initiating the depressurisation, the fluid inside the vessel and the pipe is in saturated conditions.

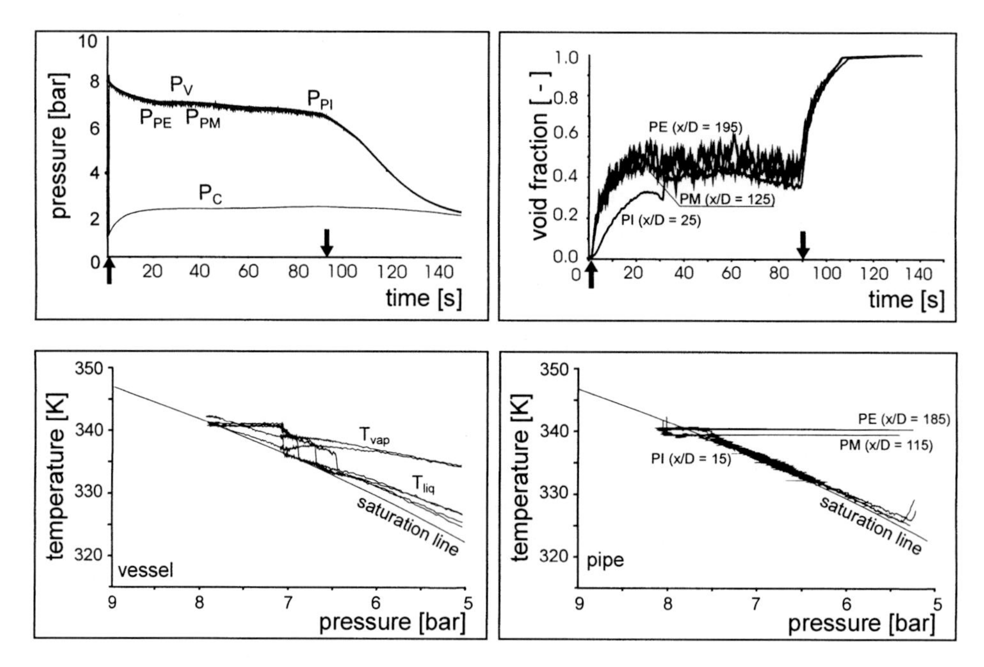

Figure 3.4: Time histories of pressure and void fraction for the blowdown of a pure fluid (butane), thermodynamic state of the fluid inside the vessel and along the pipe ($L/D = 240$, orifice 3 mm).

In the vessel: The depressurisation caused by the discharge of mass and energy is superposed with the evaporation and the increase in volume of the vapour phase in the vessel. With the beginning of the blowdown (upward showing arrow), the pressure in the vessel drops. With an increasing superheat of the liquid, the evaporation process starts to work against the pressure drop. The second period of the blowdown is characterised by an almost constant superheating of the liquid inventory. The evaporation rate in the vessel nearly balances the outflow and the pressure decrease is small. In the last stage of the release, the vessel is almost emptied, vapour enters the pipe, and, from this moment on, the pressure decrease is accelerated again (downward showing arrow).

In the pipe: The superheated liquid enters the pipe and starts flashing in the very first pipe section. The non-equilibrium of the fluid is reduced by the evaporation process and from the first sensor section ($x/D = 25$) up to the pipe end, thermal equilibrium is observed.

Due to the pressure drop in the pipe, the fluid continues to be exposed to flashing on its way downstream to the outlet. A data analysis for propane based on a thermal equilibrium assumption showed that the drift-flux model predicts the development of the void fraction downstream of the first sensor section quite well [15]. If the moderate mechanical non-equilibrium (slip $s = 1.4 \ldots 1.8$) is neglected, the level of void fraction is only dependent on the pressure distribution along the pipe and the superheat of the liquid in the vessel.

3.5 Experimental Results

3.5.1 Influence of the pipe length and diameter of orifice on the depressurisation

The mass flow rate was varied by mounting orifices of different diameters at the pipe end. For the open pipe end, the highest depressurisation rates and non-equilibrium states of the liquid in the vessel were observed due to the high mass flow rates. The large superheat of the liquid entering the pipe, the high depressurisation rate caused by the acceleration of the fluid linked with the high flow velocities favour the onset of flashing at the pipe entrance. The fluid is exposed to flashing along its way through the pipe due to the frictional pressure drop, which limits the mass flow rate.

The enrichment of the less volatile component in the liquid caused by the evaporation process inside the vessel with increasing duration of the blowdown is linked with a rise of the boiling point temperature: In the region of lower pressures, the measured temperature at the pipe entrance ($x/D = 25$) deviates from the line of boiling points which is based on the initial concentration (*T*-*p*-plot in Figure 3.5).

The preferential evaporation of the lighter component along the pipe causes a depletion in the liquid phase and an increase of the boiling point. This is illustrated by the elevated course of the temperature at the pipe end in comparison to the pipe entrance in Figure 3.5 (*T*-*p*-plot, orifice 10 mm).

The throttling of the flow by the orifices results in a lower mass flow rate, a reduced depressurisation velocity and a lower non-equilibrium of the liquid inside the vessel. Since thermal equilibrium is reached because of the flashing in the first pipe section, the level of the void fraction is reduced. The changes of the quality of the two-phase mixture in the consecutive pipe are decreased because the lower flow velocities result in a lower frictional pressure drop.

A further reduction of the outlet cross-section results in a retarded onset of the flashing process in the first pipe section because of the decreased driving forces for the nucleation-liquid superheat, depressurisation rate, and flow velocity. The longer nucleation zone lowers the vapour generation at the pipe entrance and thermal equilibrium is reached downstream of the first sensor section.

The influence of the pipe length on the history of the pressure and the void fraction is of minor importance as demonstrated by Tong [16] for a pure

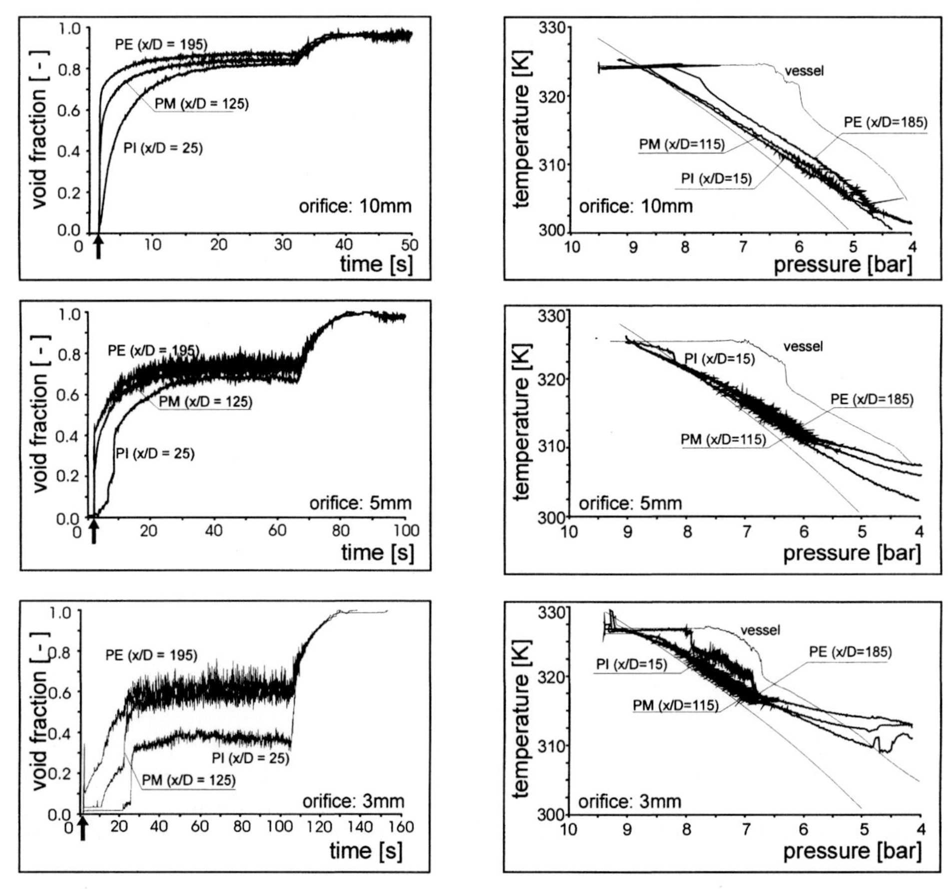

Figure 3.5: Influence of the depressurisation velocity (controlled by diameter of orifice) on the development of void fraction and thermal non-equilibrium for a binary mixture (propane/n-pentane), $L/D=240$.

substance. The pressure difference across the orifices caused by the throttling of the flow predominates the pressure drop along the pipe due to friction and acceleration. A reduction of the mass flow rate with increasing pipe length is only seen for the open pipe end.

3.5.2 Influence of the concentration on thermal non-equilibrium

Three different mixtures (1:3, 1:1, 3:1) of the binary propane/n-butane system were chosen to investigate the influence of the concentration on the flashing pipe flow during the depressurisation. Besides the increase of the superheat of the liquid inside the vessel with higher mass flow rate, the thermal non-equilibrium (Figure 3.6) for the binary mixtures is higher than for the pure fluids due to a reduced nucleation and a slower growth of the rising bubbles. The mass

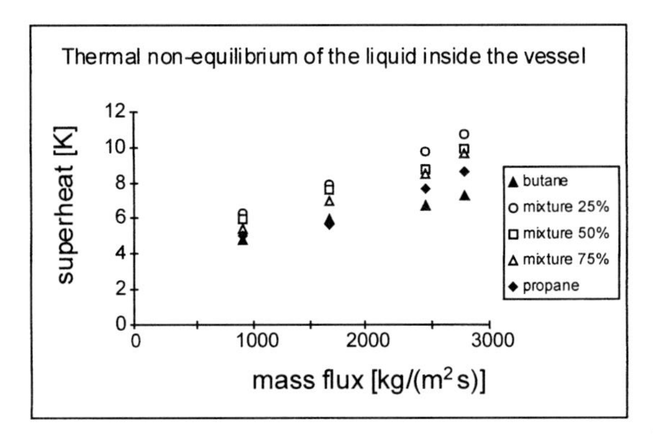

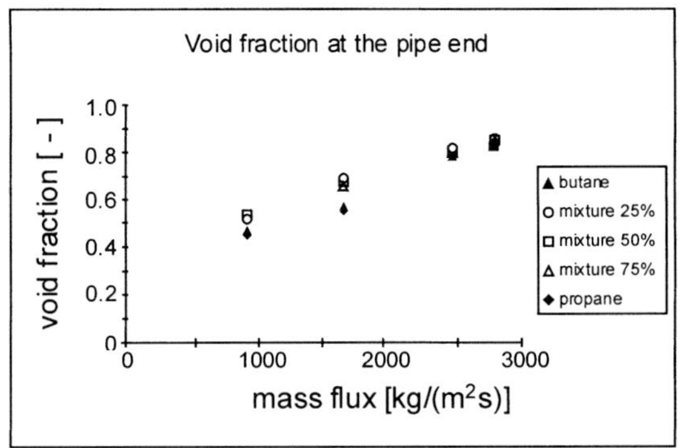

Figure 3.6: Influence of the depressurisation velocity (mass flux at time period when the pressure in the vessel is almost constant) and the concentration of the more volatile component on the thermal non-equilibrium of the liquid inside the vessel and the void fraction at the pipe end for the propane/n-butane system.

transfer of the more volatile component towards the bubble interface is the limiting factor for the bubble growth and in consequence for the evaporation process inside the vessel. With increasing difference of the concentration of the more volatile component in the liquid and the vapour phase, the non-equilibrium in the vessel is rising.

Since the thermal equilibrium in the pipe is reached by flashing, the level of the void fraction is rising with increasing superheat of the liquid inside the vessel.

3.5.3 Influence of the components

A comparison of blowdowns performed with binary mixtures of different composition shows the influence of the boiling range of the mixture components. The difference of the concentration in the vapour and the liquid phase is increasing with the rise of the boiling point temperature of the less volatile component. For binary mixture, the flashing of the liquid caused by a depressurisa-

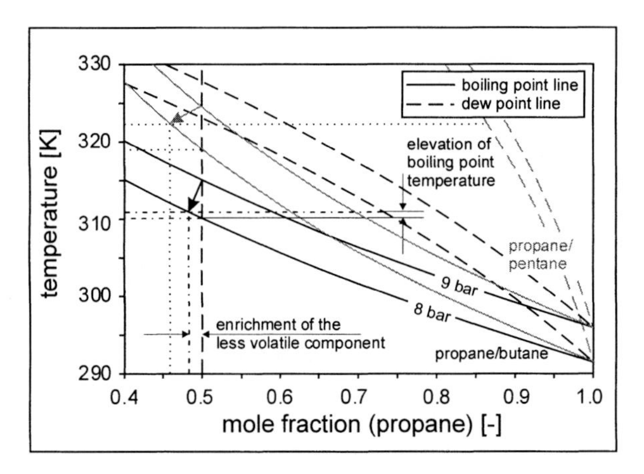

Figure 3.7: Enrichment of the less volatile component in the liquid and elevation of the boiling point temperature for binary mixtures (propane/n-butane, propane/n-pentane) caused by flashing.

tion results in an enrichment of the less volatile component in the liquid due to the preferential evaporation of the lighter component. The increasing enrichment of the heavier component in the liquid phase and the steeper slope of the line of boiling points for the wider-boiling-range mixtures results in a higher rise of the boiling point with respect to the initial concentration (Figure 3.7).

The influence of the boiling range on the separation and the rise of the boiling point caused by the evaporation process inside the vessel is given by the *T*-*p*-plots in Figure 3.8, which demonstrate the deviation of the temperature history from the line of boiling points based on the initial composition.

The temporal course of the void fraction histories (Figure 3.8) shows the retarded vapour generation at the pipe entrance for the lowest mass flow rates. With increasing the boiling point temperature of the less volatile component, the flashing inception point moves downstream and, at the first sensor section, the void fraction is lower than the void fraction that corresponds to the equilibrium quality. According to Stephan and Körner [17], the work of formation of a vapour bubble in a binary mixture is greater than for an equivalent pure fluid and depends on the concentration difference between the vapour and the liquid phase. Therefore, the activation of nucleation sites is impeded with increasing difference of the boiling point temperatures and the onset of flashing is retarded.

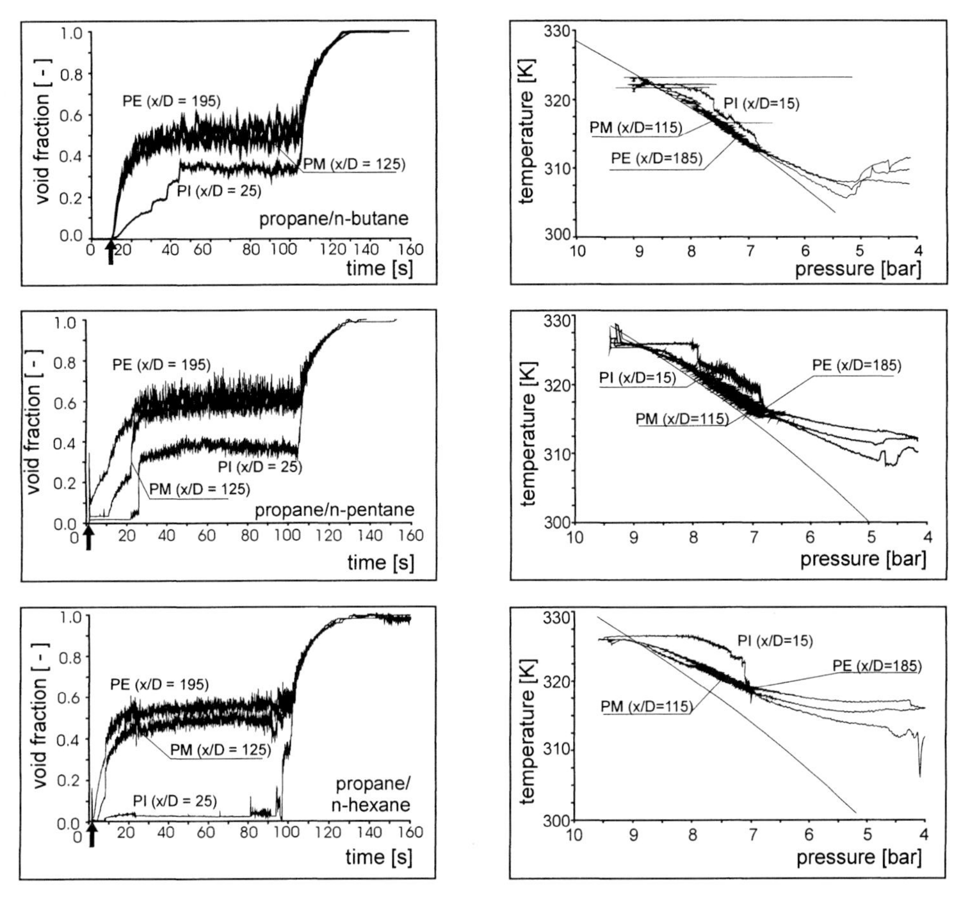

Figure 3.8: Development of the void fraction and thermal non-equilibrium for mixtures with varying boiling range, $L/D=240$, orifice 3 mm.

3.5.4 Effect of a third component

The behaviour of a multi-component mixture during a depressurisation was studied by the means of the ternary propane/n-butane/n-pentane system (molar ratio 1 : 1 : 1). Due to the presence of a third component, which flattens the line of boiling points, the difference of the concentration of the lightest component in the vapour and the liquid phase is smaller than for the binary propane/ n-pentane system. The mass transport of the lightest component is the limiting factor for the evaporation process in the vessel, because the concentration differences of the less volatile components are smaller. For the ternary mixture, lower superheats of the liquid inside the vessel in comparison to the equivalent binary mixture are observed (Figure 3.9).

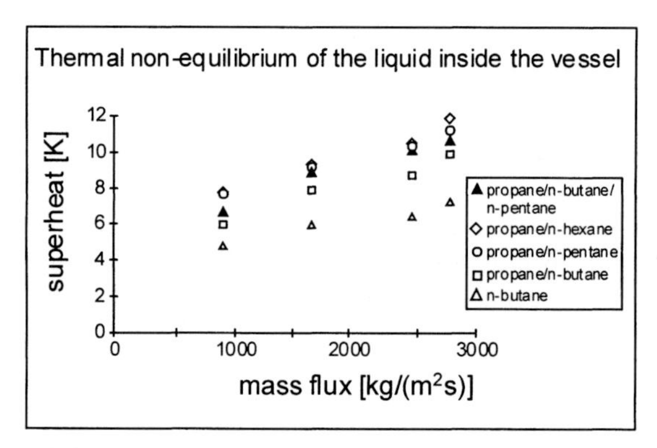

Figure 3.9: Influence of the boiling range and the depressurisation velocity on the thermal non-equilibrium of the liquid inside the vessel.

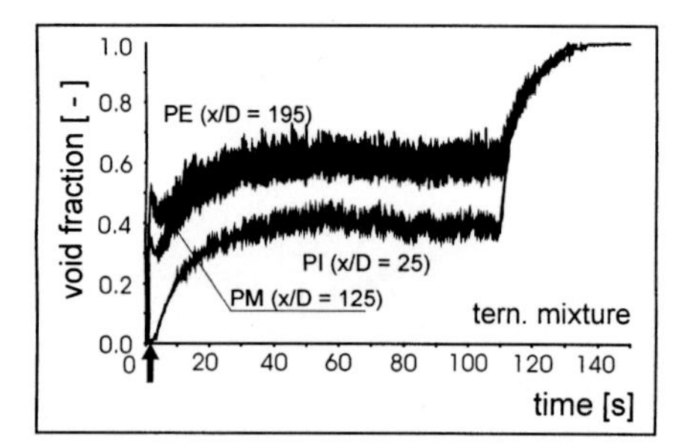

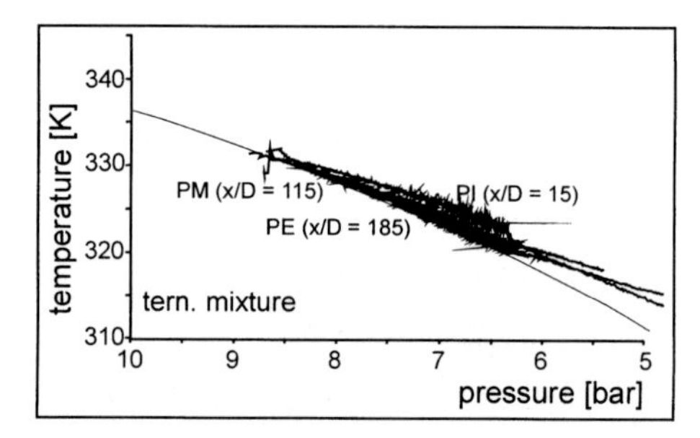

Figure 3.10: Development of the void fraction and the thermal non-equilibrium for a ternary mixture (propane/n-butane/n-pentane), L/D=240, orifice 3 mm.

Similar to the evaporation process in the vessel, the non-equilibrium phenomena at the pipe entrance are reduced. Due to the impeded nucleation in comparison to the pure fluid, the inception of flashing is retarded. In comparison to the binary mixture, the non-equilibrium at the pipe entrance is weakened by the addition of the third component (Figure 3.10).

3.6 Data Analysis

3.6.1 Conservation equations

In order to model the non-equilibrium phenomena in the flashing flow, the effects of delayed bubble nucleation and the diffusive mass transfer mechanism on bubble growth for multi-component mixtures have to be considered. The conservation laws for

- mixture mass

$$\frac{\partial \varrho_m}{\partial t} + \frac{1}{A}\frac{\partial}{\partial z}(\varrho_m w_m A) = 0 \,, \tag{2}$$

- gas mass

$$\frac{\partial}{\partial t}(\varepsilon\varrho_g) + \frac{1}{A}\frac{\partial}{\partial z}(\varepsilon\varrho_g w_m A) + \frac{1}{A}\frac{\partial}{\partial z}\left[\frac{\varepsilon(1-\varepsilon)\varrho_g\varrho_f w_r}{\varrho_m}A\right] = \Gamma \,, \tag{3}$$

- mixture momentum

$$\frac{\partial}{\partial t}(\varrho_m w_m) + \frac{1}{A}\frac{\partial}{\partial z}(\varrho_m w_m^2 A) + \frac{1}{A}\frac{\partial}{\partial z}\left[\frac{\varepsilon(1-\varepsilon)\varrho_g\varrho_f w_r^2}{\varrho_m}A\right] + \frac{\partial p}{\partial z} + \varrho_m g_z + \tau = 0 \,, \tag{4}$$

- mixture energy

$$\frac{\partial}{\partial t}(\varrho_m h_m) + \frac{1}{A}\frac{\partial}{\partial z}(\varrho_m h_m w_m A) + \frac{1}{A}\frac{\partial}{\partial z}\left[\frac{\varepsilon(1-\varepsilon)\varrho_g\varrho_f w_r(h_g - h_f)}{\varrho_m}\right] - \frac{\partial p}{\partial t} + \tau w_m = q \,, \tag{5}$$

- component mass

$$\frac{\partial}{\partial t}(\varrho_m y_{m,i}) + \frac{1}{A}\frac{\partial}{\partial z}(A\varrho_m w_m y_{m,i}) + \frac{1}{A}\frac{\partial}{\partial z}\left[\frac{\varepsilon(1-\varepsilon)\varrho_g\varrho_f(y_{g,i} - y_{f,i})w_r}{\varrho_m}A\right] = 0 \,, \tag{6}$$

- and for bubbly flow the conservation equation for the bubble number density

$$\frac{\partial n_b}{\partial t} + \frac{1}{A}\frac{\partial}{\partial z}(n_b w A) = \phi \tag{7}$$

are solved numerically.

3.6.2 Constitutive equations

For the closure of the system of equations, additional relationships are needed. These are constitutive equations to describe the pressure drop, the interfacial heat transfer, the interfacial area, the evaporation rate and the relative velocity between the phases.

3.6.2.1 Pressure drop

In two-phase flows, the drag force, which results from the viscous boundary at the wall, is usually expressed by the wall friction of the liquid multiplied with a two-phase multiplier.

Some of the available correlations for the frictional pressure drop were tested against the experimental data [15, 16]. Chisholm's formulation [18] of the two-phase multiplier, which gave the best agreement with the experiment, was selected for the calculation:

$$\frac{\Delta p_{2ph}}{\Delta p_{f0}} = 1 + (\Gamma_{fr}^2 - 1) \cdot \{B\dot{x}^{(2-n)/2}(1 - \dot{x})^{(2-n)/2} + \dot{x}^{2-n}\}\,. \tag{8}$$

3.6.2.2 Relative velocity correlation

A relation for the relative velocity accounts for the momentum transfer between the liquid and the vapour phase. In one-dimensional two-phase flow, the drift-flux formulation gives an integral information about phase and velocity distribution. The relationship for the relative velocity is given as:

$$w_r = \frac{(C_0 - 1) \cdot w_m + V_{gj}}{\left[1 - C_0\left(1 - \frac{\varrho_m}{\varrho_f}\right)\right] \frac{(1-\varepsilon)\varrho_f}{\varrho_m}}\,. \tag{9}$$

Dependent on the flow regime, the distribution parameter and the expression for the drift velocity of the vapour phase are calculated according to Nabizadeh [19] and Kolev [10].

For the assumption of mechanical equilibrium ($w_r = 0$), which was additionally tested, all terms containing the relative velocity in the set of conservation equations vanish.

3.6.2.3 Evaporation rate

To describe the rate of phase change due to nucleation and interfacial heat transfer [20]

$$\Gamma = \frac{4}{3} r_b^2 \pi \phi + \frac{h_i \cdot a_i (T_l - T_g)}{h_g - h_l}\,, \tag{10}$$

the nucleation site density and bubble departure frequency, the interfacial heat transfer and the interfacial area density – the latter depending on the flow regime – have to be taken into account.

3.6.2.4 Nucleation site density and departure frequency

Kocamustafaogullari and Ishii's model [8] for the description of the nucleation site density in subcooled boiling was modified and successfully applied to flashing pipe flow by Riznic and Ishii [7]. The wall nucleation site density is correlated in dimensionless form as:

$$N_{ns}^* = R_c^{*-4.4} \cdot f(\varrho^*)\,. \tag{11}$$

3.6.2.5 Interfacial area density and interfacial heat transfer

The dependency of the interfacial density area and the heat transfer coefficient on the flow regime was taken into account by using different formulations:
Bubbly flow ($\varepsilon \leq 0.3$):
Assuming spherical bubbles, the interfacial area density is given by:

$$a_i = \left(36\varepsilon^2 \cdot n_b \cdot \pi\right)^{1/3}\,. \tag{12}$$

The Nusselt number is derived from Scriven's bubble growth law for binary mixtures:

$$Nu = \frac{12}{\pi} \cdot Ja \cdot \left[1 - (y - x) \cdot Le^{1/2} \cdot \frac{c_p}{h_g - h_f} \frac{dT}{dx}\right]^{-1}\,. \tag{13}$$

Bubbly-slug flow ($0.3 = \varepsilon_{B_{max}} < \varepsilon < \varepsilon_{S_{max}} = 0.8$):
According to Blinkov et al. [21], it is assumed that some of the bubbles coalesce to larger Taylor bubbles, while others continue to grow. The vapour generation takes place on the surface of both kinds of bubbles and the total interfacial heat flux is given by:

$$\dot{q}_i = \dot{q}_S a_S + \dot{q}_B a_B\,. \tag{14}$$

The area density of the Taylor bubbles

$$a_S = \frac{4\varepsilon_S^{2/3}}{\varepsilon_{S_{max}}^{1/6} D} \tag{15}$$

is expressed in terms of the void fraction of the larger bubbles

$$\varepsilon_S = \frac{1}{1 - \varepsilon_{B_{max}}} \left\{\varepsilon - \varepsilon_{B_{max}} \left[1 - \frac{(\varepsilon - \varepsilon_{B_{max}})(1 - \varepsilon_{S_{max}})}{\varepsilon_{S_{max}} - \varepsilon_{B_{max}}}\right]\right\}, \tag{16}$$

and the interfacial heat flux is given by:

$$\dot{q}_S = 0.0073 \cdot \varrho_l \cdot c_{pf} \cdot (T_f - T_g) \; . \tag{17}$$

The contribution of the smaller bubbles is determined by Equations (12) and (13).

3.6.3 Comparison with the experiment

For the calculation of the thermophysical properties and namely the vapour-liquid equilibria, the NIST database REFPROP was embedded in the computer code. The numerical integration of the set of equations gives the development of the void fraction and the composition of the mixture. The boundary conditions are given by the conditions inside the vessel (pressure and superheat of the liquid), and the pressure in the observation chamber for the open pipe, or the mass flow rate for the experiments performed with an orifice mounted at the downstream end of the pipe.

Due to the vapour-liquid equilibrium assumption, the concentration of the more volatile component in the liquid is decreasing as the mixture is exposed to flashing along the depressurisation pipe. Since thermodynamic equilibrium is reached in the pipe, the enrichment of the less volatile component in the liquid phase is linked with the development of the void fraction, which depends on the boundary conditions (pressure and liquid superheat in the vessel), the pressure drop along the pipe and the velocity ratio of the vapour and the liquid phase. The computed void fractions are close to the experimental results of the blowdown with an open pipe end (Figure 3.11) and the tests with orifices (Figure 3.12).

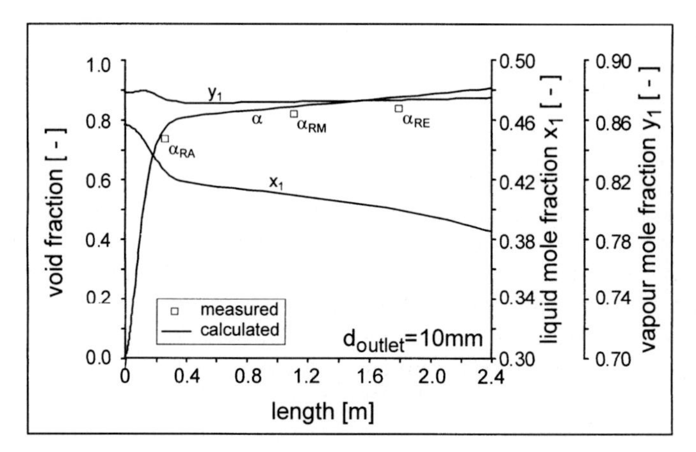

Figure 3.11: Development of void fraction and concentration along the pipe (no orifice at pipe end), binary mixture (propane/n-pentane).

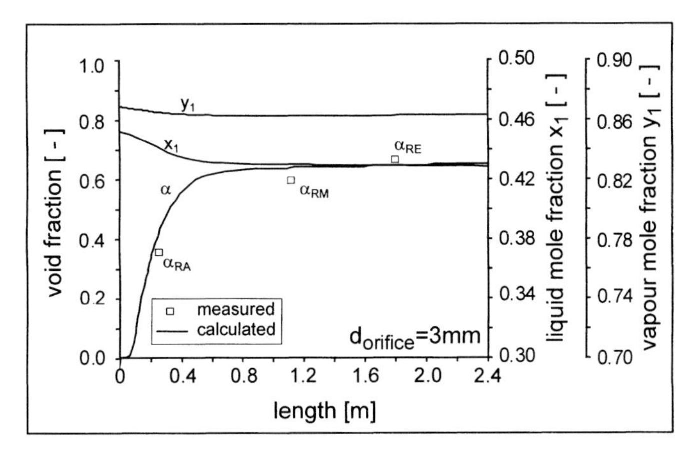

Figure 3.12: Development of void fraction and concentration along the pipe with reduced cross-section of the outlet (orifice 3 mm), binary mixture (propane/n-pentane).

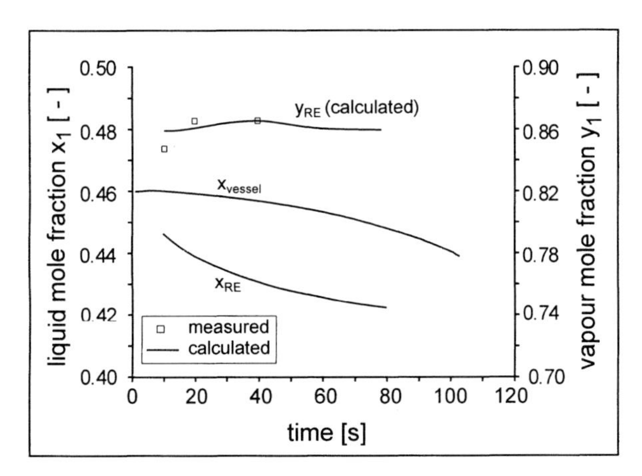

Figure 3.13: Temporal development of the concentration inside the vessel and at the pipe end, binary mixture (propane/n-pentane), $L/D=240$, orifice 3 mm.

Figure 3.13 shows the temporal development of the composition of the liquid and the vapour phase at the pipe end. The enrichment of the less volatile component at the pipe end in comparison to the liquid composition inside the vessel results from the preferential evaporation of the lighter component. The corresponding concentrations of the vapour phase are in accordance to the composition of gas samples, which were analysed in a gas chromatograph and prove the vapour-liquid equilibrium assumption.

3.7 Summary and Conclusions

Experiments on the transient pipe flow of binary, ternary mixtures and for reference of pure substances under flashing conditions were performed. At the pipe entrance, the flow is characterised by thermal non-equilibrium, which is reduced in the consecutive pipe sections by the flashing process. For binary mixtures, the thermal non-equilibrium is increased both inside the vessel participating in the depressurisation and in the first pipe section. The non-equilibrium depends on the molar composition and on the components of the mixture. The highest superheats were observed in the region of the maximum difference in the composition of the vapour and the liquid phase. With larger difference of the boiling point temperatures of the pure components, the thermal non-equilibrium effects are increasing. Non-equilibrium phenomena are weakened by the addition of a third component due to the flattened bubble point curve. The higher non-equilibrium effects and the retarded inception of flashing are considered to result from a reduction of the nucleation site density and the degradation of the interfacial heat transfer in binary mixtures. In applications (e.g. short pipes or nozzles), where the transit time through the duct is not sufficient to reach thermal equilibrium, these effects are expected to gain in importance.

References

[1] J.R. Chen, S.M. Richardson, G. Saville: *A Simplified Numerical Method for Transient Two-Phase Pipe Flow.* Chemical Engineering Research & Desing **70**(A3) (1993) 304–306.

[2] J.R. Chen, S.M. Richardson, G. Saville: *Modelling of Two-Phase Blowdown From Pipelines – I. A Hyperbolic Model Based on Variational Principle.* Chem. Engineering Science **40**(4) (1995) 695–713.

[3] E. Elias, G.S. Lellouche: *Two-Phase Critical Flow.* Int. J. Multiphase Flow **20** (Suppl.) (1994) 91–168.

[4] R. Dagan, E. Elias, E. Wacholder, S. Olek: *A Two-Fluid Model for Critical Flashing Flows in Pipes.* Int. J. Multiphase Flow **19**(1) (1993) 15–25.

[5] H.J. Richter: *Separated Two-Phase Flow Model: Application to Critical Two-Phase Flow.* EPRI-NP-1800, Palo Alto, Ca., USA, 1981.

[6] F. Dobran: *A Nonequilibrium Model for Two-Phase Critical Flow Analysis in Variable Area Ducts.* Techn. Report ME-RT-84005, Stevens Institute of Technology, Hoboken, NJ, USA, 1984.

[7] J.R. Riznic, M. Ishii: *Bubble Number Density and Vapor Generation in Flashing Flow.* Int. J. Heat Mass Transfer **32**(10) (1989) 1821–1833.

[8] G. Kocamustafaogullari, I.Y. Chen, M. Ishii: *Correlation for Nucleation Site Density and its Effect on Interfacial Area.* NUREG/CR-2778, ANL-82-32, Argonne National Laboratory, IL, USA, 1982.

[9] K.H. Ardron, R.A. Furness, P.C. Hall: *Experimental and Theoretical Studies of Transient Boiling and Two-Phase Flow During the Depressurisation of a Simple Glass Vessel.* CSNI Specialist Meeting, Toronto, 16–19 July, 1970, vol. 2, 1970, pp. 610–636.

[10] N.I. Kolev: *Transiente Zweiphasen-Strömung.* Springer Verlag, Berlin, 1986.

[11] W.H. Bednar, K. Bier: *Wärmeübergang beim Blasensieden von binären Kohlenwasserstoffgemischen.* Fortschr.-Ber. VDI Reihe 3, Nr. 357, Düsseldorf, VDI Verlag, 1994.

[12] J.R. Thome, A.W. Shock: *Boiling of Multicomponent Liquid Mixtures.* Advances in Heat Transfer **16** (1984) 5–156.

[13] L.E. Scriven: *On the Dynamics of Phase Growth.* Chemical Engineering Science **10**(1/2) (1959) 1–13.

[14] M.S. Plesset, S.A. Zwick: *The Growth of Vapour Bubbles in Superheated Liquids.* Journal of Applied Physics **25**(4) (1954) 493–500.

[15] F. Mayinger: *Zwischenbericht des DFG-Vorhabens MA 501/32,* München, 1996.

[16] K. Tong: *Dampfgehalt und Druckverlust in Rohren bei Druckentlastung.* Dissertation Technische Universität München, 1996.

[17] K. Stephan, M. Körner: *Berechnung des Wärmeübergangs verdampfender binärer Flüssigkeitsgemische.* Chemie-Ingenieur-Technik **41** (1969) 409–417.

[18] D. Chisholm: *Pressure Gradient due to Friction During the Flow of Evaporating Two-Phase Mixtures in Smooth Tubes and Channels.* Int. J. Heat and Mass Transfer **16**(2) (1973) 347–358.

[19] H. Nabizadeh: *Modellgesetze und Parameteruntersuchungen für den volumetrischen Dampfgehalt in einer Zweiphasenströmung.* Dissertation Universität Hannover, 1977.

[20] N.I. Kolev: *The code IV4: nucleation and flashing model.* Kerntechnik **60**(4) (1995) 157–165.

[21] V.N. Blinkow, O.C. Jones, B.I. Nigmatulin: *Nucleation and Flashing in Nozzles – 2, Comparison with Experiments Using a Five-Equation Model for Vapor Void Development.* Int. J. Multiphase Flow **19**(6) (1993) 968–986.

4 Transient Two-Phase Flow in Thin Pipes Caused by Degassing during Fast Pressure Drops

Sebastian Fleischer and Rainer Hampel *

4.1 Introduction

Pressure and differential pressure sensors (hydrostatic level measurement) are important parts of the standard instrumentation of pressure vessels. They are generally connected through thin measuring pipes with technical facilities. The medium in the measuring pipe must be in single-phase liquid state under high pressure and temperature conditions. It is a general requirement for correct measurements. During fast pressure drops, processes with boiling point underflow or degassing of noncondensable gases can occur. In this case, a two-phase flow will appear in the measuring pipe.

As a result of the transient two-phase flow, large faulty measurements and mal functions in the level measurement can appear and bring about safety related consequences to the process and the measuring system.

The dynamic behaviour of the measuring systems must be included in dynamic process models of pressure tanks. Then, it is possible to describe the correlation between the actual process parameters and the indicated measuring parameters.

4.2 Aims and Objectives

The aim of the theoretical and experimental investigations is the model development for the mathematical description of the gas dissolutions and the transient two-phase flow in thin pipes during fast pressure drops. Selected results of the theoretical and experimental investigations are presented in Section 4.5.

---

* Hochschule Zittau/Görlitz (FH), Institut für Prozeßtechnik, Prozeßautomatisierung und Meßtechnik, Theodor-Körner-Allee 16, D-02763 Zittau, Germany

The pressure decreasing and the transient two-phase flow in vertical and sloped pipes depend on the following variable parameters:

- pressure, temperature, initial gas concentration, void fraction, properties of the system water and the solved gas;
- surface condition, diameters and isometric parameters of the pipe.

These influence quantities were varied in experiments for single effect and complete process analysis. The results build up the data base for the verification of the theoretical models.

4.3 Modelling the Transient Two-Phase Flow

The model of degassing and transient two-phase flow in the pipe contain the following basic models:

- model of the mass and impuls conservation for the calculation of the bubble velocity,
- model of the bubble formation through degassing,
- bubble growth model (combining of bubbles, change of gas density and diffusion).

The following simplifications in the modelling were assumed:
- one-dimensional flow (thin pipes),
- adiabatic system,
- constant density of the liquid phase.

The classical Euler model with spatially constant control volumes is unsuitable for the description of the individual gas bubbles (fixed number and length of the nodes). The nodalization of the flow channel with the Lagrange model is variable (number and length of the nodes). For this reason, a combination of Lagrange and Euler models was chosen. The nodalization of the flow channel and the allocation of a gas bubble to a node will be calculated after each time step (Figure 4.1).

Description of model equations:

- mass conservation [1, 2]
  for the liquid phase control volume

$$\frac{\partial L_L}{\partial \tau} + \frac{\partial}{\partial z}[u_L \cdot L_L] = 0\,, \tag{1}$$

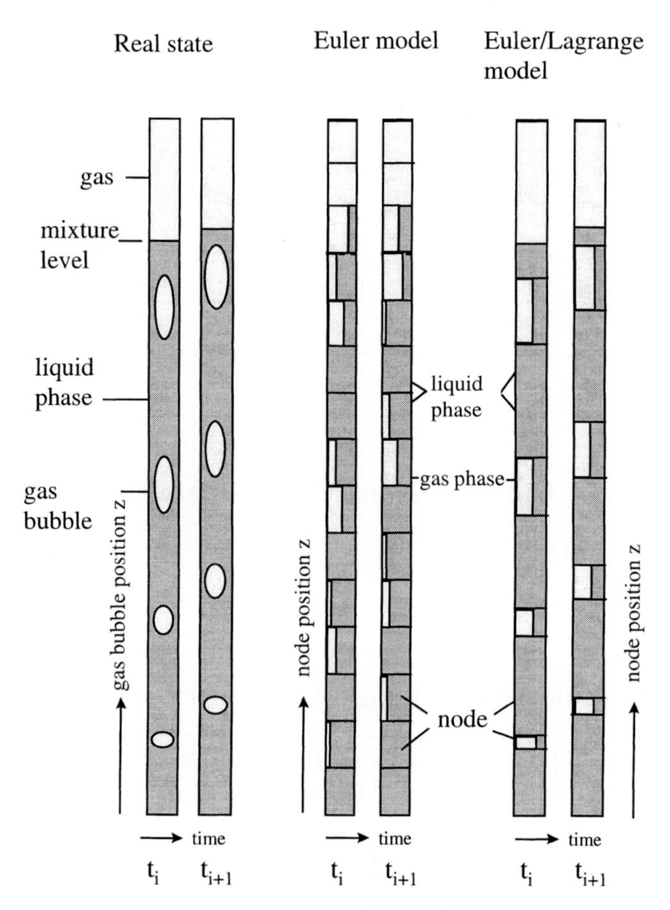

Figure 4.1: The nodalization of the flow channel according to Euler and Lagrange.

for the two-phase control volume

$$\frac{\partial}{\partial\tau}[\varepsilon\cdot\rho_G\cdot L_G]+\left\{\frac{\partial}{\partial z}[\varepsilon\cdot\rho_G\cdot u_G\cdot L_G]\right\}+\frac{\Gamma_{LG}}{A_c}=0\,, \tag{2}$$

$$\frac{\partial}{\partial\tau}[(1-\varepsilon)\cdot\rho_L\cdot L_G]+\left\{\frac{\partial}{\partial z}[(1-\varepsilon)\cdot\rho_L\cdot u_L\cdot L_G]\right\}-\frac{\Gamma_{LG}}{A_c}=0\,, \tag{3}$$

- impulse conservation [1, 2]
  for the liquid phase control volume

$$\frac{\partial}{\partial\tau}[u_L\cdot L_L]+\frac{\partial}{\partial z}[u_L\cdot u_L\cdot L_L]+\frac{L_L}{\rho_L}\frac{\partial p_L}{\partial z}+L_L\cdot g\cdot\sin a=0\,, \tag{4}$$

pressure gradient

$$\left(\frac{\partial p_L}{\partial z}\right) = \frac{32 \upsilon_L \cdot \rho_L \cdot u_L}{D_c^2}\,, \tag{5}$$

for the two-phase control volume

$$\frac{\partial}{\partial \tau}[\varepsilon \cdot \rho_G \cdot u_G \cdot L_G] + \left\{\frac{\partial}{\partial z}[\varepsilon \cdot \rho_G \cdot u_G \cdot u_G \cdot L_G]\right\} + \varepsilon \cdot L_G \cdot \frac{\partial p_{(G)}}{\partial z}$$

$$+\varepsilon \cdot \rho_G \cdot L_G \cdot g \cdot \sin\alpha + \frac{F_R^i}{A_C} + \frac{\Gamma_{LG}}{A_C}(u_L - u_G) + \frac{F_{Drag}}{A_C} = 0\,, \tag{6}$$

calculation of the pressure gradients [3]

$$\frac{\partial p_{(G)}}{\partial z} = \left(\frac{\partial p_L}{\partial z}\right) - g \cdot \rho_L \cdot \sin\alpha\,, \tag{7}$$

$$\left(\frac{\partial p_L}{\partial z}\right) = \left(\frac{\partial p_L}{\partial z}\right)_{Cap} + \left(\frac{\partial p_L}{\partial z}\right)_{Acc,1} + \left(\frac{\partial p_{L,F}}{\partial z}\right) + \left(\frac{\partial p_L}{\partial z}\right)_{Acc,2}\,, \tag{8}$$

pressure gradient for bubble flow

$$\left(\frac{\partial p_{L,F}}{\partial z}\right) = 2 \cdot f_{F,LF} \cdot \frac{\dot{m}^2}{\overline{D} \cdot \rho_L} = 2 \cdot f_{F,LF} \cdot \frac{u_L^2 \cdot \rho_L}{\overline{D}}\,, \tag{9}$$

$$f_{F,LF} = 0{,}079 \cdot Re_{LF}^{-0{,}25}\,, \tag{10}$$

pressure gradient for slug flow (Taylor bubble) [4]

$$\left(\frac{\partial p_{L,F}}{\partial z}\right) = \frac{(\tau^i - \tau^w) \cdot P_C}{A_c \cdot L_{LF}}\,, \tag{11}$$

- calculation of the frictional force

$$F_R^i = c_{w,Cap}\frac{\pi}{8}D_B^2 \cdot c^2 \cdot \rho_L + \tau^i \cdot A_{LF} + (K_{Acc,1} - K_{Acc,2})\frac{\rho_L}{2}A_c \cdot c \cdot |c|\,, \tag{12}$$

- additional bubble acceleration according to Taitel et al [5]

$$F_{Drag} = K_{Drag,1} \cdot \rho_L \cdot u_L^2 \cdot (1 - \tanh(K_{Drag,2} \cdot L_L))\,, \tag{13}$$

- calculation of the medium flux (diffusion) in the bubble for $\delta \ll R$

$$\Gamma_{LG} = \rho_L \cdot \pi \cdot D_c \cdot \int_{y=0}^{\delta} [u_L(y) \cdot (\xi_{In} - \xi_{Out}(y))dy\,, \tag{14}$$

- bubble diameter for homogenous bubble formation

$$r_B := \frac{3\sigma}{p_L} \cdot \left(\xi_G \frac{1}{M_G\left(\frac{\xi_G}{M_G} - \frac{(1-\xi_G)}{M_L}\right)} - \frac{p_L}{K_H} \right)^{-1}, \tag{15}$$

$$n_B = \varepsilon_0 \frac{p_{0,L}}{p_L} \cdot \frac{T_0}{T_{0,L}} \cdot \frac{3}{4\pi} [r_B]^{-3}\,. \tag{16}$$

4.4 Test Facility

In order to create a data base for the verification of the model, a test facility for the pressure decreasing experiments was constructed (Figure 4.2). The facility consists of two components:

- facility for setting the solution concentration, and
- experimental pipe with variable isometry.

The test medium is air-saturated water with 70 bar maximal solution pressure. The facility instrumentation for setting the gas solution concentration consists of a pressure, level and an oxygen measuring system. The experimental pipe is equipped with 12 needle-shaped conductivity probes and 3 capacitivity probes as well as pressure and differential pressure measuring. The probes are installed in the vertical and sloped pipes along the experimental piping. The gas bubbles are detected in the transient two-phase flow with the help of the conductivity and capacitivity probes. The gas bubble size (length and diameter), the gas bubble velocity and the mixture level are measured. In addition, Figure 4.2 shows selected signal courses for a pressure drop experiment.

As shown in Figure 4.3, the bubble velocity and length resulting from the probe signals are calculated with the help of Equations (17) and (18):

$$v_B = \frac{L_S}{t_L}\,, \tag{17}$$

$$l_B = v_B \cdot t_K\,. \tag{18}$$

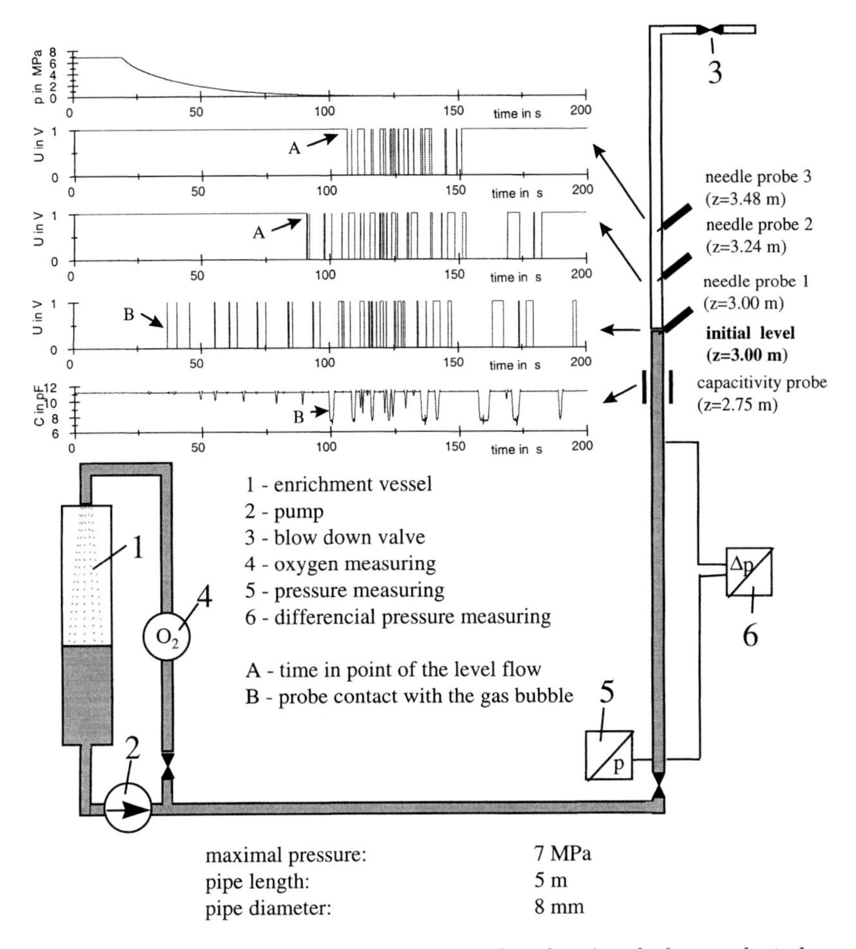

Figure 4.2: Test facility and the temporal course of probe signals for a selected pressure drop experiment.

4.5 Results

A mathematical model was developed on the basis of the equations described in Section 4.3. The model is able to calculate the transient two-phase flow in the vertical and sloped flow channels including degassing. Selected results of the experiments and simulations will be presented in the following sections. The experimental data were analysed by statistical methods. They are used for the model verification.

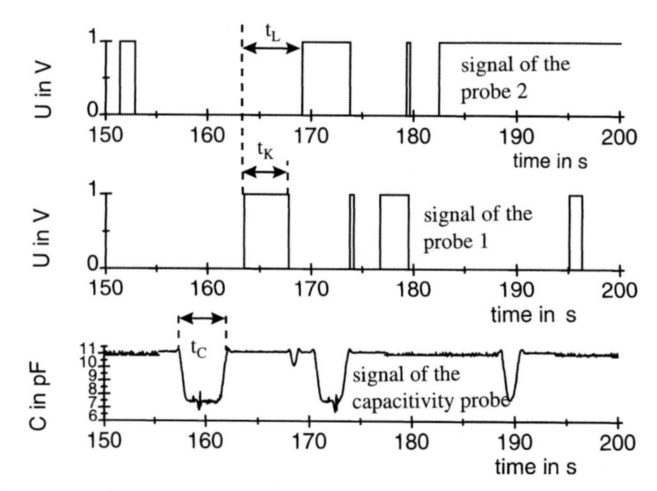

Figure 4.3: Signal course of the conductivity and capacitivity probe gas bubble for the gas bubble parameter calculation.

4.5.1 Bubble ascent velocity

The verification of the flow dynamic part was carried out through experiments under atmospheric pressure. Gas bubbles were blown in with a nozzle/valve into the experimental pipe. The pipe material ($D = 0.008$ m) is transparent. The analysis of experiments was carried out with the help of an optical measurement system (video camera). The comparison of experimental and simulated bubble velocities v_B for different bubble lengths l_B is demonstrated in Table 4.1.

The statistical deviation of experimentally determined and calculated bubble velocities is c. 10%. The bubble ascent velocities are computed accurately by the model.

Table 4.1: Measured and calculated bubble ascent velocities.

Pipe slope 15°			Pipe slope 30°			Pipe slope 45°			Pipe slope 90°		
l_B in m $\cdot 10^{-3}$	Exp. v_B in m/s $\cdot 10^{-2}$	Model v_B in m/s $\cdot 10^{-2}$	l_B in m $\cdot 10^{-3}$	Exp. v_B in m/s $\cdot 10^{-2}$	Model v_B in m/s $\cdot 10^{-2}$	l_B in m $\cdot 10^{-3}$	Exp. v_B in m/s $\cdot 10^{-2}$	Model v_B in m/s $\cdot 10^{-2}$	l_B in m $\cdot 10^{-3}$	Exp. v_B in m/s $\cdot 10^{-2}$	Model v_B in m/s $\cdot 10^{-2}$
6	5.5	4.1	5.1	8.3	7.6	4.7	10.0	11.0	5.2	7.9	8.8
15.1	1.9	1.7	11.1	5.3	4.2	12.7	7.4	6.4	15.7	4.9	5.8

4.5.2 Degassing

The verification of the degassing was much more difficult. The temporal dissolved gas mass is not directly measurable. Indicators for the dynamic of degassing are the temporal course of the mixing level, the moment for degassing (differential pressure signal), and the moment for first detection of bubbles on the probes. The dynamic of degassing is influenced by the flow. The integrated equations for the degassing in the model represent the equations known from the literature [2, 6, 7].

For the dynamic of degassing, the following cases appearing in combinations are to be considered:

1. homogeneous degassing,
2. heterogeneous degassing,
3. degassing only by diffusion.

In cases 1 and 2, additional micro bubbles in the oversaturated medium can occur.

In the model, the degassing through gas formation of bubbles is presupposed and pure diffusion is excluded. In addition, in practice and in the experiments, finely distributed micro bubbles are located in the saturated water. In the course of the pressure drop, these micro bubbles have a great influence on the dynamic of the degassing. Presently, the modelling of the micro bubbles in the model is not yet solved satisfactorily. To limit the number of nodes (c. 500), these bubbles are modelled with an initial diameter of c. 1 mm. Therefore, the bubbles reach a higher velocity than the real micro bubble velocities do. The bubbles are removed quicklier from the system. As a result, gas mass output is higher so that the mixture level is too small.

The dissolution pressure can be detected using following effects:

1. first bubble formation (not measurable) $\rightarrow p_{Diss.\_1}$,
2. first detection of bubbles $\rightarrow p_{Diss.\_2}$,
3. large pressure impuls oscillations in the bottom pressure $\rightarrow p_{Diss.\_3}$.

A characteristic ratio for beginning of degassing was defined as follows:

$$K_{Diss.} = \frac{p_{Sol.}}{p_{Diss.}} \, . \tag{19}$$

For all experiments accomplished, we obtained $K_{Diss.\_3} = 10$. This ratio corresponds to the supersaturation grade for homogeneous bubble formation.

As a result of registration first detectable bubble the dissolution pressure $p_{Diss.\_2}$ is higher than $p_{Diss.\_3}$. There are two causes for this:

1. additional micro bubbles build larger bubbles,
2. heterogeneous bubble formation at exposed points in the pipe.

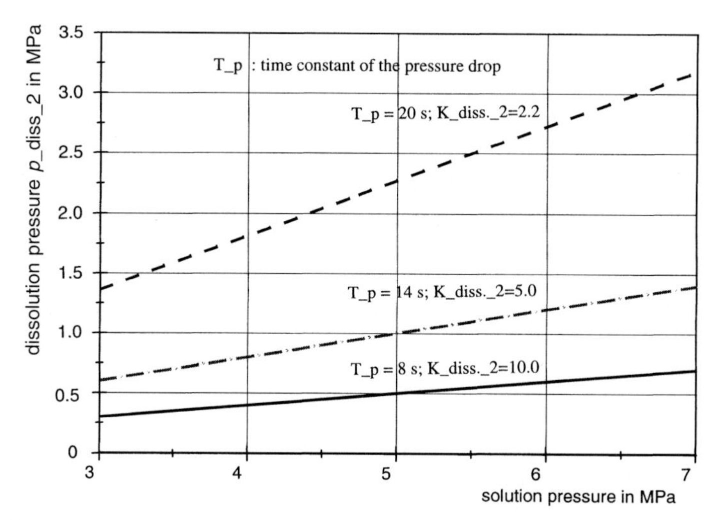

Figure 4.4: Dissolution pressure $p_{\mathrm{Diss.\_2}}$ in dependence on the solution pressure and pressure gradient.

The experimentally determined dissolution pressures for the system water/air are presented in Figure 4.4. According to the set pressure gradients, different dissolution pressures arose. The larger the pressure gradient the more the dissolution pressure $p_{\mathrm{Diss.\_2}}$ shifts in direction of $p_{\mathrm{Diss.\_3}}$.

4.5.3 Mixture level

The temporal courses of the mixture levels are given in Figure 4.5. The experimental conditions are described in Tables 4.2 and 4.3.

The gas bubble velocity decreases if the slope angle is reduced. The mixed level increases correspondingly. The continuity of mixture level was determined in fixed positions using needle-shaped conductivity probes. The dynamic behaviour of the bubble-up is nearly independent of the slope angle of the flow channel. The dynamic behaviour of the decrease depends on the slope angle conditioned by different bubble ascent velocities.

4.5.4 Results of selected simulations/calculations

The computational results of the experiments 1, 2 and 3 for different mathematical models of degassing process are given in Figure 4.6 a.

The level course with characteristic ratio $K_{\mathrm{Diss}}=2.2$ is represented in Figure 4.6 b. The calculated level did not reach the level determined experimentally because micro bubbles were not considered. The same calculation with $K_{\mathrm{Diss}}=10$ is represented in Figure 4.6 c. The mixture level (Figure 4.6 c) is

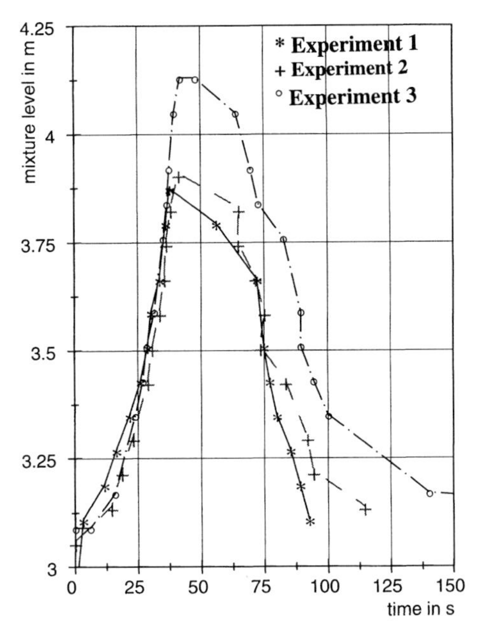

Figure 4.5: Dynamic behaviour of the mixture levels for different experiments.

Table 4.2: Parameter settings.

Experiment nr.	Initial level in m	System pressure in MPa	Solution pressure (air) in MPa	Pressure gradient (PT1) T in s	Pipe diameter in m
1,2,3	3.0	3.0	3.0	14.0	0.008

Tabelle 4.3: Isometry of the flow channel.

Exp. nr.	Isometry of the flow channel					
	X1 (90°)	X2 (45°)	X2 (30°)	X2 (15°)	X3 (90°)	X3
1	0–1 m	1–3 m			3–5 m	X2
2	0–1 m		1–3 m		3–5 m	
3	0–1 m			1–3 m	3–5 m	X1

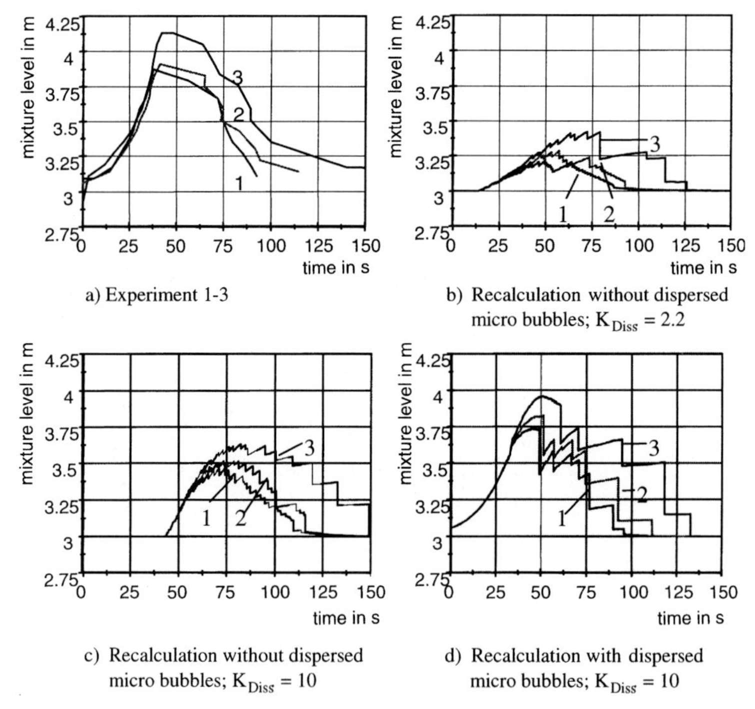

Figure 4.6: Recalculation of the dynamic behaviour of mixture levels (experiments 1, 2 and 3).

higher than the mixture level (Figure 4.6 b) because the degassing started later. For small slope angles of the pipe, mixture level signal jumps occur about 20 cm during the level decreasing. These jumps results from exhausting of gas slugs, which were also observed in experiments.

Figure 4.6 d demonstrates the results of the calculation with micro bubbles at the beginning of the pressure drop. The dynamic behaviour and the position of the maximum value correspond to the experiments.

Comparing experiments and computations, we conclude that the initial void fraction of micro bubbles must be accounted for calculation.

Figure 4.7 demonstrates the dynamic behaviour of the level, position, and diameter of gas bubbles in a vertical pipe during a pressure drop with degassing.

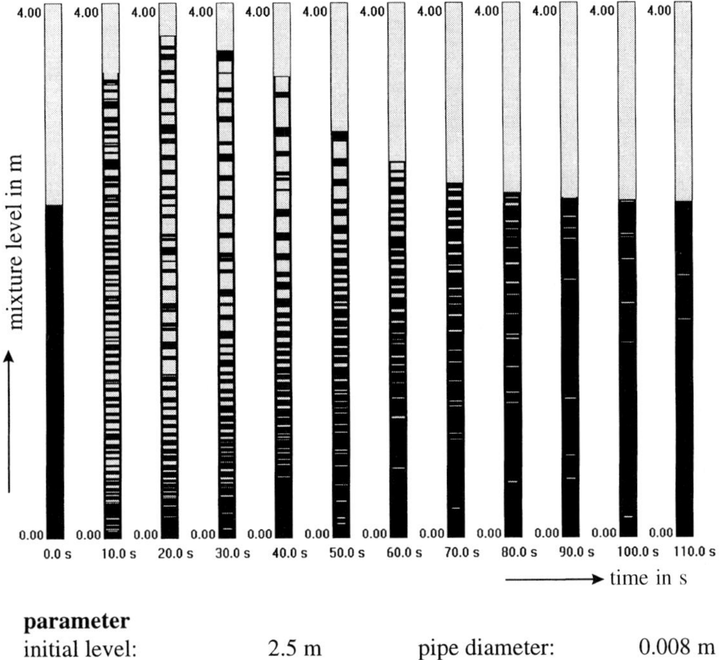

parameter

initial level:	2.5 m	pipe diameter:	0.008 m
solution pressure (air):	5 MPa	pipe length (vertical):	2.50 m
degassing time constant:	15 s	time constant for the	
dissolution pressure:	0.5 MPa	pressure drop:	5 s

Figure 7: Graphical representation of the transient two-phase flow for the pressure drop with degassing (simulation).

4.6 Summary

The developed model is able to supply the following information for pressure drops with degassing in thin pipings:

- local bubble formation as function of the system pressure and solution conditions,
- temporal course of the single bubble movement,
- bubble growth and coalescence,
- time-dependent local pressure distribution,
- dynamics of the degassing processes and the medium exchange.

The verification of the model parameters was carried out with help of experimentally determined bubble volume and velocities as well as with the mixture level course. The experiments supplied a statistically reliable data base.

Through variation, calculation and comparison with the experimentally determined mixture levels, it is possible to determine the initial gas bubble concentration and the temporarily dissolved gas mass.

List of Symbols

L_L liquid node length
L_LF liquid film length
u_G bubble velocity
c relative velocity
R pipe radius
D_C pipe diameter
D_B bubble diameter
ρ_L liquid density
τ time
$c_{\mathrm{w,\,cap}}$ friction coefficient
$\dot{m}_\mathrm{L}$ liquid mass flow
M_L liquid mol mass
ξ_G gas concentration
n_B bubble density
τ^i interfacial area shear stress
σ surface stress
p_L liquid pressure
p_Diss dissolution pressure
K_Diss dissolution ratio
l_B bubble length
z progress coordinate
$\dfrac{\partial p_{\mathrm{L,F}}}{\partial z}$ friction pressure gradient in the liquid
$\left(\dfrac{\partial p_\mathrm{L}}{\partial z}\right)_{\mathrm{Acc,1}}$ acceleration pressure gradient before the bubble
$\dfrac{\partial p_\mathrm{L}}{\partial z}$ pressure gradient in the liquid
F_R^i frictional force
t_K, t_C bubble contact time with the probe
$K_{\mathrm{ACC,\,1}}$ acceleration coefficient before the bubble
$K_{\mathrm{DRAG,\,1}}$ correction factor for the suction force

L_G two-phase node length
L_S probe distance
u_L liquid velocity
v_B bubble velocity
$\overline{D}$ middle diameter
A_C pipe cross section area
P_C pipe perimeter surface
ρ_G gas density
α pipe slope
Re_LF Reynold's number in the film
g gravitational acceleration
M_G gas mol mass
K_H Henry constant
r_B bubble radius
τ^w wall shear stress
A_LF perimeter surface of the film
p_Sol solution pressure
ε void fraction
Γ_LG medium flux (liquid/gas)
T time constant or temperature
δ film thickness
$\left(\dfrac{\partial p_\mathrm{L}}{\partial z}\right)_{\mathrm{Cap}}$ form pressure gradient
$\left(\dfrac{\partial p_\mathrm{L}}{\partial z}\right)_{\mathrm{Acc,2}}$ acceleration pressure gradient after the bubble
$\dfrac{\partial p_{(\mathrm{G})}}{\partial z}$ pressure gradient in the gas phase
F_Drag suction force
t_L bubble course time between two probes
$K_{\mathrm{ACC,\,2}}$ acceleration coefficient after the bubble
$K_{\mathrm{DRAG,\,2}}$ correction factor for the suction force

References

[1] S. Benerjee, G. Hetsroni, G. F. Hewitt, G. Jadigaroglu: *Modelling and Computation of Multiphase Flow.* Notes for Short Courses at the ETH Zurich, Zürich 1989.

[2] N. I. Kolev: *Transiente Zweiphasenströmung.* Springer-Verlag, Berlin Heidelberg New York Tokyo, 1986.

[3] D. J. Tritton: *Physical Fluid Dynamics.* 2nd Edition, Clarendon Press, Oxford, 1987.

[4] G. F. Hewitt, N. S. Hall-Taylor: *Annular Two-Phase.* 1st Edition, Pergamon Press, Oxford, 1970.

[5] Y. Taitel, D. Bornea, A. E. Dukler: *Modelling Flow Pattern Transitions for Steady Upward Gas-Liquid Flow in Vertical Tubes.* AIChE Journal **26**(3) (1980) 345.

[6] H.-H. Möbius, W. Drüselen: *Chemische Thermodynamik.* VEB Deutscher Verlag für Grundstoffindustrie, Leipzig, 1989.

[7] E. Molter: *Druckentlastung von Gas-/Dampf-Flüssigkeitsgemischen.* Dissertation TU Dortmund, 1991.

5 Investigation of the Transient Behaviour of Cavitation Effects in Liquid Injection Nozzles

Peter Roosen*

5.1 Introduction

In most technical injection systems, liquids are propagated with partially high velocities. Due to the hydrodynamic pressure reduction, a spontaneous evaporation, called cavitation, may take place that has to be described as a dual phase phenomenon. Contemporary injectors are rotationally symmetric. Therefore, the cavitation phenomena exist in three dimensions, obscuring the central part of the flow field, especially the interesting interface of undisturbed liquid and surrounding gas bubbles and/or dual phase flow. By putting a planar transparent nozzle to work, the elimination of the third dimension for cavitation induction allows a good access to those core regions of the nozzle canal. Its design will be discussed in Section 5.2.

During the series of research projects, a number of structural effects with respect to the formation of cavitating areas were observed. The following sections will show some examples of them, with a special stress of the influence of the flow field determining variables (geometry, choice of liquids, effect of counter pressure) on the development and dynamics of the dual phase regions. Further important information is gained from the observation of the velocity distribution in the non-cavitating regions of the nozzle canal.

* RWTH Aachen, Lehrstuhl für Technische Thermodynamik, Schinkelstr. 8, D-52062 Aachen, Germany

5.2 Planar, Optically Accessible Nozzle and Counter Pressure System

The optical investigation of cavitating structures requires the respective access to the nozzle canal. This was only partially possible in recent designs – glass nozzles with realistic axisymmetric shape – as e.g. reported in [1, 2, 3]. To circumvent this problem and gain a better view into the individual phenomena, a planar injection nozzle was developed (Figure 5.1). It is contracted only in one geometric direction, so, no obscuring phase boundaries inhibit the optical access to the core of the partially cavitating flow field.

Two 3 cm thick quartz glass plates (3, 14) enclose a metal lamella (13 and rightmost figure), together forming a rectangular nozzle canal. This sandwich construction is slightly stressed by two halves of a surrounding steel corpus (8, 11). So, the streaming liquid can only propagate through the canal to the counter pressure vessel attached at (16).

The lamella is manufactured by laser cutting from stainless steel. In the reported experiments, its thickness was chosen in the range of 0.1 to 0.3 mm. The width of the injection canal was set to 0.2 mm, but the larger injection pressures eventually tend to spread the lamella to a wider value. Releasing the pressure does not lead to a return to the intended value but leaves the lamella in the spreaded state until the next dismantling. So, successive experiments with lower pressure values without dismantling the nozzle are performed in the spreaded state. Its actual value may be readily assessed by the image processing, so, no effort was made to refrain the lamella from spreading. Two large recesses in the steel corpus allow the optical vista through the nozzle hole and the application of various optical measurement techniques.

The liquid is fed through a hole in one of the quartz plates. Therefore, a rather simple rubber gasket (6, 9, 10) around the plates and the feed tube (5) is

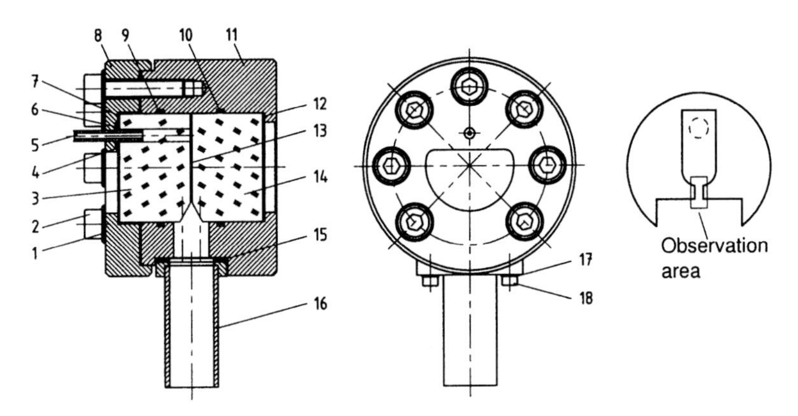

Figure 5.1: Schematic of the planar nozzle.

sufficient to seal the high pressure domain (maximum operating pressure 200 bar). The injection pressure/counter pressure ratio may be chosen quite freely due to the installation of a counter pressure system. It consists of an interception vessel with an electric release valve operated by a magnetic swimmer level monitor. The insertion of an orifice into the release system limits the liquid release to a value yielding only a slow change of the counter pressure value due to the level change in the vessel. For security reasons, the vessel had to be chosen rather small. To prevent large amplitudes in counter pressure variation, an additional gas buffer volume was added, resulting in a total counter pressure volume of roughly 20 l. This leads to a counter pressure variation of about 1.5% across collection and release intervalls, which is regarded as sufficiently constant for the performed experiments.

5.3 Fluorescence Pulse Illumination

The small size of the bore hole and the high velocities of the ejected liquid necessitate a very short image acquisition interval. As a standard video equipment was to be used to easily sample larger amounts of pictures for statistical evaluation, a very short flashing time, together with a non-interfering double exposure technique was necessary.

This is achieved by using two cuvettes containing fluorescent laser dyes as illumination sources, being irradiated by two separate excimer lasers with pulse durations of 10 to 25 ns (see Figure 5.2). Since the fluorescence of the dyes decreases in a similar period, we achieve an overall pulse duration of about 50 ns each. As we do not use the lasers directly for scene illumination

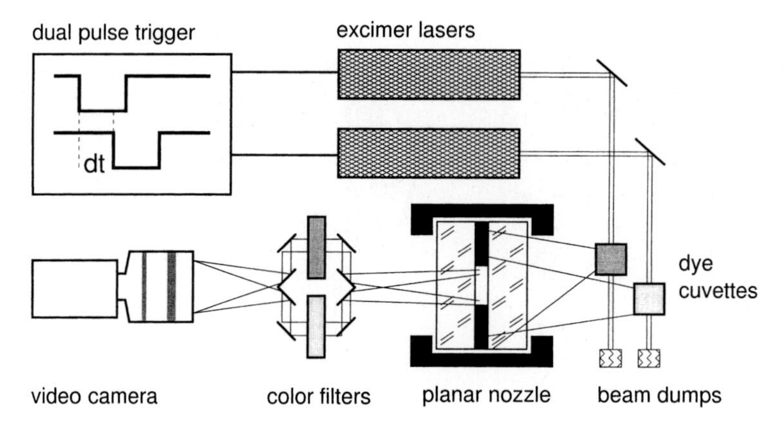

Figure 5.2: Optical set-up of the cavitation investigation system.

but exploit a rather broad fluorescence spectrum of several ten nanometers per dye, we do not experience problems of speckles or interferences. The primary light source lasers may be triggered at any time, so, the delay between them can be chosen freely, yielding a very versatile double pulse system.

The light of the dyes is shone through the nozzle canal and illuminates it homogeneously. As the wavelength ranges of the applied dyes do not overlap, their light may be filtered individually by respective band pass filters: the transmitted light is fed into a dual path mirroring system, which is equipped with such filters. Both paths feed a black & white camera at slightly different positions on the frame, thus forming an efficient double exposure image acquisition system.

Due to the rather high velocity of the liquid travelling in the nozzle hole, time delays of about 2 μs between the first and the second fluorescence pulse set an upper limit to the double exposure interval for prosecutable image details, whereas delays of less than 150 ns usually do not exhibit any significant changes in the picture. In any case, these intervals are very short in comparison to the video timing (40 ms per frame), so, we observe the two images as doubly exposed video frames. The video signal is digitized via a PC based digitizer board with an 8 bit intensity resolution. The images are post-processed on an unix workstation with the Khoros image processing library to obtain motion and surface dynamics information. A similar experimental set-up is used to determine the flow field in the non-cavitating regions of the nozzle flow. The liquid to be investigated is seeded with $\phi = 10$ μm tracer latex spheres of the same specific weight. Comparing the unseeded flow with the seeded one with respect to emergence and dynamics of the cavitation process showed, within the bounds of statistical certainty, no change of the observable effects. The velocity distribution of the non-cavitating liquid was finally carried out with a rather simple shadowgraphy set-up without color coding, where the latex particles appear as small dark dots in the homogeneously lit liquid when observed with a standard video camera. Existing cavitation regions also appear dark on the pictures, but can easily be distinguished from the particles by their location and geometrical appearance. There is no possibility to distinguish particles closely surrounded by cavitation bubbles, though. Therefore, the investigation is limited to the region of undisturbed liquid, existing in both frames of each double exposure. The number density of the PIV particles was chosen very low in order to minimize the influence on the cavitating flow. Typically only two or three particles are to be observed in any video frame. Therefore, the usual automatic correlation evaluation techniques could not be applied. Instead, a larger number of single shot frames was evaluated by hand both to eliminate staining effects and to yield a sufficient characterization of the mean flow field conditions.

5.4 Transient Behaviour of Individual Cavitation Bubbles

The described dual pulse technique allows the investigation of the transient behaviour of individual cavitation bubbles and ligaments in the planar nozzle canal.

Figure 5.3 shows on the left side a cavitation distribution in Decaline at an ejection pressure of 30 bar against atmospheric pressure. The high definition of the picture is immediately visible, as is the general distribution of cavitating regions in the nozzle canal: beginning at the sharp canal inlet corners regions of layered cavitation, to be described as dense dual phase region, protrude into the canal. They are stabilized by the hydrodynamic flow detachment at those points. After about two thirds of the canal length, these layers disintegrate into distinguishable individual bubbles or swarms of bubbles.

In numerous pictures taken for identical flow conditions, a differing cavitation distribution is to be seen on the different hole sides. Considering pictures taken in short order, a symmetric switching of flow behaviour with respect to the hole center is detected. The assumption that the switch is caused by tiny stochastic changes in the inflow conditions could only be verified partially by comparing results with those gained in the investigation of asymmetric nozzle shapes, as will be discussed below (see Section 5.5).

The observation of single bubble dynamics is possible due to the sub-microsecond resolution: The relatively large bubble near to the nozzle exit is undergoing a form change and shear, while a little one farther upstream shrinks significantly. The right part of Figure 5.3 shows a typical ejection of Decaline

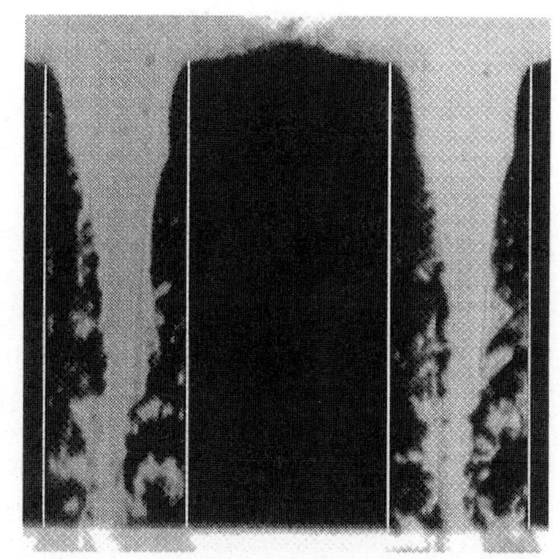

Figure 5.3: Double exposure of the form variation of cavitation structures in the planar nozzle canal. Ejection with $p_e = 30$ bar (left) and 100 bar (right) against atmospheric pressure. $T = 295$ K, liquid: Decaline, streaming top down. Temporal interval of the respective single frames $\Delta t = 750$ ns resp. 680 ns. Additional bright lines clarify the borders of the canal.

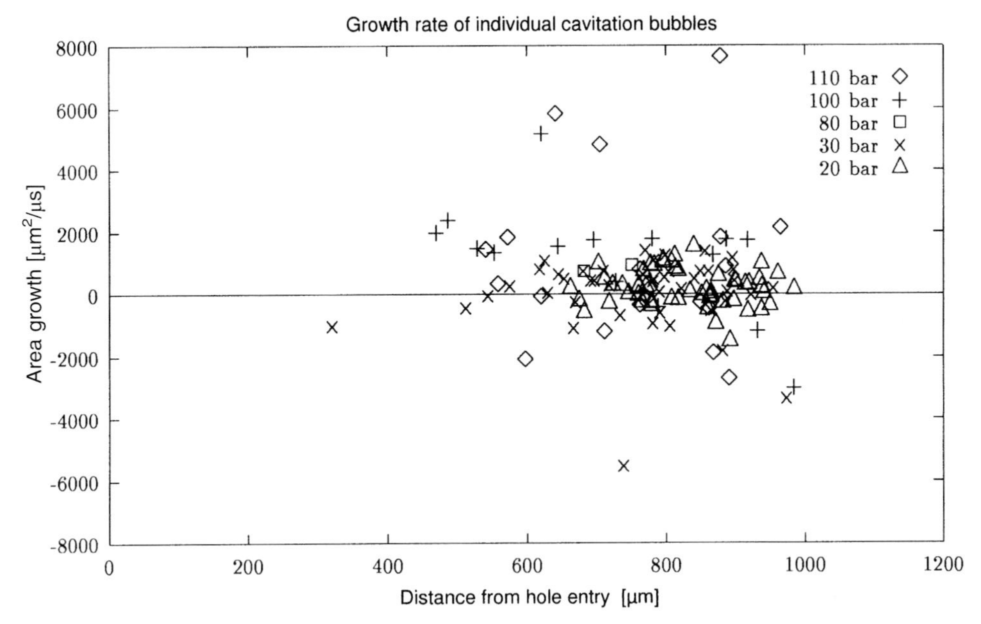

Figure 5.4: Dependence of the visible rate of growth on the ejection pressure and axial position in the nozzle canal. Liquid: Decaline. Counter pressure: 1 bar.

with an ejection pressure of $p_e = 100$ bar against atmospheric conditions. The kind of homogenuous layered cavitation induced by the sharp inlet bends disintegrates further downstream into individual ligaments being separated from each other by areas of clear liquid. Only in the latter regions, the dynamics of individual structures may be monitored with respect to determination of growth rates or velocities.[1]

The large number of acquired images has been post-processed with respect to a statistical evaluation of the cavitating flow behaviour. As an example, the size changes of individual bubbles depending on the axial position in the nozzle hole shall be discussed (Figure 5.4). The unit of the ordinate was chosen deliberately not as a volume since there is no information available in the direction of view. The change of the dectable area was chosen instead.

At ejection pressures of several ten bars, mainly size fluctuations around the bubble size observed in the respective first pictures are to be found. At elevated pressures (approx. 100 bar), significant volume increases are monitored for discernible cavitation structures. It has to be kept in mind, though, that in this statistical evaluation growth rates are compared regardless of their lateral position in the cross section, regardless that relevant structures near to the center of the canal are observed at rather high pressure ratios only.

[1] Only a few examples can be given here; more detailed discussions are found in the papers [4] to [8] and in the annual reports of the project.

A locally resolving comparison is very difficult as at higher pressure ratios, the areas showing cavitation at lower ones are mostly obscured. Using a very limited part of images for the characterization of those low pressure drop, cavitation phenomena would yield a risk of biased information as that set might be pre-selected in an adultered way.

5.5 Instabilities of the Flow Field in Asymmetric Nozzles

In the symmetric nozzles, a switch of the cavitation structures was observed as discussed above. After discussion with modellers recent experiments, focusing on the effect of the inlet corner sharpness, have therefore been performed with asymmetric nozzles. A sketch of the respective set of lamellas, together with one photographed example, is shown in Figure 5.5. An in-depth discussion of results may be found in [8].

Besides the geometrical starting point, the general behaviour and distribution of the cavitation regions was monitored. The pressure/counter pressure ratio was varied in the feasible range for each lamella by changing the ejection pressure value for a set of fixed counter pressures, the latter being indicated by the dashed lines in Figure 5.6. While raising and lowering the ejection pressure, numerous images of the cavitation distribution are acquired, leading to the identification of characteristic ratios, at which flow conditions change significantly. Ratios yielding similar flow field changes at different counter pressure values were connected, and thus produce areas of comparable flow states in the diagram. If a general flow field structure is not found when changing to another counter pressure value, the separating lines end in between the dashed lines. The exact positioning of the boundary line may of course not be assessed in those cases.

At very low pressure drops, the water in the nozzle hole is completely translucent and without a noticeable structure. This undisturbed state is found

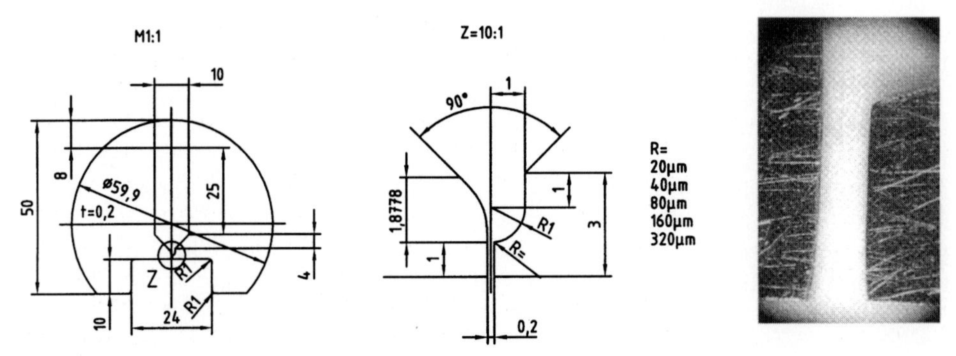

Figure 5.5: Schematic of the asymmetric nozzle lamella and actual shape (growing magnification from left to right).

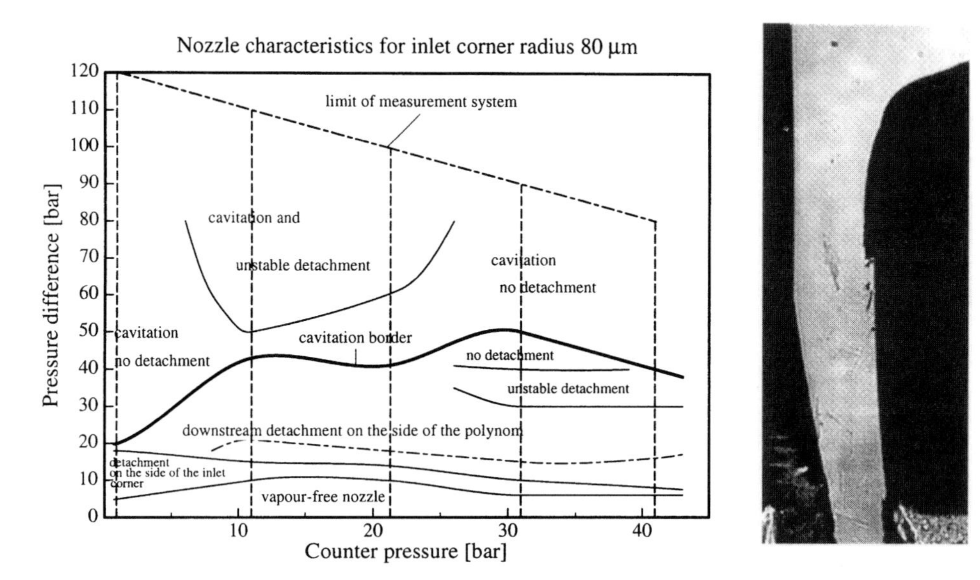

Figure 5.6: Chart of characteristic cavitation distributions (left) in non-degassed water for a flat asymmetric nozzle with an 80 µm inlet edge radius; typical image for p_e=80 bar, p_c=11 bar (right).

in a narrow strip along the bottom side in the diagram of Figure 5.6. At slightly elevated pressure differences, the flow recesses from the wall in the orifice zone first from the wall opposite the sharp inlet bend – contrarious to other investigated inlet edge radii. An oszillating interface may be watched between translucent liquid and gas/vapour phase whose amplitude grows with raising ejection pressure. At a certain point of its value, the flow recess jumps to the side of the inlet bend. This region is further separated by a dashed line: The recess shows two clearly discernible modes. A slow raise of the ejection pressure leads to a sudden switch between them.

There is region above the inlet cavitation boundary, in which the flow recess from the wall is completely suppressed by cavitation with changing properties. At larger counter pressures, a short cavitation tube with a split-off of individual bubbles at its top is found. The length of the tube slowly grows with raising ejection pressure. At larger pressure drops, a gas/vapour zone with an unstable surface structure straddles the complete length of the nozzle canal. At lower counter pressures, the length of the cavitation tube is larger at its onset, and it grows faster with raising ejection pressure. At even lower counter pressures, a smooth flow recess sets in at the same point and reaches to the orifice. Raising the ejection pressure at this state, the vapour/liquid boundary becomes unstable and changes into a state, at which individual bubbles separate from the coherent vapour zone. With the counter pressures 11 and 21 bar, this flow recess becomes unstable again. Figure 5.6 shows an individual picture example of a flow recess in its full size.

5.6 Flow Field of Non-Cavitating Liquid Regions

The flow field in the liquid fraction in the nozzle hole is investigated with an adapted version of particle imaging velocimetry (see Section 5.3). As a larger number of single shot frames has to be evaluated to yield a sufficient characterization of the mean flow field conditions, and only an averaged flow field picture is shown. Comparing the unseeded flow with the seeded one with respect to emergence and dynamics of the cavitation process showed, within the bandwidth of statistics, no change of the observable effects. No spontaneous cavitation caused by the addition of the tracer particles has been observed. As in the experiments without them, the onset always was induced at the inlet bend.

Most averaged pictures show also particle tracks in areas overlaid with cavitation structures. They stem from individual shots without cavitation at those positions, thus allowing an unambiguous assessment.

Figure 5.7 shows flow velocities for an ejection pressure of 80 bar and atmospheric counter pressure. The growth of the velocities at the inlet is obvious, as is a slight asymmetry of the canal, being caused by the manufacturing process of the nozzle lamella.

The center region of the flow shows a rather smooth flow region with undisturbed liquid. It should be caused by an only weak coupling of the homogeneous liquid zone to the dual phase flow interacting with the walls. The existence of particle traces overlapping each other depict the statistical character of the flow, it is not stable but fluctuates slightly relative to the average picture even in the center part.

Discernible particle tracks deeply in the averaged cavitating regions indicate that the cavitation bubble distribution in individual frames is strongly structured. The cavitation region frequently switches between a long and a short one. At the discussed pressure ratio, it collapses or expands remarkably

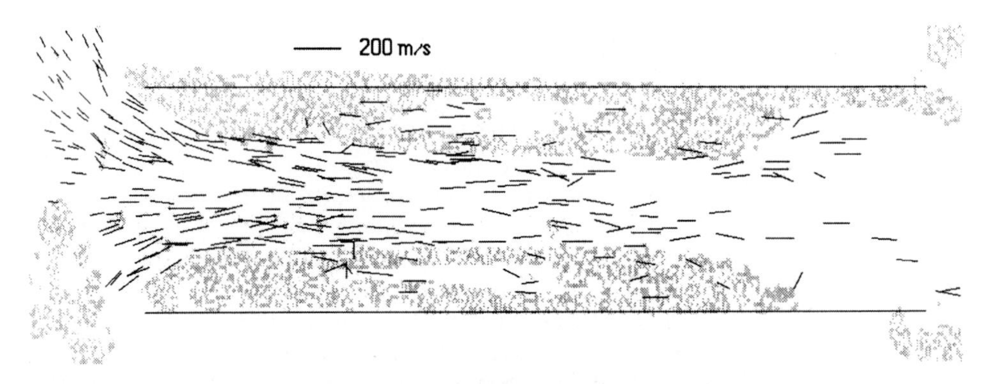

Figure 5.7: Tracer particle velocities for an ejection pressure of 80 bar and a counter pressure of 19 bar. Liquid: non-degassed water, streaming from left to right. Shadowed regions indicate areas with statistically observed cavitation.

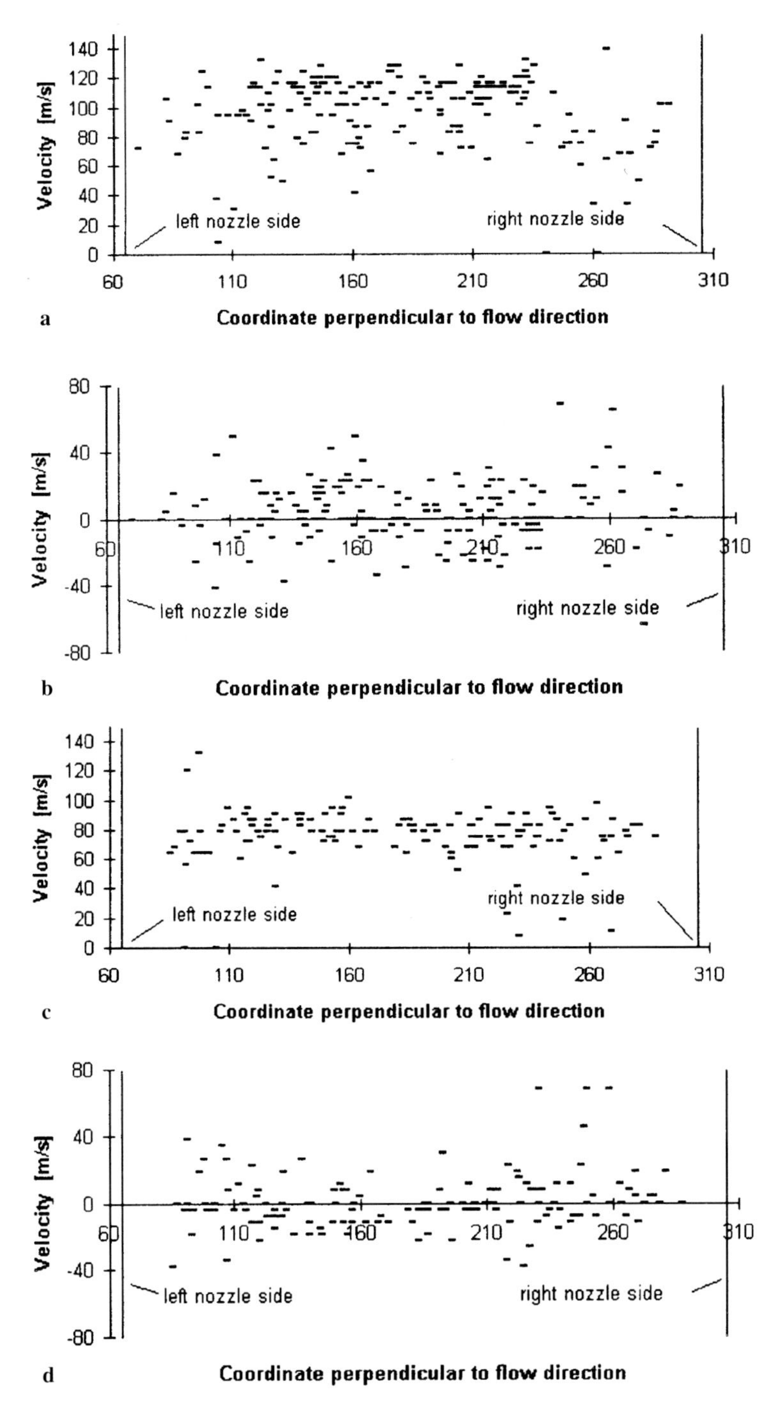

Figure 5.8: Statistical evaluation of the velocity components for ejection pressure 80 bar and counter pressures of 19 bar (a, b) and 21 bar (c, d). Liquid: non-degassed water.

fast. At both cavitation breakdown positions, additional non-axial particle tracks indicate the existence of recirculation regions.

The turbulence-inducing effect of the cavitation collapse is even more evident when comparing statistical evaluations of the velocity components at 19 bar and 21 bar counter pressure (Figure 5.8): While at 19 bar, a relatively wide distribution of values is evident for both lateral and radial values, it is narrowed significantly at the more stable operating point with 21 bar counter pressure. In the latter case, the collapse already takes place upstream the area taken for statistical evaluation. Similar results are obtained for significantly different pressure/counter pressure ratios as 80/6 or 80/26 (see also [7]).

5.7 Summary and Outlook

The investigation of the transient properties of cavitation areas in planar translucent nozzle systems has yielded relevant information for the modellers on the flow field behaviour to support the development of simulation codes. For the near future, there is no hope, though, that the observed abundance of details will be represented in such models. Especially, the simulation of single bubble dynamics in interaction within the surrounding cavitating flow field will take some time. The most recent developments with respect to global flow field and cavitation area prediction are quite encouraging nevertheless [9].

The parametric investigations, especially the hole geometry variations, have shown that, at least within closer limits, the actual distribution of the cavitating zones may be influenced. This may lead to targeted preparations, e.g. of fuel-air-mixtures, in combustion systems.

In an adjacent, new research project, this idea is prosecuted by investigating this possibility, in spite of the very limited insight into the intricate interactions of the various small-scale hydrodynamic effects, by an evolutionary adaption of the nozzle geometry to a pre-defined goal and an a-posteriori explanation of the evolved structures.

References

[1] J. Bode, H. Chaves, F. Obermeier, T. Schneider: *Influence of Cavitation in a Turbulent Nozzle Flow on Atomization and Spray Formation of a Liquid Jet.* Proc. Sprays and Aerosols, ILASS Europe, 1991.

[2] J. Bode, H. Chaves, A.M. Kubitzek, W. Hentschel, F. Obermeier, K.P. Schindler, T. Schneider: *Fuel Spray in Diesel Engines.* Proc. 3rd Int. Conf. "Innovation and Reliability in Automotive Design and Testing", Firenze, **2** (1992) 749–759.

[3] W. Eifler: *Untersuchungen zur Struktur des instationären Dieselöleinspritzstrahles im Düsennahbereich mit der Methode der Hochfrequenz-Kinematografie.* PHD Thesis, Kaiserslautern, Germany, 1990.

[4] P. Roosen, O. Unruh, M. Behmann: *Investigation of cavitation phenomena inside fuel injector nozzles.* ISATA, Florenz, Juni 1997.

[5] P. Roosen, M. Behmann, O. Unruh: *Optische Untersuchung von Kavitationserscheinungen in einer schnell durchströmten, quasizweidimensionalen Düse.* Tagungsband zum 3. Workshop über Sprays, Erfassung von Sprühvorgängen und Techniken der Fluidzerstäubung; Lampoldshausen, Oktober 1997, ISBN 3-89100-029-4.

[6] P. Roosen, S. Kluitmann: *Entwicklung und Einsatz eines Laser-Doppelpulsverfahrens zur optischen Untersuchung der Dynamik von Kavitationserscheinungen in einer schnell durchströmten, planaren Düse.* Workshop „Meßtechnik für stationäre und transiente Mehrphasenströmungen"; FZR-204, Forschungszentrum Rossendorf, Dezember 1997.

[7] P. Roosen, O. Unruh: *Cavitation-induced flow field inside fuel injector nozzles.* ISATA, Düsseldorf, Juni 1998.

[8] P. Roosen, O. Genge: *Optische Untersuchung von Kavitationsfluktuationen in einer planaren, nicht-achsensymmetrischen Düse.* Spray 98, Essen, Oktober 1998.

[9] D.P. Schmidt, C.J. Rutland, M.L. Corradini, P. Roosen, O. Genge: *Cavitation in Two-Dimensional Asymmetric Nozzles.* SAE Technical Paper 1999-01-0518.

Flow Pattern

6 Phase Distribution and Bubble Velocity in Two-Phase Slit Flow

Lutz Friedel and Steffen Körner*

6.1 Status and Problem

For the prediction of the establishing two-phase mass flux for a given pressure difference besides the knowledge of the occurring boiling delay respectively thermodynamic non-equilibrium between the phases, also the magnitude of the mean fluiddynamic non-equilibrium in form of the slip is needed. Referring to John et al. [1] or Westphal [2], the highest prediction accuracy for the two-phase mass flow through narrow rectangular slits is obtained by using the Homogeneous Equilibrium Model according to Pana [3]. Besides others, it includes the assumption of a homogeneous flow respectively a negligibly small slip, an immediate adjustment of the thermodynamic equilibrium between the phases and a pressure loss submodel. The suitability of the model by Pana [3] seems not to be logical because following the common expert opinion with the Homogeneous Equilibrium Model for given conditions in comparison to other flow models, the minimum (critical) mass flux is predicted. When taking intentionally a boiling delay and/or a heterogeneous flow into account, a significantly larger mass flux is predicted. In view that each submodel includes assumptions, which in reality may not be fully met, it can not reasonably be excluded that the appropriate prediction of the establishing two-phase mass flux is caused only by the mutual compensation of simplifying assumptions in the model. Hence, the extrapolatability of the model by Pana [3] can not be taken for granted.

* Technische Universität Hamburg-Harburg, Arbeitsbereich Strömungsmechanik, D-21071 Hamburg, Germany

6.2 Aim and Approach

For validation of the assumption, concerning the ignored slip, dedicated experiments were conducted. Using high-speed cinematography, the occurring phase distribution in flashing water and water/air bubbly flow were recorded. After processing the results by object-detection algorithms and computer-based picture analysis, the velocities of the bubbles were obtained. These have then been compared to the calculated mean homogeneous mixture velocity. Additionally, the state change of the air in the bubbles was estimated on the basis of the volume increase due to the pressure decrease along the flow path.

6.3 Narrow Slit Test Section

In the experiments, an adiabatic model slit in the form of a channel with constant rectangular cross section limited by two borosilicat glass or plexiglass walls was used so that the movement of the bubbles could be observed optically along the total flow path (Figure 6.1). In accordance to the large cross section upstream of the test section there, the mean velocity is negligibly small

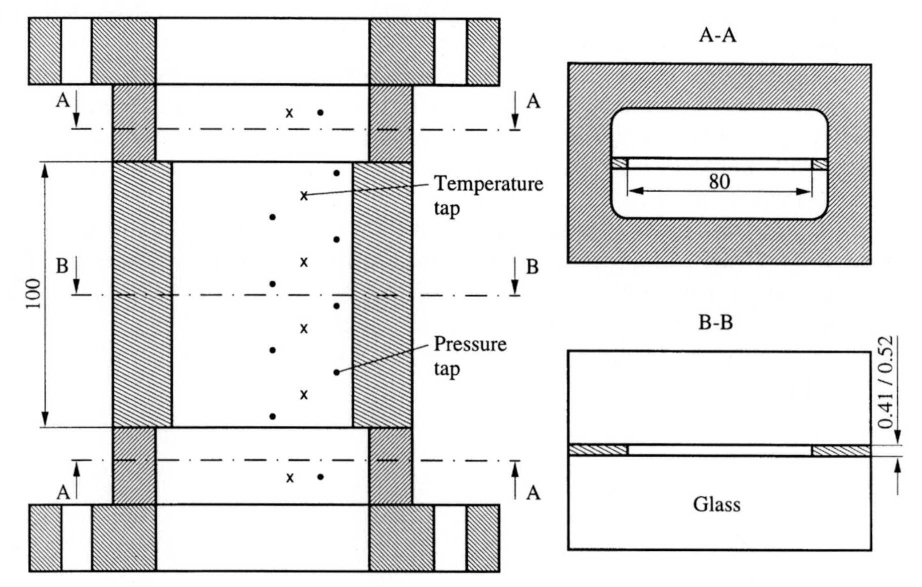

Figure 6.1: Geometrical dimensions of the model slit and arrangement of the pressure and temperature taps.

compared to that in the slit, and so, the recorded wall pressure equals practically with the stagnation pressure, and a thermodynamic equilibrium respectively the same temperature for both nearly stagnant phases can be presumed. The aspect ratio of the test section amounts to more than 100. As a consequence of the small channel opening in the range of some 0.5 mm, the pressure loss due to friction should mainly contribute to the total pressure drop in relation to that due to acceleration and the negligibly small geodetical lift.

6.4 Phase Distribution in Water/Air Bubbly Flow

The bubbles in the flow seem to be relatively homogeneously distributed across the flow cross section (Figure 6.2). The equivalent diameter of the establishing bubbles varies in a wide range even at an equal longitudinal position in the slit, the bubble volumina being between 0.5 and some 100 mm^3. An effect of the used static device for mixing the air and the water on the distribution and the size of the bubbles could not be observed. This may be attributable to the distance of more than 200 hydraulic diameters between the mixer and the slit inlet, and especially to the negligibly small velocity of the mixture upstream of the slit. Both features induce in total a relatively long mean residence time of the bubbles in the water before entering the slit.

In the test section, the (specific) air volume increases mainly because of the pressure decay due to friction along the flow path. As a consequence, the individual bubbles expand, at the same time the corresponding water volume has to be displaced. As a result of the relatively steep local pressure gradient in downstream flow direction and the less effort requiring removal of the cocurrent flowing liquid, the bubbles expand mainly at the top, herewith forming streaks and indentations. The reason for this behaviour can be seen in local random interactions respectively metastable equilibria between the forces in the bubble top surface and those due to the turbulence and the inertia of the surrounding water. This effect is especially observed for large diameter air bubbles. In this case, due to the great curvature radius, the surface forces only exhibit a relatively small impact on the resulting structure of the interface. The consequence is an irregular and fissured, large specific surface area of the bubbles at the top.

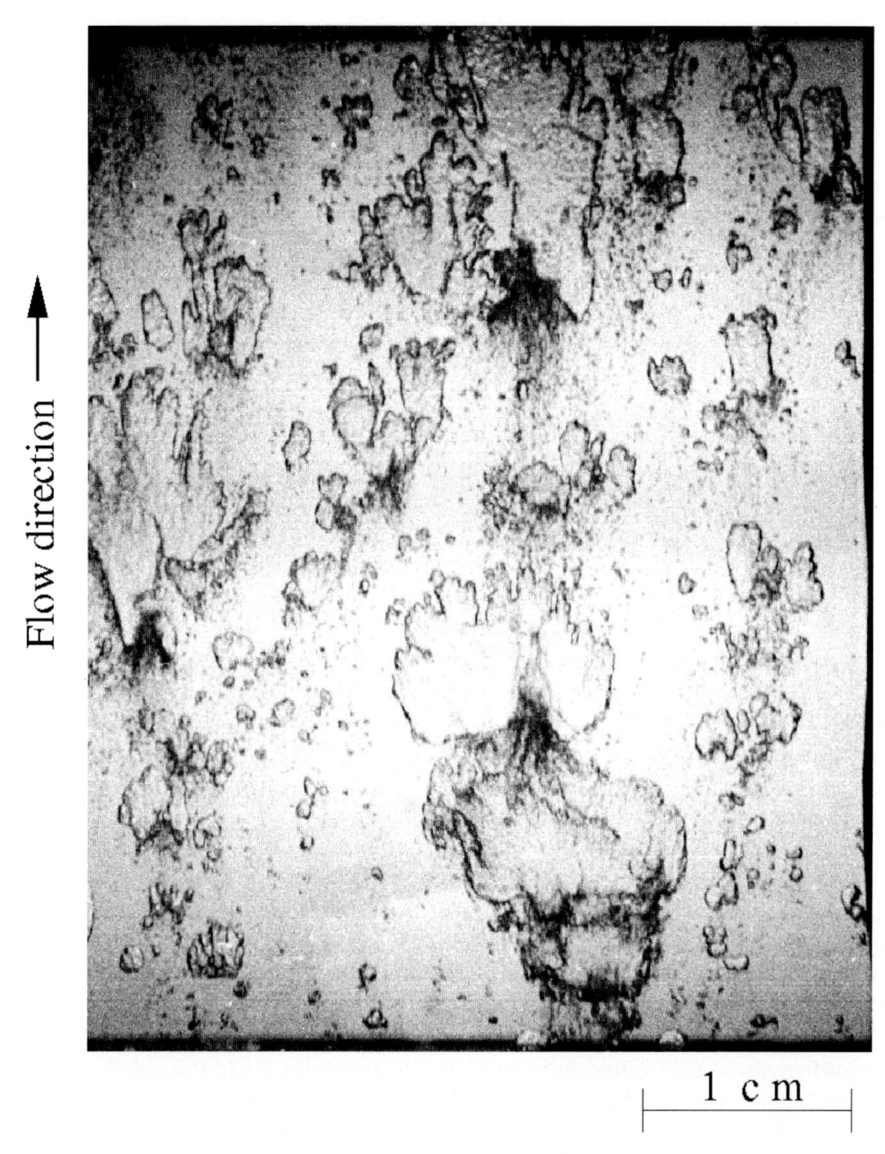

Figure 6.2: Picture of a vertical upward (adiabatic) water/air slit flow (opening 0.52 mm, length 100 mm) for a pressure difference of 4 bar and a mass flow quality of 0.1%.

6.5 Phase Distribution in Flashing Water Flow

Considering a subcooled water flow at the inlet of the adiabatic slit due to the pressure drop caused by friction, acceleration and geodetical change, the saturation pressure corresponding to the mean local fluid temperature can further downstream be attained or even fall below so that in this latter case the water still in the slit becomes superheated and finally flashing will set in (Figure 6.3). In this experiment, a pressure distribution along the downward flow path establishes in such a way that some 20 mm upstream of the slit outlet, the saturation pressure is just exceeded. With a view to the actual definition of the place of boiling inception in technically pure or clean superheated liquids, flashing can only start on the surface of existing activable nuclei. These are normally present in form of small bubbles or microbubbles adsorbed on crevices in the walls of the channel respectively on impurities in the water. On account of these nuclei, small (spherical) steam bubbles form averaged over the time rather homogeneously distributed in the superheated water[1]. Along with the further decrease of the pressure along the flow path, the bubbles expand, while in parallel, in consequence of the progressing phase change due to flashing, their volumes steadily increase. Herewith, due to the necessary large nucleus formation energy, the number of bubbles remains nearly constant. Despite of the remarkable volume increase, the radii of the steam bubbles are here considerably smaller and within a narrower diameter range than those of the formerly shown air bubbles for quasi identical density ratios. This can be referred to the different origin of the air respectively steam bubbles in the surrounding water.

Further on, considering again an initially subcooled water inlet flow but now with a larger mass flow rate respectively a higher mean velocity in consequence of the abrupt change of the cross section at the slit inlet, bubbles can develop due to local vortex cavitation caused by a local pressure undershot in the vena contracta [4] (Figure 6.4). To emphasize this effect, the slit inlet was modified by installing a sharp edged short orifice with a diameter of about 50% of the slit width shortly ahead of the test section so that in this special case an inlet velocity distribution is enforced with a remarkably higher fluid velocity in the center than at the periphery of the slit inlet. In consequence of the smaller local velocity and therefore the modest pressure change at the periphery, no cavitation bubbles could be created, or they are (at least) invisible in the picture. Downstream of the slit inlet, the center flow is strongly decelerated due to the re-attachment at the wall and the establishment of a fully developed velocity profile. Correspondingly, the local pressure increases and the cavitation

[1] Because of the smooth surface of the borosilicat glass walls, the boiling nuclei are here mainly adsorbed on impurities supposed to be rather homogeneously distributed in the flowing liquid. The uneven distribution of the bubbles can be attributed to the random nucleation process, in this special recording instant predominantly at the right side of the slit flow.

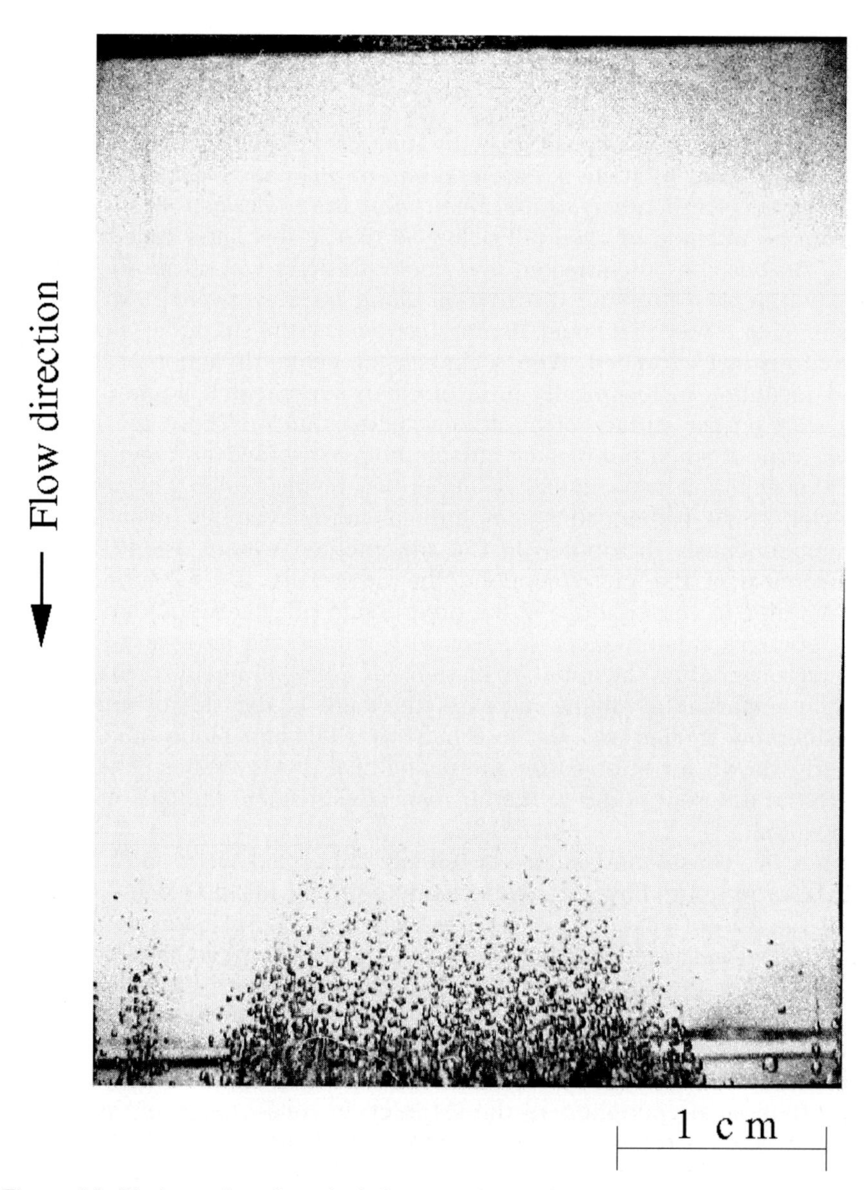

Figure 6.3: Photography of vertical downward (adiabatic) flashing water slit flow (opening 0.41 mm, length 100 mm) for an inlet stagnation pressure of 1.3 bar and an initial subcooling of 3 K.

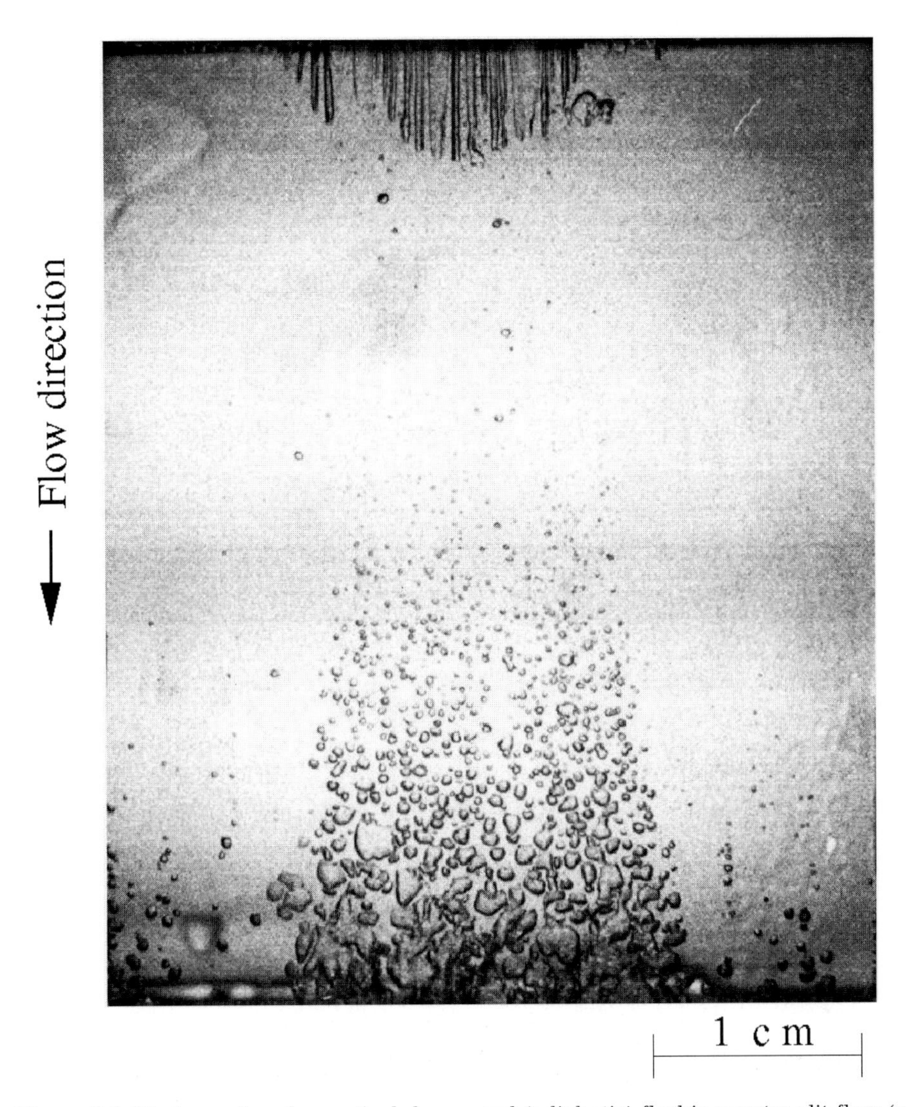

Figure 6.4: Photography of a vertical downward (adiabatic) flashing water slit flow (opening 0.41 mm, length 100 mm) coupled with an inlet orifice for a modified velocity profile for an inlet stagnation pressure of 5 bar and an initial subcooling of 5 K.

bubbles collapse. However, a substantially large number of microbubbles consisting of non or only slowly condensable gases are supposed to remain in the water. Because of their tiny diameter in the range of just 1 μm, these bubbles are hardly to detect and almost invisible in the shown photography [5]. If further downstream, again the saturation pressure of the water is significantly undershot compared to that in the still partly undisturbed water at the periphery, a larger number of microbubbles can immediately be reactivated as boiling

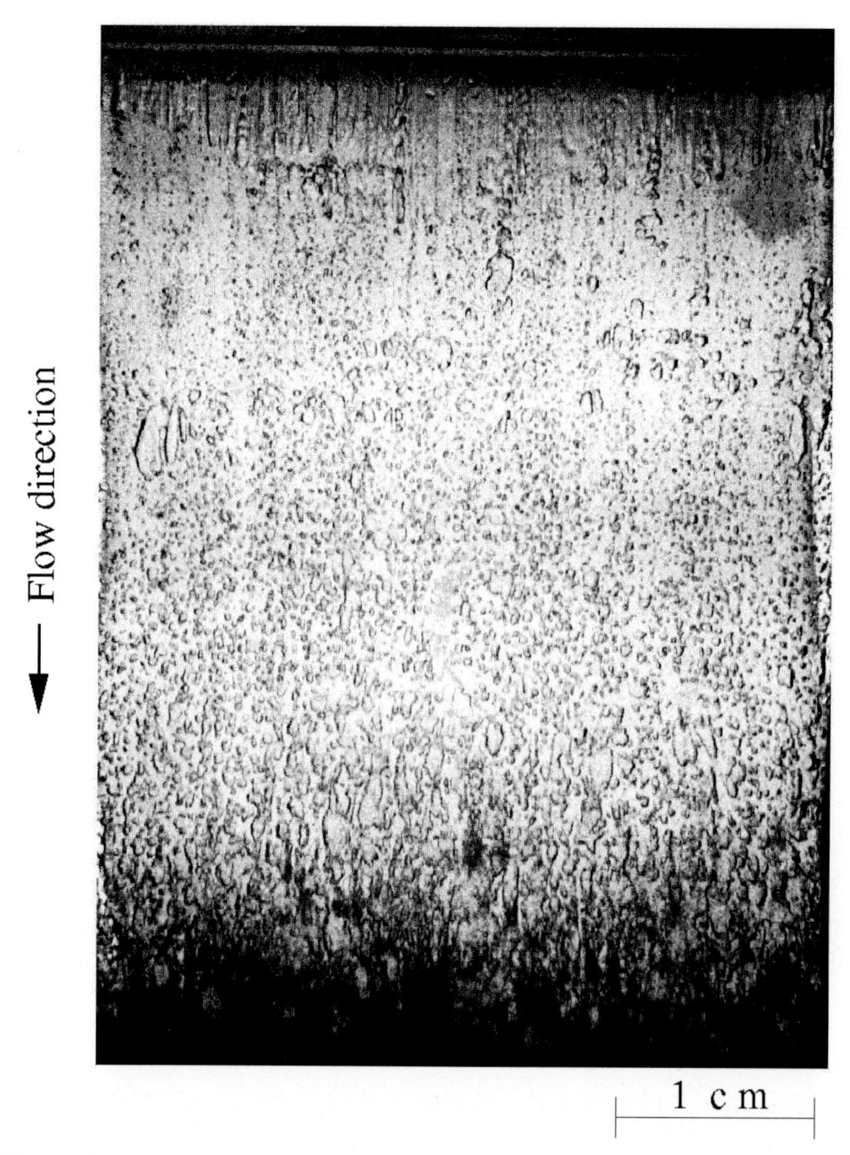

Figure 6.5: Photography of a vertical downward (adiabatic) flashing water slit flow (opening 0.41 mm, length 100 mm) for an inlet stagnation pressure of 12 bar and an initial subcooling of 1 K.

nuclei for the phase change. In consequence, subsequently now, a more intensive flashing occurs and a quasi homogeneous bubble distribution in the flashing water jet is attained. Besides this, the higher fluid velocity causes a more intensive turbulence and, thus, a larger heat flux from the superheated water to the bubble surface, and so, the boiling delay is less significant. In conclusion, it is obvious that in case of cavitation bubbles being created at the sharp edged inlet, a smaller water superheat develops despite of the higher mean velocity.

In case of an initially nearly saturated inlet water flow and high mean velocities as typical for fluiddynamic critical flow conditions at the slit outlet, the cavitation bubbles induced at the inlet do not collapse or vanish in total, supposed the time for the condensation being too short respectively the initial subcooling being too small. In contrast, a considerable part of the bubbles remain till the saturation pressure is again undershot and then act as nuclei so that intensive and early flashing occurs (Figure 6.5). As a consequence of the now enlarged interphase area and the enhanced evaporation, only a negligibly small water superheat can establish, and the local pressure and temperature of the mixture along the flow path are in reasonable agreement with the static equilibrium values. Again, caused by the decreasing pressure and the evaporation, the bubble volume increases. But, though coalescence of neighbouring bubbles occurs, the recorded diameters are significantly smaller and within a narrower diameter band than those observed in water/air bubbly flow in the same range of density ratio. In total, the bubbles are homogeneously distributed in the transversal direction.

6.6 Change of State of the Air in the Bubbles

For a relatively large bubble with an air mass of 0.3 mg, the change of the volume along the slit length compared to that at the inlet was opposed to the calculated volume increase assuming an isothermal respectively an adiabatic expansion (Figure 6.6). Obviously, the experimental results at the slit inlet are closer to the lower curve based on an adiabatic change of state. With increasing distance respectively residence time of the bubbles in the channel, a more isothermal expansion behaviour is occurring. In detail, for distances of more than about 50 mm downstream of the inlet, no longer a significant deviation between the experimental results and the calculated curve, assuming an isothermal change of state, can be observed. Indeed, the isothermal behaviour seems logical as the temperature decrease induced by the pressure profile along the flow path is slow compared to the (rapid) change at the sharp edged inlet. With respect to the air streaks developed during the bubble (top) expansion, the relatively large specific phase contact area allows for a high specific heat flux and, thus, a small resulting temperature difference between the phases will establish. On transferring this idea to the behaviour of smaller bubbles especially in case of flashing flow due to their higher specific surface area and the minor heat capacity, the thermal equilibrium will be attained earlier so that with a higher degree of physical evidence an isothermal behaviour can be attributed.

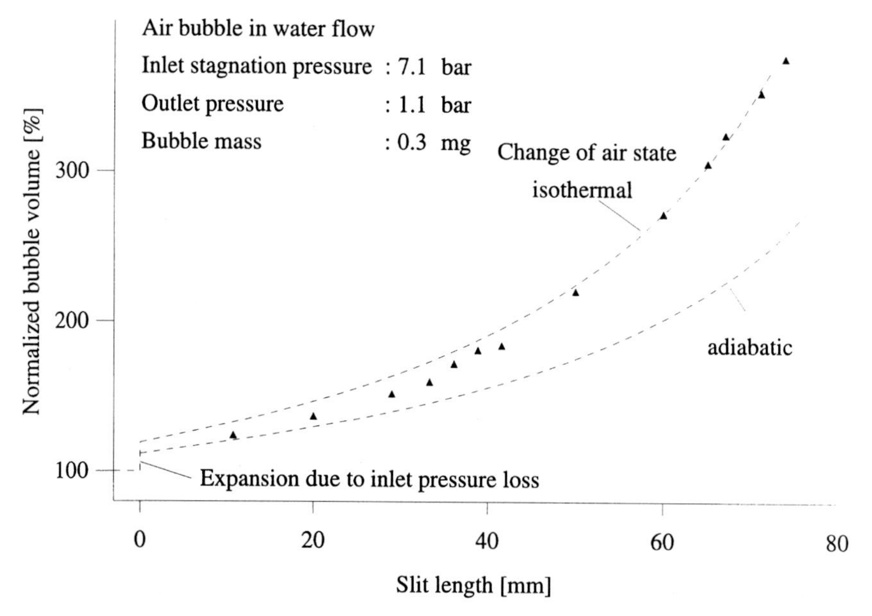

Figure 6.6: Measured normalized bubble volume and calculations, assuming an isothermal and an adiabatic change of state of the air in the bubble, as a function of the slit length.

6.7 Bubble Velocity and Slip

The measured local (absolute) velocities of the bubbles in an (adiabatic, non-flashing) air/water flow with a mass flow quality of 0.05% are depicted in Figure 6.7. Bubbles with a volume between 3 and some 40 mm^3 are encountered in transversal as well as longitudinal direction. With enlarging bubble volume, a degressive increase of the measured velocity is evident. This trend is well known from literature, valid for a single bubble rising in stagnant water, and should also be conclusive at least for upward flowing two-phase mixtures. In this context, the relatively higher velocities of the larger bubbles can be related to the smaller specific interphase area. As a result, a less intensive momentum exchange between the phases occurs. Further on, depending on the position in the slit, the (static) pressure decreases significantly between the inlet and the outlet. As a consequence, the air density and the calculated homogeneous mixture density increase. In this environment, the large variation of the absolute bubble velocity at an equal volume is on the one hand coupled with the local position in the slit in the instant of recording, the lower values corresponding to the inlet position with a smaller liquid velocity, while in the outlet due to the inferior pressure larger velocities prevail. This trend is also reflected by the calcu-

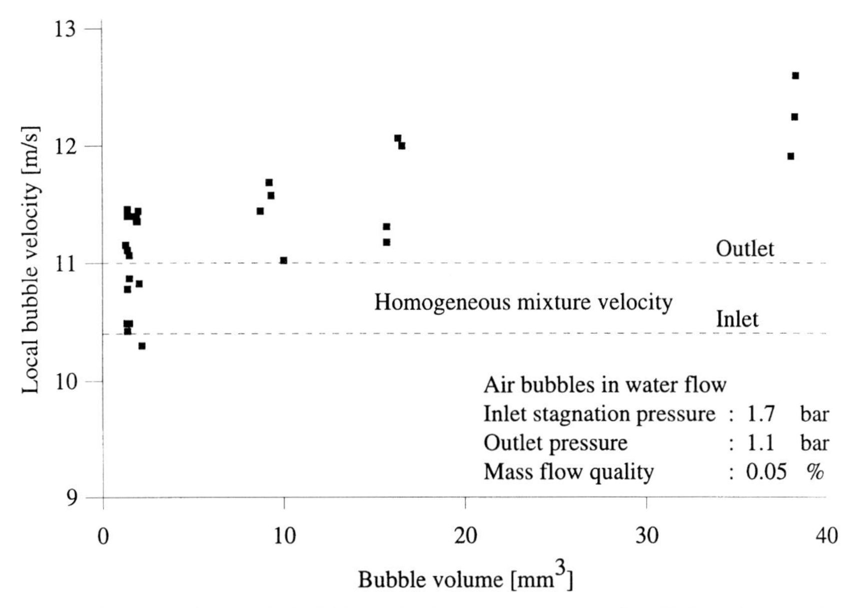

Figure 6.7: Measured local air bubble velocities in an upward (adiabatic) water/air flow (opening 1.1 mm, length 100 mm) as a function of the bubble volume for an inlet stagnation pressure of 1.7 bar, an outlet pressure of 1.1 bar and a mass flow quality of 0.05%.

lated homogeneous mixture velocity; the velocities being here between some 10.4 and 11 m/s. An additional variation in the measurements is introduced by the irregular top side expansion of the bubbles, the interactions between their movement in a swarm and the turbulence forces in the surrounding water flow. Indeed, only the mean value of the velocities of small bubbles would be quite close to that calculated, assuming an one-dimensional homogeneous flow. For larger bubbles, nevertheless, significant deviations between the actual velocity and that of the homogeneous mixture are evident, even at an equal longitudinal position in the slit.

On the basis of the individual mass fluxes, the local pressure respectively the air and water density as well as the mean bubble velocity, the mean water velocity as well as the (local) slip ratio can be calculated with respect to the longitudinal and transversal position in the slit. It is evident from Figure 6.8 that, in parallel to the trend seen for the bubble velocity, also the slip ratio increases with larger bubble volumes as well as bubble position in the slit being largest at the outlet. For the smallest bubble size, the derived mean slip ratio is nearly unity, in any case even for the largest bubbles, here the maximum experimental (local) slip ratio is less than 1.2, allowing in a first approximation for applying the homogeneous mixture flow assumption in transversal and longitudinal direction.

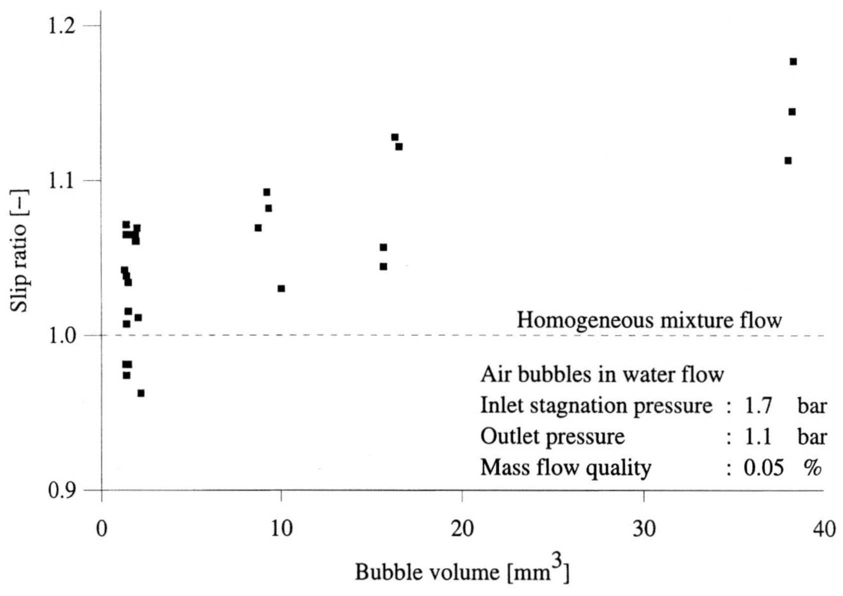

Figure 6.8: Calculated local slip ratio in an upward (adiabatic) water/air slit flow (opening 1.1 mm, length 100 mm) as a function of the bubble volume for an inlet stagnation pressure of 1.7 bar, an outlet pressure of 1.1 bar and a mass flow quality of 0.05%.

6.8 Recalculation of the Bubble Velocity

The final drift velocity of a single air bubble in a stationary water flow can be estimated by using the (static) equilibrium between the driving pressure force and the resistance force due to friction. The pressure force follows from integrating the local pressure distribution on the bubble surface:

$$F_{\Delta p} = \frac{\varrho_{liq} - \varrho_g}{\varrho_{liq}} \int_O p \, dO_{Bubble} \, . \tag{1}$$

On assuming a single spherical gas bubble, the resistance force can be calculated by using the drag coefficient of bubbles:

$$F_R = c_W \frac{\pi}{4} D^2_{Bubble} \frac{\varrho_{liq}}{2} (\bar{u}^2_{Bubble} - \bar{u}^2_{liq}) \tag{2}$$

with

$$c_W = \frac{21}{Re} + \frac{6}{\sqrt{Re}} + 0.28 \quad \text{and} \quad Re = \frac{D_{Bubble}(\bar{u}_{Bubble} - \bar{u}_{liq})}{\nu_{liq}} \leq 4000 \, . \tag{3}$$

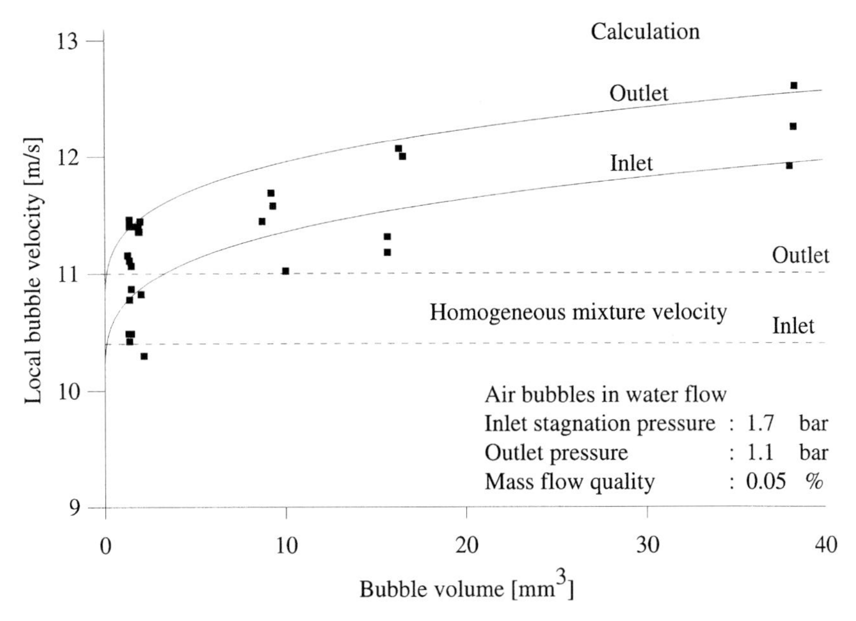

Figure 6.9: Calculated and measured local air bubble velocity in an upward (adiabatic) water/air slit flow (opening 1.1 mm, length 100 mm) as a function of the bubble volume for an inlet stagnation pressure of 1.7 bar, an outlet pressure of 1.1 bar and a mass flow quality of 0.05%.

The stationary single bubble velocity in the liquid after rearrangement results from:

$$\bar{u}_{\text{Bubble}} = \sqrt{\bar{u}_{\text{liq}}^2 + \frac{\varrho_{\text{liq}} - \varrho_{\text{g}}}{\varrho_{\text{liq}}} \frac{4}{c_{\text{W}} \pi D_{\text{Bubble}}^2} \int_{\text{O}} p \, \mathrm{d}O_{\text{Bubble}}} \,. \tag{4}$$

For a given bubble diameter and pressure profile along the flow path, which can be calculated i.e. by integrating the Bernouili equation for compressible fluids (at subcritical flow conditions) after introducing a dissipative term, the drift velocity of the bubbles depending on their size can be estimated. It is herewith assumed that the interactions respectively the momentum exchange between the bubbles are negligibly small compared to that with the surrounding water, as the mean liquid phase velocity exceeds by far the free ascending bubble velocity.

The predicted and the former already quoted experimental bubble velocities in a water/air flow are given in Figure 6.9. Due to the pressure decay and the increasing water velocity along the slit length, the bubble velocities for different sizes are calculated for the inlet and outlet condition. With the assumption of a negligibly small momentum exchange between the bubbles compared to that between the phases, in principle, all experimental values should be arranged between these two limiting graphs. For larger bubble volumes, this is true, whereas

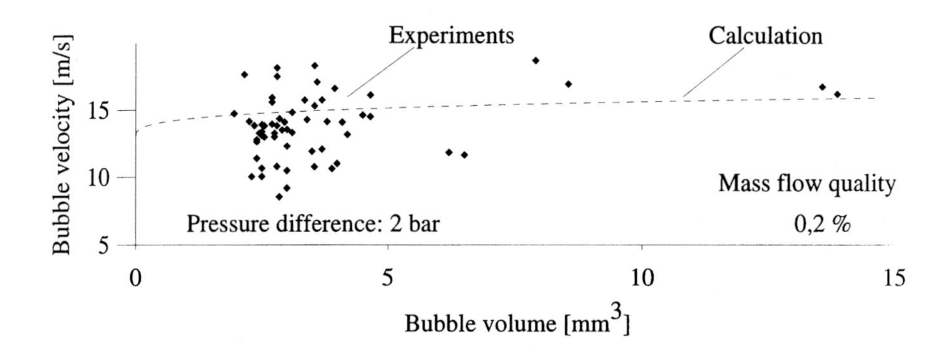

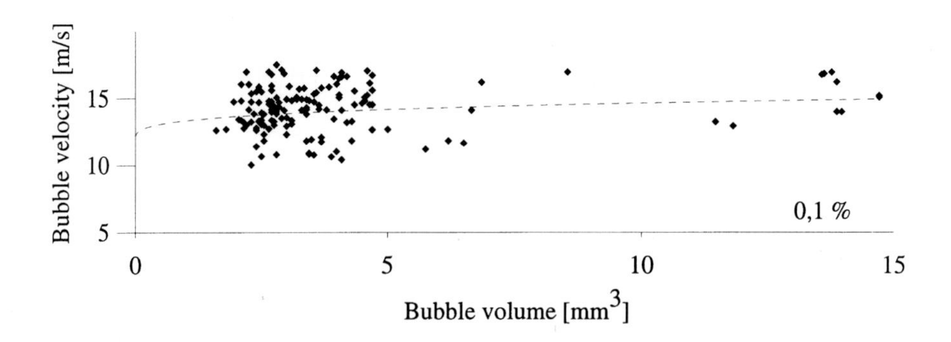

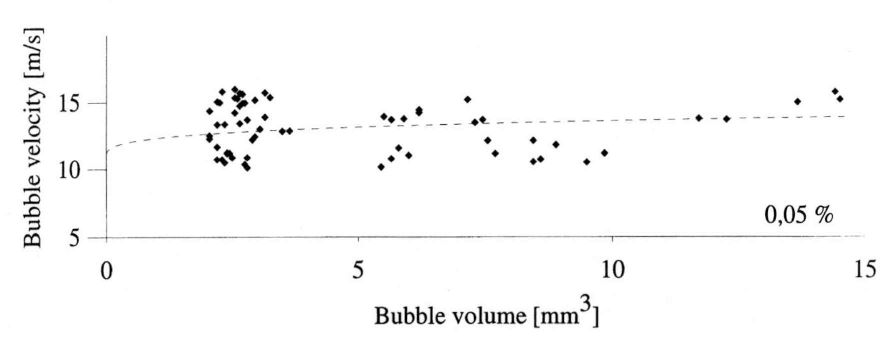

Figure 6.10: Calculated and measured local air bubble velocity in an upward (adiabatic) water/air slit flow (opening 1.1 mm, length 100 mm) for a pressure difference of 2 bar and different mass flow qualities.

some of the smaller bubbles are actually slower than predicted. This deviation could be nevertheless referred to interactions between the movement of the bubbles in a swarm and the (turbulent) water flow, the deceleration of the liquid flow caused by the higher velocity of large bubbles as well as especially to the incompleteness of the one-dimensional mixture flow assumptions.

The local bubble velocity predictions are further validated for water/air flow for a constant pressure difference of 2 bar and different mass flow qualities (Figure 6.10). For the sake of clarity, the calculated bubble velocities were not depicted for the inlet and outlet conditions but only for a mean (linear) value of the homogeneous mixture velocity in the slit. Again, in all cases, the measured and the predicted velocity increases with a larger bubble volume. With a higher mass flow quality, the air expansion due to the pressure decay becomes more distinct, and, in consequence, the velocity of the homogeneous mixture respectively that of the bubbles waxes significantly between the inlet and outlet of the slit. In parallel, for higher mass flow qualities due to the greater increase of the mixture volume between the inlet and outlet of the slit, a larger fluctuation in the measured bubble velocity at an equal bubble volume is prevailing compared to that at smaller mass flow qualities. In concluding, the calculated and the mean value of the measured velocities exhibit the same trend with increasing bubble volumes and do not significantly deviate from each other.

6.9 Drift Velocity in Flashing Water Flow

The establishing mass flow quality in flashing two-phase flow can not yet be accurately predicted due to the occurring boiling delay. As a consequence, an unambiguous slit velocity is not calculable from the measured bubble velocities. In view of the fact that a statement about the steam bubble drift velocity would be helpful, it is offered here the thesis that the (measured) drift velocities in water/air flow should not significantly differ from that in flashing water flow if an equal bubble diameter range, an identical local pressure gradient and a similar density ratio are prevailing. It bases on the evidence that, in comparison to the bubble sizes in water/air mixtures, the bubbles in flashing water flow are in a more narrow diameter band and smaller and, therefore, more spherically. This property would already allow for a more uniform radial expansion and lower bubble drift velocities and, therefore, a proper representation of the homogeneous mixture flow assumption. Further on, in view of the quasi identical thermodynamic (boiling) conditions for all nuclei at a certain longitudinal position in the slit, the size of the steam bubbles in the same distance downstream of the inlet are (again) in the same range. Therefore, also due to this, the velocities should not differ significantly against each other, especially when comparing to the variations prevailing in water/air bubbly flow. In view of all that, a quasi homogeneous mixture flow assumption seems reasonable.

6.10 Conclusion

The phase distribution in (adiabatic) flashing water and water/air bubbly flow is relatively homogeneous. Significant differences especially with respect to the transversal bubble size distribution and the structure of the interphase area, however, prevail. Due to the pressure undershot in the sharp edged slit inlet, cavitation bubbles can be formed even in slightly subcooled water. In consequence, a larger number of microbubbles can be activated for flashing in the slit resulting in this case in a significant smaller liquid superheat.

Apart of the local adiabatic expansion of the air in the bubble in a water/air flow caused by the rapid pressure decrease at the slit inlet further downstream, an almost isothermal expansion behaviour is occurring. The mean velocity of the bubble does not significantly differ from that of the homogeneous mixture; it only slightly increases with higher bubble volume. The local slip ratio was in all cases less than 1.2 for the largest bubbles, so here, a quasi homogeneous one-dimensional two-phase flow can be assumed. The isothermal expansion at the later stage and the identity between the homogeneous mixture velocity and that of the bubbles may be referred to the high specific interphase area caused by the irregular bubble top expansion and resulting air streaks. The drift velocity of the air bubbles can be recalculated by using the equilibrium between the pressure and friction forces on the bubble surface. In principle, the assumption of a homogeneous mixture flow in the slit used by Pana [3] is adequate.

References

[1] H. John, J. Reimann, G. Eisele: *Kritische Leckströmung aus rauhen Rissen.* Bericht KfK 4192, 1987.

[2] F. Westphal: *Berechnungsmodell für die Leckraten aus Rissen in Wänden druckführender Apparate und Rohrleitungen.* Diss. Univ. Dortmund, 1991.

[3] P. Pana: *Eine modifizierte Bernoulli-Gleichung für die Berechnung der Strömungsvorgänge im unterkühlten Wassergebiet.* IRS-W-18, 1975.

[4] W.-H. Isay: *Kavitation.* Schiffahrts-Verlag, 1989.

[5] N. Westphal: *Keimverteilungsmessung mit dem Laser-Streulichtverfahren bei Tragflügel- und Propeller-Strömungen.* Bericht Nr. 408 Institut für Schiffbau, Univ. Hamburg, 1981.

7 Transient Behaviour of Two-Phase Slug Flow in Horizontal Pipes

Kathrin Grotjahn and Dieter Mewes *

Abstract

Two-phase co-current flow is experimentally and numerically investigated in horizontal pipes in the transition region between plug and slug flow. Single plugs and slugs are accelerated in a pipe with the length of 44 m and the inner diameter of 59 mm. The measurements are performed with air and water at atmospheric conditions. Superficial gas velocities are chosen between 0,6 m/s and 3,9 m/s. The obtained results are corresponding to pressure drop, slug front velocities and slug length. Tomographic measurements are performed on the liquid hold-up profile within the flow. The results of the experimental investigations are compared to results from calculations made for the acceleration of single plugs and slugs with a simple numerical approach.

7.1 Introduction

Gases and liquids are often transported along horizontal or inclined pipes within chemical and energetic processes. For example in the oil- and gas industry, multiphase flows of gas, vapour and liquid are transported on the way from the well to reservoir tanks or separators in pipes over several miles. Different phase distributions can be observed during the transportation [1, 2]. The phase distribution is a function of the volumetric flow rates of the phases, the boundary conditions of the flow field (inclination and pipe diameter) and the physical properties of the phases [3–6].

---

* Universität Hannover, Institut für Verfahrenstechnik, Callinstr. 36, D-30167 Hannover, Germany

Plug and slug flow as well as the transition between those two flow regimes occur most frequently. Besides the integral parameters like superficial velocities, the pressure drop is the characterizing parameter of the flow and of the transition region. It is most influenced by the volumetric liquid hold-up and the distribution of the phases [7]. Corresponding to the phase distribution, different forces are exhibited by the flow onto the pipe walls and connected fittings, separators or pumps. Alternate unknown stresses especially in the transition region are caused. For the calculation of the pipe diameter or the pressure drop, the local velocity field and phase distribution inside every cross-sectional area have to be known.

In this paper, experimental investigations are obtained for the phase velocities, phase distribution and the pressure drop of single plugs and slugs. Measurements of single plugs or slugs moving along the pipe over a stratified flow enable the detailed description of the flow phenomenon and physical behaviour. Interference of preceding or following plugs or slugs according to entrance or outcome effects is avoided. The local phase distributions are measured with a conductive tomographic system developed by Reinecke et al. [8]. Experimental results will be compared with numerical results calculated with a numerical approach for the acceleration of single plugs and slugs.

7.2 Slug Flow in Horizontal Pipes

According to Figure 7.1, four major flow regimes can be described in co-current horizontal two-phase flow. These are dependent on the volumetric flow rates of the gas and the liquid phase, the stratified flow, the plug flow, the slug flow and the annular flow. Further flow regimes as well as the regime boundaries or transition regions between the major flow regimes have been discussed by several authors [7, 9]. For low volumetric flow rates, the stratified flow occurs. For increasing gas flow rates, Kelvin-Helmholtz-Instabilities occur, and the transition to the plug flow regime takes place. Alternating liquid and gas zones can be observed. Increasing the gas flow moreover, bubbles will be entrained into the front of the liquid zones. In accordance to this gas entrainment, the transition from plug to slug flow is defined.

In Figure 7.2, the characteristic flow parameters of the slug flow are depicted. The liquid zone is not aerated for plug flow and aerated with the gas hold-up ε_S for slug flow. The slug length is defined by l_S. Several authors published data for a stable slug length l_S of about $12d$ until $40d$ [10, 11]. During the motion of a slug, liquid is picked-up from the film with the height h_{film} in front of the slug. The liquid is transported through the liquid zone of the slug and is distributed at the end to the liquid film behind the slug. For increasing gas flow rates, simultaneous to the pick-up of liquid, gas is entrained in the liq-

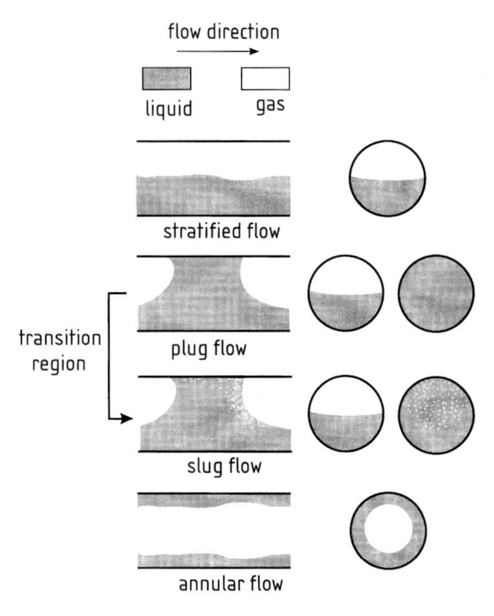

Figure 7.1: Flow regimes of two-phase cocurrent flow in horizontal pipelines.

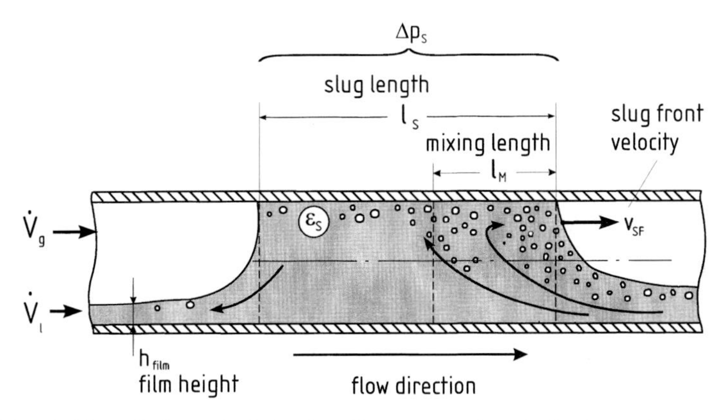

Figure 7.2: Definition of characteristic slug flow parameters in horizontal pipes.

uid zone. The gas is swirled in the mixing zone with the length l_{M}. Due to the lower density, gas bubbles rise up to the top of the pipe. They are transported through the liquid zone until they coalescence with the gas zone following the slug. The slug front moves with the slug front velocity v_{SF}. It depends on the superficial velocities of the gas and the liquid phase. Linear and exponential dependencies are described by Jepson [12] and Gregory and Scott [5].

Plug and slug flow are furthermore characterised for stationary volumetric flow rates by the pressure drop Δp_{S}. For the calculation of the pressure drop, several mathematical correlations are given by Kordyban [13] and Spedding et

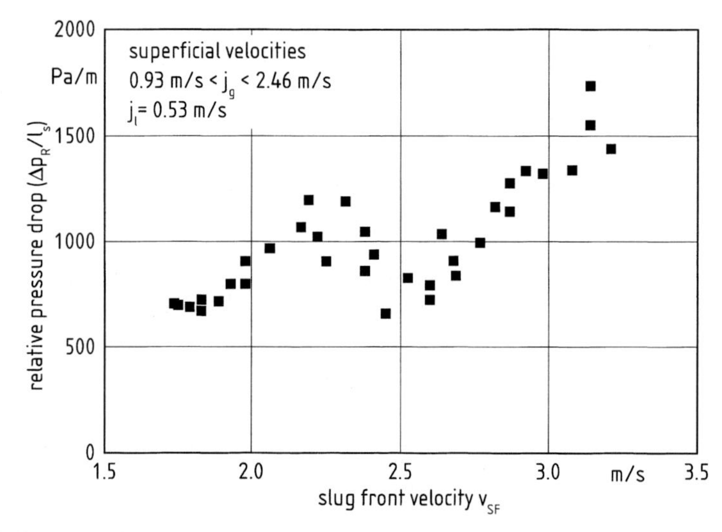

Figure 7.3: Pressure drop of individual slugs as a function of the slug front velocity found by Reinecke et al. [8].

al. [14]. Dukler and Hubbard [10] and Nickolson et al. [11] describe the physical properties of the slug flow in detail for calculating the pressure drop. According to Dukler and Hubbard [10], the pressure drop Δp_S can be devided into the acceleration pressure drop Δp_B due to the accelerated liquid picked-up from the film and transported in the mixing zone and into the frictional pressure drop Δp_R. Azzopardi et al. [15] compare experimental results with calculations. The calculated pressure drop is dramatically influenced by the empirically correlated gas hold-up within the used equations.

Ozawa et al. [16] measure the pressure drop for two-phase flow in capillaries in dependence on the relative gas hold-up. Similar measurements are conducted by Reinecke et al. [8] in horizontal gas-liquid slug flow and shown in Figure 7.3. The relative pressure drop is plotted as a function of slug front velocity. The measurements are conducted in a horizontal pipe with the inner diameter of 59 mm and with air and water for atmospheric condition.

The pressure drop decreases for increasing gas flow in the transition region between plug and slug flow. This occurs due to the decreasing density and viscosity of the mixture within the slug body for increasing gas hold-up. Therefore, the velocities of the plug or slug body increase.

Experimental investigations on the dispersion of gas into the front of the slug are described by several authors. Nydal and Andreussi [17] measure the void fraction of slugs, using flush mounted ring-electrodes. The dispersion of the gas into the liquid is only observed when the velocity of the slug reaches a critical value and sufficient liquid in front of the liquid zone in form of the liquid film is present.

For the calculation of the gas hold-up in a slug, different methods are developed. The equations can be classified into three groups. Equations in the first group are investigated empirically and without relation to a flow regime. A view of these empirical correlations is given by Spedding et al. [18]. A second group of equations consists of simple physical models using experimental results for correction. Frequently employed equations include the one from Gregory et al. [19]. From their experimental investigation in pipes with 25,8 mm and 51,2 mm inner diameter, they conducted the gas hold-up only to be dependent on the superficial velocities of the phases. The equations of the third group are based on physical models of the dispersion and distribution of the gas within the liquid. The model developed by Barnea and Brauner [20] is based on the assumption that the dispersed phase behaves like ideal bubbles.

7.3 Description of the Experimental Set-Up

In Figure 7.4, the experimental set-up used in this work is schematically depicted. It can be divided into two sections. These are the entry section for the accumulation of single plugs or slugs, and the measurement section with the measurement technique. The inner diameter of the pipe is $d = 59$ mm and the total length is $l = 44$ m. For making the flow visible, a transparent PVC is used as pipe material. The experimental investigations of single plugs or slugs are carried out during atmospheric conditions with air and water. Before the measurement starts, the part of the entry section is filled with a defined amount of water. In the measurement section, a stratified flow of continuous flowing air and water is produced. The film height h_{film} of these stratified flow is detected at different pipe length with two vertical, parallel wires using the conductance measured between them. The volumetric flow rates of the phases are measured with orifice meters.

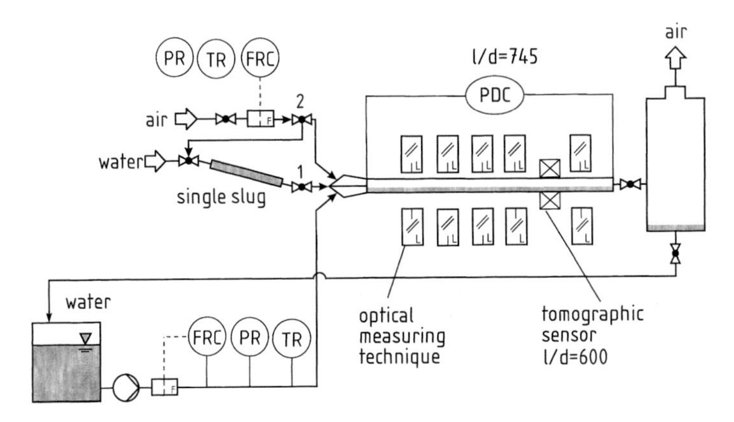

Figure 7.4: Schematic representation of the experimental set-up.

Starting the measurements, valve 1 is opened and valve 2 is switched from the measurement pipe to the inlet of the entry pipe. Thereby, the gas flow accelerates the water plug out of the entry section into the measurement section. During the experiments, the characteristic plug and slug parameters are measured time- and location-dependent. The slug front velocities v_{SF} are measured every meter. When the plug or slug passes optical barriers mounted directly on the pipe circumference, a volt signal dependent on the phase distribution in the barrier is transmitted.

7.3.1 Conductive tomographic measurement technique

At the relative pipe length of $l/d=600$, the tomographic sensor developed by Reinecke et al. [8] is installed. The sensor for the conductive tomography is depicted schematically in Figure 7.5. In three parallel planes (3 mm apart), sets of 29 parallel wires are fixed across the cross-sectional area of the pipe. The wire diameter is 100 μm, and the wires are fixed between the planes with a distance of 2 mm so that the arbitrary cross-sectional area is about 95%. Thus, the flow and the pressure drop are not strongly influenced by the sensor.

The conductivity is measured between two wires and directly depends on the electrical properties of the phases and thus of the phase distribution. By the continuous measurement of the conductivity between pairs of wires, for each plane, 28 linear independent values are conducted. The measurement is started shortly before the plug or slug enters the measurement plane of the sensor, and its duration is 1.8 s with an image refresh rate of 110 pictures per second. After the measurement of the conductivity field, the phase distribution

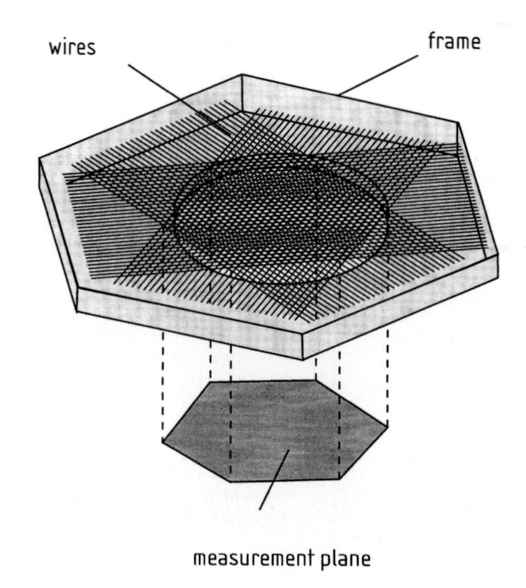

Figure 7.5: Schematic representation of the tomographic sensor.

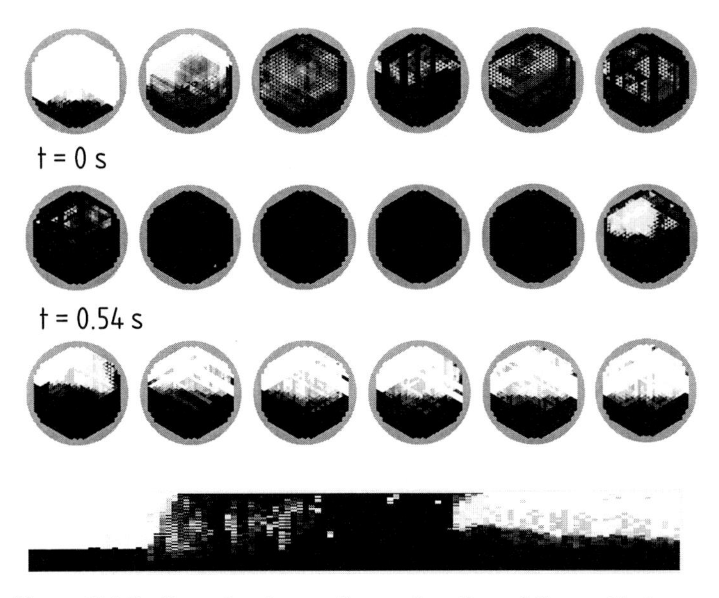

Figure 7.6: Phase distribution of a slug unit as a function of time with temporal distance of 0.09 s found by Reinecke et al. [8].

in the measurement plane has to be reconstructed. For the reconstruction of the phase distribution, an iterative method called ART is used. The reconstruction method was optimised with regard to the quality, i.e. by the time correction, proposed by Reinecke et al. [21].

Typical reconstructed phase distributions of a slug body for different times are given by Reinecke et al. [8] and are depicted in Figure 7.6. Every tenth image of the reconstruction is given. Thus, the temporal distance between the images is 0.09 s. The liquid phase is represented in dark shading and the gas phase in bright one. The slug front velocity was 3 m/s. By the continuous adding of the single cross-sectional areas, three-dimensional pictures can be obtained. Cutting these three-dimensional pictures in flow direction, the two-dimensional image at the bottom of Figure 7.6 is obtained.

7.4 Experimental Results

The acceleration behaviour of single plugs and slugs is detected in dependence on different superficial velocities of the phases. Thus, the liquid film thickness of the stratified flow in front of the plugs and slugs is varied. The existence of stationary and instationary movement is dependent on the covered distance along the pipe. The slug growth and the decay of slugs dependent on the film thickness of the liquid layer in front of the slug is investigated. In the

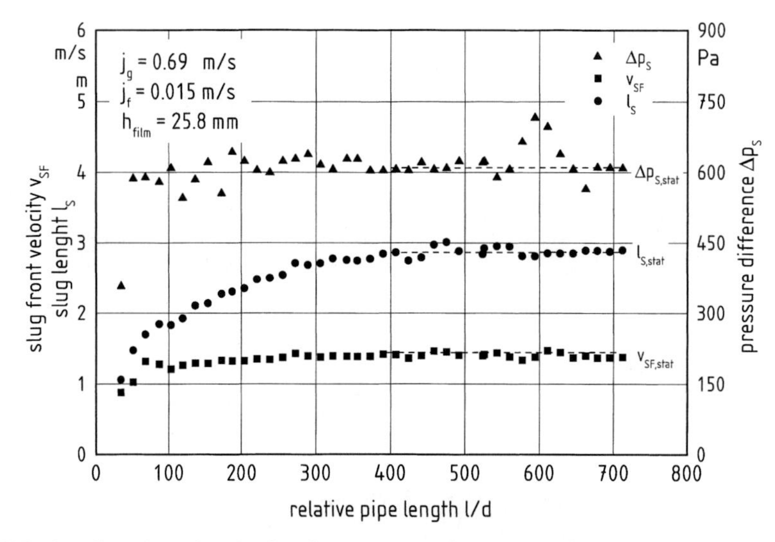

Figure 7.7: Acceleration of a single plug moving along a pipeline.

present paper, only plugs or slugs from measurements with stationary conditions at the end of the pipe are conducted. The superficial gas velocity is varied between j_g=0,69 m/s and j_g=3,59 m/s and the superficial liquid velocity between j_l=0,005 m/s and j_l=0,02 m/s. The detected film heights of the stratified flow in front of the plugs and slugs thus varied between h_{film}=15 mm and h_{film}=30 mm. The plug feed in the entry section has a length of l_S=1 m for starting conditions.

One typical measurement result for a not aerated plug with stationary conditions at the end of the pipe is shown for superficial velocities j_g=0,69 m/s and j_l=0,015 m/s in Figure 7.7. The film height of the stratified flow conducts h_{film}=25,8 mm. After starting the measurement, the plug accelerates with rapidly increasing slug front velocity along the pipe. Averaged stationary values for the slug front velocity $v_{SF,stat}$=1,40 m/s are measured after the relative pipe length of l/d=400. Correspondingly, the slug length and the pressure difference increase until the stationary values of $l_{S,stat}$=2,92 m and $\Delta p_{S,stat}$=620 Pa are reached. For l/d=600, the negligible influence of the sensor on the measured pressure difference Δp_S can be obtained. Due to the reduced cross area of about 5%, the pressure difference increases about 100 Pa.

In Figure 7.8, the tomographically measured liquid hold-up (1–ε) is depicted for several single plugs and slugs. The liquid hold-up is plotted as a function of length for three different values of the superficial gas velocity. For this, the measured time plots of the liquid hold-up are correlated with the different slug front velocities to enable the comparison. For the plug presented in Figure 7.7, typical liquid hold-up values are investigated in Figure 7.8. The plug moves with the stationary front velocity of v_{SF}=1,40 m/s through the sensor. In front of the plug, the stratified flow with a very smooth surface equiva-

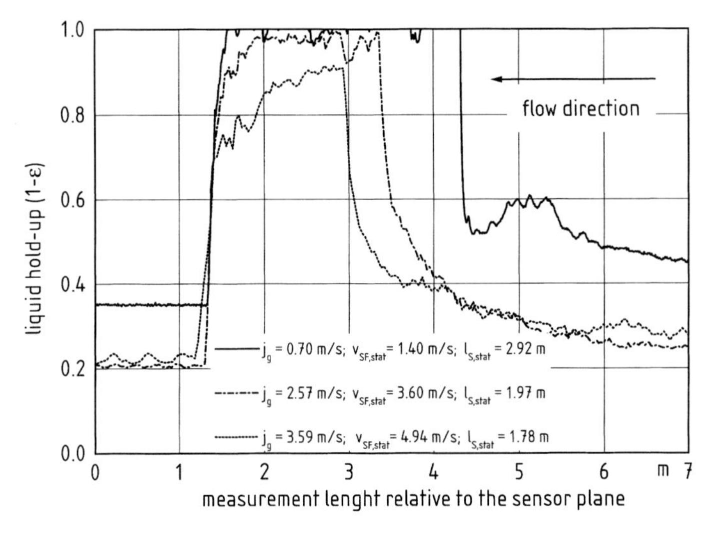

Figure 7.8: Liquid hold-up in the slug as a function of the measurement length and different inlet conditions.

lent to the low superficial gas velocity is conducted due to a liquid hold-up of $(1-\varepsilon)=0{,}35$. The front of the plug is only slightly aerated and so, the liquid hold-up in the plug zone rapidly increases to values of $(1-\varepsilon)=1$. In the liquid zone, gas bubbles at the top of the pipe can be detected as small releases in the liquid hold-up. After the plug length of $l_{\mathrm{S,stat}}=2{,}92$ m, the liquid hold-up decreases and the stratified flow follows with waves on the surface. Increasing the superficial gas velocities up to values of about $j_{\mathrm{g}}=3{,}59$ m/s, significant different profiles of the liquid hold-up are detected. Due to the increased superficial gas velocities and the appropriate slug front velocities, the slugs are now highly aerated in the mixing zone of the slug. Along the liquid zones of the slug, the liquid hold-up increase up to values of 0,9 for very fast moving slugs. The front profiles of the slugs are less sharper than the profile of the plug due to increased gas entrainment. After the liquid zone, the liquid hold-up decreases slowly down to values of the stratified flow conducted in front of the slugs. From this representation, the different heights of the liquid film in the stratified flow in front of and behind the slug can be seen. The surface behind the slug is wavier than in front of the slug. Due to the transferred liquid, picked-up in front and deliver at the end of the slug, the liquid level behind the slug is higher than in front, compared with the results obtained by Cook and Behnia [22].

In Figure 7.9, the slug front velocity for stationary movement of single slugs is plotted as a function of the mixture superficial velocity for all performed measurements. A linear correlation for the data can be found as:

$$v_{\mathrm{SF}} = v_0 + C_0\, j_{\mathrm{m}} = 0{,}55 + 1{,}144\, j_{\mathrm{m}}\,. \tag{1}$$

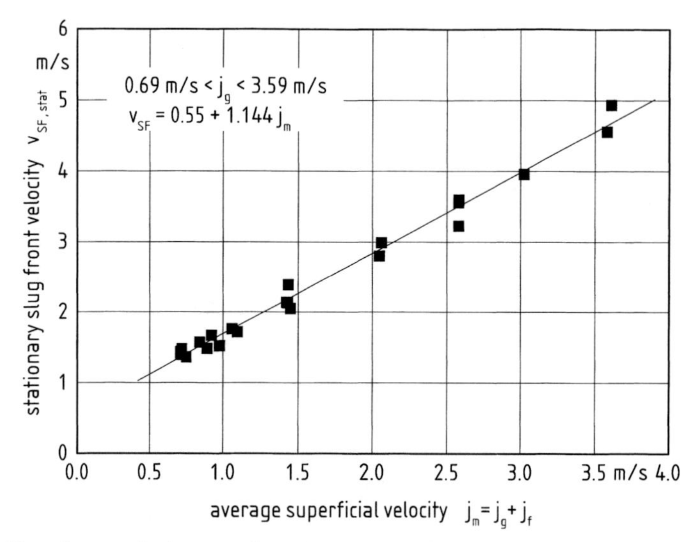

Figure 7.9: Slug front velocity as a function of superficial mixture velocity.

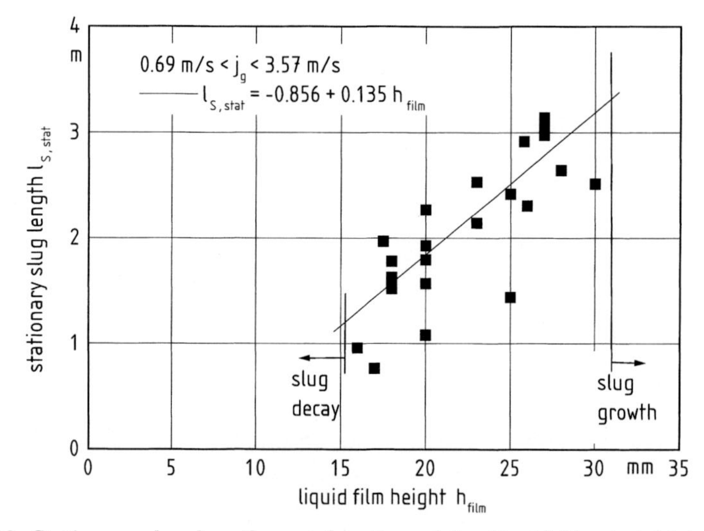

Figure 7.10: Stationary slug length as a function of the liquid film height in front of the slug.

The result sustained from the experimental investigations of single plugs and slugs in this paper is in accordance to the data investigated on slug flow with several slugs found by Reinecke et al. [8] (v_0=0,13 m/s; C_0=1,14). They are also comparable to correlations found by other authors, for example Gregory and Scott [5] (C_0=1,35), Dukler and Hubbard [10] (C_0=1,24–1,3), and Jepson [12] (C_0=1,4–1,5). For further investigations on the calculation of single plug and slug movement, the simplified linear correlation $v_{SF}=1{,}41\ j_g$ is used as will be shown below.

Investigations on the slug growth and decay dependent on the liquid film height of the stratified flow in front of the slug are depicted in Figure 7.10. The stationary slug length $l_{S,stat}$ is plotted as a function of the liquid film height h_{film} for measurement data with superficial gas velocities between $j_g=0{,}69$ m/s and $j_g=3{,}57$ m/s. During the investigations, no stationary slug length at the end of the pipe can be found for a liquid film height in front of the slug less than 15 mm or above 30 mm. For film heights over 30 mm, the accelerated slugs growth over the whole pipe length. For film heights less than 15 mm, no acceleration until stationary conditions is possible because the feeded slug disappears during the measurement. These results are in agreement to the results found by Woods and Hanratty [23] on the stability of slugs. They obtained their results for an air and water flow at atmospheric condition in a horizontal pipe with 0,0953 m inner diameter. The relative minimum liquid height in front of the slugs, which enables a stable slug lenght are found to be $h/d \sim 0{,}2$ ($h=19$ mm).

7.5 Calculation of the Transient Behaviour of Single Plugs and Slugs

For the calculation of the transient movement of one plug or slug, two boundary conditions are considered. Once the acceleration through a constant pressure difference Δp_S, and twice the acceleration through a constant mass flow rate of the gas phase. In this work, the acceleration due to constant mass flow will be presented. This occurs corresponding to the experiments performed with constant superficial gas velocities.

7.5.1 Equation of motion for a single plug

In Figure 7.11 a, the forces acting on a single plug are depicted schematically. The plug will be accelerated by the constant mass flow rate $\dot{M}_g$ of the gas (Figure 7.11 b). Starting point is the local parameter x=0 m. The pressure p_1 depends on the mass flow rate of the gas and the storage volume V_P due to the free gas volume behind the slug. During the acceleration of the single plug through the pipe, the free volume behind the slug increases. Thus, the pressure p_1 will be changed. The pressure p_0 is assumed to be the same as environment. By summing up all forces, the equation of motion

$$M\ddot{x} = \left(p_1 - p_0 - \left(\frac{\Delta p_R}{l_S}\right) l_S\right) A \tag{2}$$

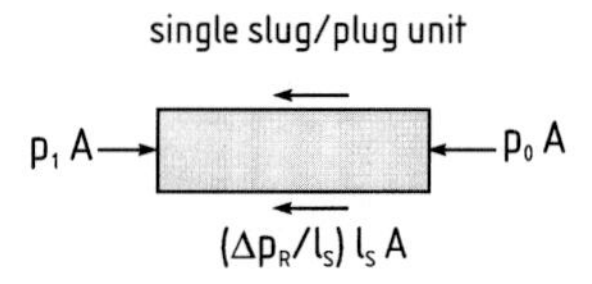

Figure 7.11a: Forces acting on a single plug or slug unit.

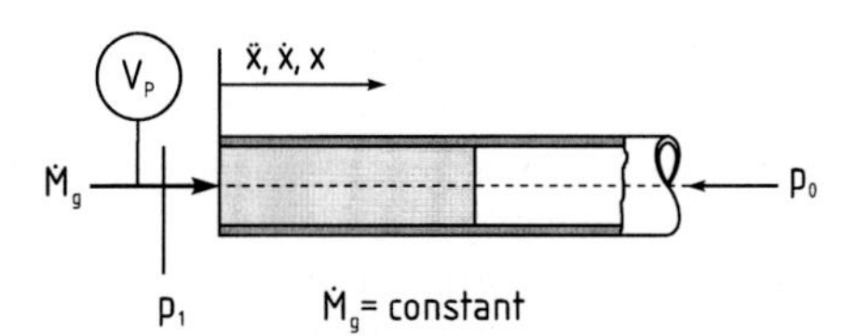

Figure 7.11b: Acceleration due to a constant mass flow rate with the gas storage volume V_P.

with the acceleration $\ddot{x}$ follows. The frictional pressure drop referring to the slug length l_S is defined to Δp_R. The cross area of the pipe with the diameter $d=0{,}059$ m is A. The mass of the water plug with the water density ρ_W is defined to:

$$M = \frac{\pi d^2}{4} l_S \rho_W \,. \tag{3}$$

The pressure p_1 will be calculated by means of the ideal gas law to:

$$p_1 = \frac{(M_P + \dot{M}_g t) R_g T}{V_P + xA} \tag{4}$$

with the mass M_P of the storage volume for the time $t=0$ s. T is the atmospheric temperature and R_g the ideal gas constant. x is the covered way through the pipe.

For the equation of motion, the inhomogeneous differential equation of second order follows to:

$$\ddot{x} = \left(\frac{(M_P + \dot{M}_g t) R_g T}{V_P + xA} - p_0 - \left(\frac{\Delta p_R}{l_S} \right) l_S \right) \frac{A}{M} \,. \tag{5}$$

The analytical solution of Equation (5) does not exist. Thus, the numerical Runge-Kutta-Nystroem algorithm with constant time-steps of 0,0001 s will be applied. The calculation starts for known initial conditions for the time $t=0$ s. Then, x is zero and the velocity $\dot{x}$ of the plug is also zero. The acceleration, velocity and the way passed by the plug will be calculated for each time-step with the values of these data at the time-step before. The calculation ends for the length of x=44 m equivalent to the measurement set-up. Before starting one calculation, the input parameters $\dot{M}_g$, the plug length l_S and the storage volume V_P for the time $t=0$ s must be chosen. For constant mass flow rates of the gas, $\dot{M}_g$ will be calculated for a given superficial gas velocity with the following equation for stationary end velocities:

$$\dot{M}_g = \frac{(\Delta p_{S,stat}(j_g) + p_0)1{,}41 j_g A}{R_g T} \,, \tag{6}$$

using $\dot{V}_g = v_{SF} A = 1{,}41\, j_g A$ and $p_1 = \Delta p_R(j_g) + p_0 = \Delta p_{S,stat}(j_g) + p_0$.

This accords the above presented empirical correlation for the slug front velocity $v_{SF,stat}$ equal to the velocity $\dot{x}_{stat}$ at the end of the calculation. For stationary final velocities, the pressure p_1 is equivalent to the frictional pressure drop Δp_R and thus to the measured pressure difference $\Delta p_{S,stat}$ shown in the experiments above. The pressure depends on the superficial gas velocity j_g.

In Figure 7.12, the frictional pressure drop $\Delta p_R/l_S$ used for the calculations is depicted as a function of the superficial gas velocity for a moving single slug or plug. The frictional pressure drop is calculated by different parts of polynomials. The polynomials are chosen in accordance to the measurement data obtained by Reinecke et al. [8] shown in Figure 7.3. For low superficial gas velocities, a single not aerated plug is moving through the pipe with a stationary end velocity in the subcritical range up to $j_g = 1$ m/s. For increasing velocity, the slug will be entrained as shown above, and the pressure drop decreases in the transition region. The average densities and viscosities are decreasing because of increasing gas entrainment in the liquid zone.

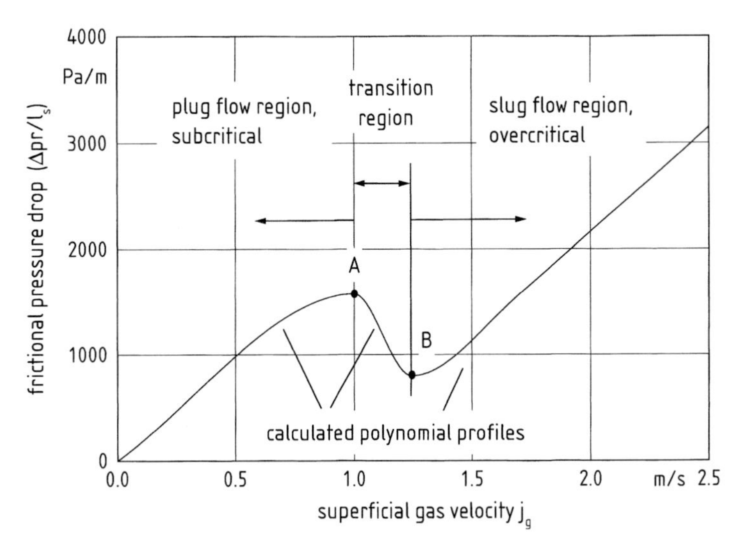

Figure 7.12: Frictional pressure drop as a function of the superficial gas velocity.

7.6 Comparison between Measured and Calculated Results

In Figure 7.13, experimental results for the slug front velocity and calculated results for three different inlet conditions are plotted as a function of the relative pipe length l/d. The storage Volume V_P for the time $t=0$ s is chosen to be 0,1 l equivalent to the free pipe gas volume at the beginning of the measurements behind the entry section. The mass flow rates are calculated from the conditions used for the experimental investigations depicted in Figure 7.13. For all calculated profiles, the slug velocities oscillate until the stationary velocity $v_{SF,stat}$ is reached at difference relative pipe length. For decreasing inlet mass flow rates, the stationary condition will be reached faster. The period of the oscillations increases for all calculations with length because of the increasing volume of gas behind the liquid volume. The oscillations are caused by the inertia of the liquid slug. For all depicted cases, the measurement data for the stationary slug front velocity are reproduced in accordance by the calculated data.

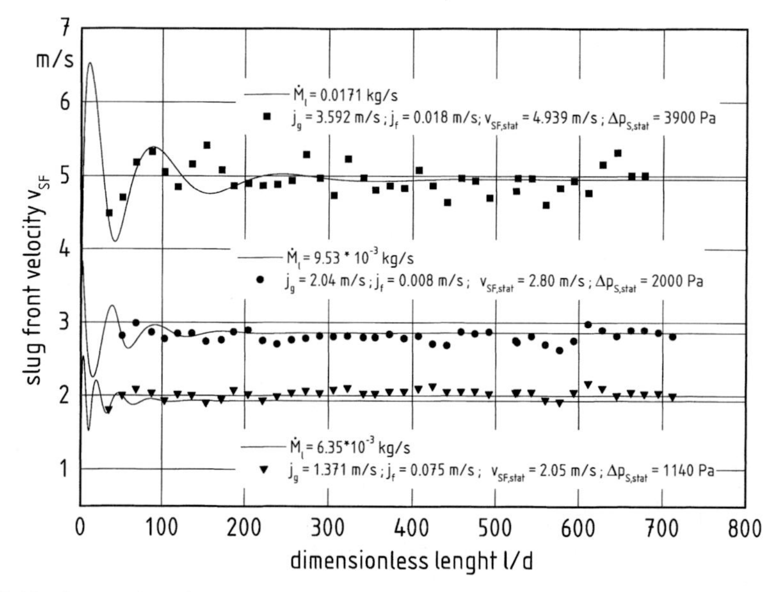

Figure 7.13: Comparison between calculated and measured data for different inlet conditions.

7.7 Conclusion

Experimental and numerical investigations for the acceleration of single plugs and slugs are described. The results for stationary slug front velocities are in accordance to each other. In further works, the measurements and the calculations have to be extended especially on the liquid hold-up in the transition region, which is not reported in this work (see [8]). Furthermore, the calculations have to be performed for several plugs and slugs, using the physical behaviour obtained on single plugs and slugs. Using the results, precise calculation of velocity fields and phase distributions of plug and slug flow can be evaluated

References

[1] F. Mayinger: *Strömung und Wärmeübergang in Gas-Flüssigkeits-Gemischen*. Springer Verlag, 1982.

[2] Y. Taitel: *Flow pattern transition in two-phase flow*. 9th Int. Heat Transfer Conference, Jerusalem, 1990.

[3] M. Nädler, D. Mewes: *The effect of gas injection on the flow of immiscible liquids in horizontal pipes*. Chem. Eng. Technology **18** (1995) 156–165.

[4] M. Nädler, D. Mewes: *Effects of liquid viscocity on the phase distribution in horizontal gas-liquid slug flow*. Int. J. Multiphase Flow **21** (1995) 253–266.

[5] G. Gregory, D.S. Scott: *Correlation of the liquid slug velocity and frequency in horizontal cocurrent gas liquid flow*. AIChE J. **15** (1969) 933–935.

[6] H. Herm-Stapelberg, D. Mewes: *The pressure loss and slug frequency of liquid-liquid-gas slug flow in horizontal pipes*. Int. J. Multiphase Flow **20** (1994) 285–303.

[7] G.F. Hewitt: *Phenomena in horizontal two-phase flow*. Proc. Int. Conf. on Multiphase Flows, Tsukuba, Japan, 1991.

[8] N. Reinecke, G. Petritsch, M. Boddem, D. Mewes: *Tomographic imaging of the phase distribution in two-phase slug flow*. Int. J. Multiphase Flow **24** (1998) 617–634.

[9] O. Baker: *Designing for simultaneous flow of oil and gas*. Oil a. Gas J. **53** (1954) 185–195.

[10] A.E. Dukler, M.G. Hubbard: *A model for gas-liquid slug flow in horizontal and near horizontal tubes*. Ind. Eng. Chem. Fund. **14** (1975) 337–347.

[11] M. Nicholson, K. Aziz, G.A. Gregory: Intermittent two-phase flow in horizontal pipes: Predictive models. *Can. J. Chem. Engng.* **56** (1978) 653–663.

[12] W. Jepson: *The flow characteristics in horizontal slug flow*. Paper F2, 3rd Int. Conf. on Multiphase Flow, The Hague, 1987.

[13] E.S. Kordyban: *Flow model for two-phase slug flow in horizontal tubes*. J. Bas. Engng. **83** (1961) 613–618.

[14] P.L. Spedding, J.J. Chen, V.T. Nguyen: *Pressure drop in two-phase gas-liquid flow in inclined pipes*. Int. J. Multiphase Flow **8** (1982) 407–431.

[15] B.J. Azzopardi, A.H. Govan, G.F. Hewitt: *Two-phase slug flow in horizontal pipes*. Preprints Vol. 2, Symposium Pipelines, Institution Chem. Engineers European Branch, Utrecht, 1985, pp. 213–225.

[16] M. Ozawa, K. Akagawa, T. Skagucchi: *Flow instabilities in parallel-channel flow systems of gas-liquid two-phase mixtures.* Int. J. Multiphase Flow **15** (1989) 639–657.

[17] O.J. Nydal, P. Andreussi: *Gas entrainment in a long liquid slug advancing in a near horizontal pipe.* Int. J. Multiphase Flow **17** (1991) 179–189.

[18] P.L. Spedding, K.D. O'Hare, D.R. Spence: *Prediction of holdup in two-phase flow.* Particulate Phenomena and Multiphase Transport, Vol 1, Ed.: T.N. Veziroglu: Hemisphere Publishing Corp., London, 1988.

[19] G.A. Gregory, M.K. Nicholson, K. Aziz: *Correlation of the liquid volume fraction in the slug for horizontal gas-liquid slug-flow.* Int. J. Multiphase Flow **4** (1978) 33–39.

[20] D. Barnea, N. Brauner: *Holdup for the liquid slug in two-phase intermittend flow.* Int. J. Multiphase Flow. **11** (1985) 43–49.

[21] N. Reinecke, M. Boddem, D. Mewes: *Improvement of tomographic reconstructions by time-correction of sequential measurements.* Proc. ECAPT '94, Oporto, Portugal, Process Tomography – A Strategy for Industrial Exploitation –, Eds.: M.S. Beck et al., 1994, pp. 381–392.

[22] M. Cook, M. Behnia: *Film profiles behind liquid slug in gas-liquid pipe flow.* AIChE J. **43** (1997) 2180–2186.

[23] B.D. Woods, T.J. Hanratty: Relation of slug stability to shedding rate *Int. J. Multiphase Flow* **22** (1996) 809–828.

8 Experimental Investigations of Transient Two-Phase Bubbly Flow in Vertical Pipes

Frank Tillenkamp and Ralf Loth *

8.1 Introduction

This work studies the transient behaviour of a two-phase pipe flow during a rapid change from one flow condition to another due to a rapid increase of the gas flow rate at the bottom of a vertical pipe. To characterize the transient flow behaviour, it was intended to measure and determine local flow parameters for several transients within the bubbly flow regime. Shown are those results, which could be achieved during the sponsored time period 1995 until 1996. Parameters, which were measured respectively determined with time, are the local void fraction, the local phase velocities and the local bubble chord lengths.

The ensemble averaging method was used, and the experiments had to be planned accordingly for many repetitions of the same transient.

8.2 Experimental Facility

To investigate transient two-phase flows in a vertical pipe, a double-pipe test section was integrated in an air-water test loop. A schematic diagram of the facility is given in Figure 8.1. One of the two optically clear pipes is used as the measuring pipe with six measuring levels, the other pipe serves as hydraulic compensator for the loop during a rapid exchange of the different flows between the two pipes. This flow exchange is initiated via a quick working valve system and defines a transient. It was experimentally proven that the same

* Technische Universität Darmstadt, Fachgebiet Energietechnik und Reaktoranlagen, Petersenstr. 30, D-64287 Darmstadt, Germany

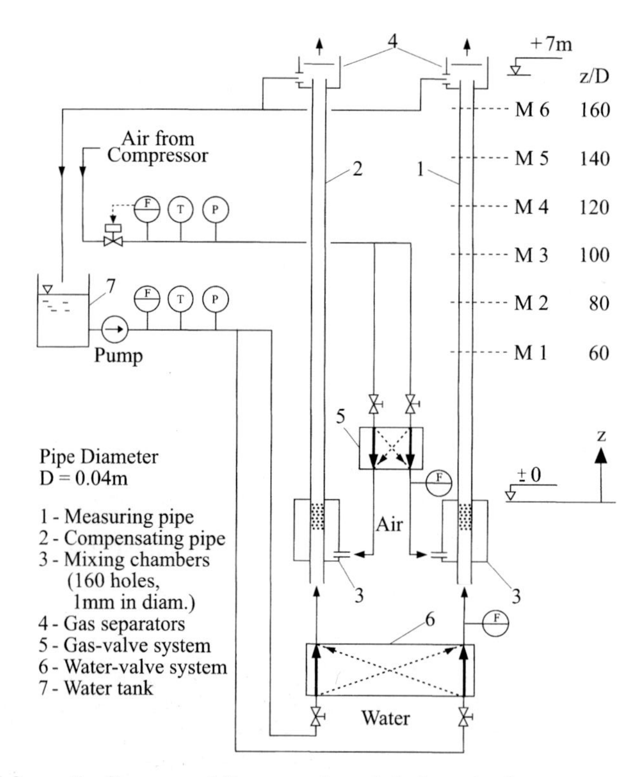

Figure 8.1: Schematic diagram of the experimental air-water-loop.

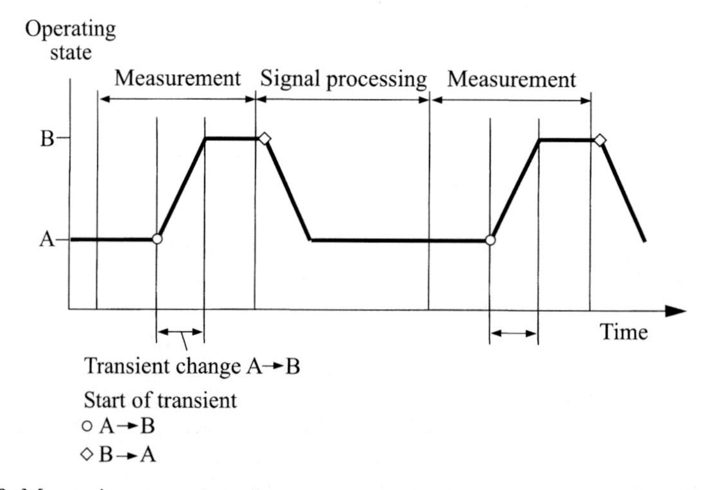

Figure 8.2: Measuring sequence.

transient could be almost identically repeated many times. To fulfil the needs for ensemble averaging, the same transient was 200 times repeated. Figure 8.2 shows the measuring sequence on principle. At the bottom of each pipe, air and water flow is injected via a mixing chamber.

8.3 Measuring Technique

For measuring the two-phase parameters, the following sensors were used: For the local void fractions rectangular fiber-optic sensors with a tip diameter less than 0.25 mm, for the local gas velocities fiber-optic double-sensors whose tips were 5 mm apart, and for the local water velocities a conical miniature hot-film probe, applied to a constant temperature anemometer (CTA). Each of these probes can be traversed radially to achieve local measurements at several radial positions.

As measuring system, the transputer-based measuring system from Schmitt et al. [1], was used, but in a further developed design in such a way that simultaneously 12 channels can be used with a sampling frequency of 1 MHz.

Definition of the measured local two-phase flow parameters:

ε_G: local void fraction (time-averaged phase density function),

$<\varepsilon_G>$: cross-section averaged void fraction under the assumption of rotational symmetry,

v_G: gas-phase velocity measured via interface displacement time,

v_L: liquid-phase velocity measured CTA signal where bubble signals are cut out,

c: bubble chord length determined via the bubble detection time and gas-phase velocity.

8.4 Experimental Results

Local measurements at different heights (M1 through M6) were performed for three different transients, which are characterized by a rapid increase of the gas and water flow rate at the bottom of the measuring pipe via the mixing chamber. The investigated transients I, II and III are defined in Table 8.1 and shown in a flow pattern map by Taitel et al. [2] (see Figure 8.3).

Transient I: Change from a pure water flow (S1) to a bubbly two-phase flow (S2) with a cross-section averaged void fraction $<\varepsilon_G>$ of approx. 3%.

Table 8.1: Superficial velocities of operating points.

Point	Beginning of transient	End of transient	j_G [m/s]	j_L [m/s]
S1	I	–	0	0.82
S2	–	I	0.03	0.88
S3	II	–	0.02	0.96
S4	–	II	0.03	1.02
S5	III	–	0.03	0.41
S6	–	III	0.20	0.70

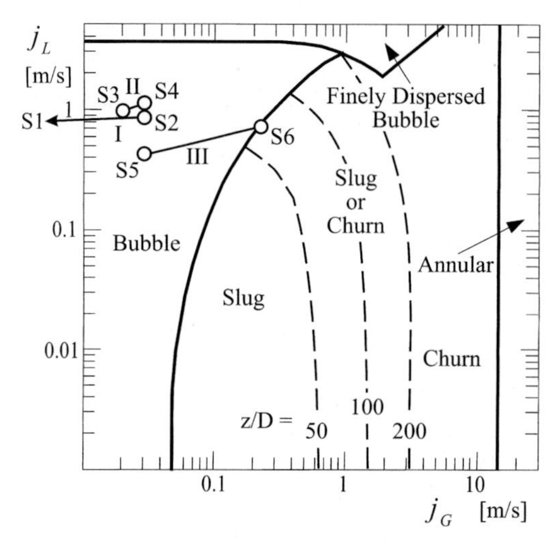

Figure 8.3: Flow pattern map for vertical upflow air-water in pipe [2]. Shown are the flow operating points S1...S6 and the measured transients I, II and III.

Transient II: Change from a bubbly flow (S3), where the radial gas profile shows a wall peak and $<\varepsilon_G>$ is approx. 1% to a bubbly flow (S4) with $<\varepsilon_G>$ of approx. 2.5%.

Transient III: Change from a bubbly flow (S5) with $<\varepsilon_G>$ of approx. 6% to a flow (S6) of almost slug character with $<\varepsilon_G>$ of approx. 18%.

Figure 8.4 shows for the three transients in an overview the variation of local void fraction with time for different radial positions and heights. One sees very clearly the timely change of that flow parameter.

From now on, we only show measurements for pipe height 3.20 m (M2), 4.00 m (M3), and 4.80 m (M4).

In Figure 8.5, the time histories of the cross-section averaged void fractions for the three transients at different pipe heights are shown. In Figure 8.6,

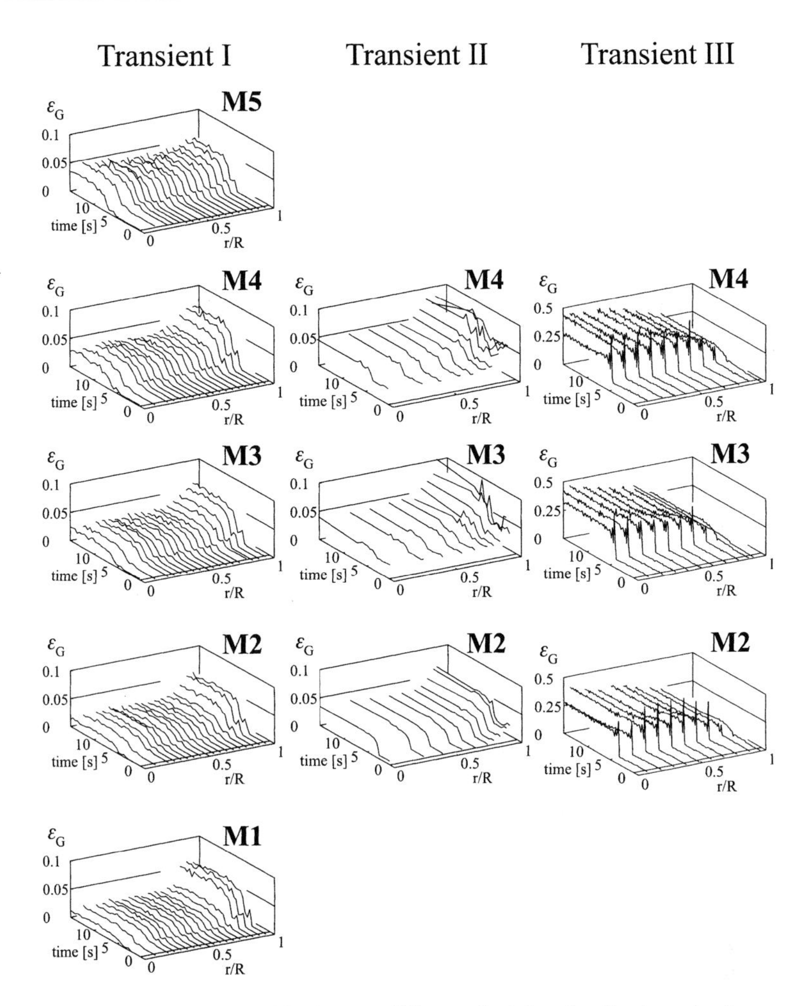

Figure 8.4: Local void fraction with time at different heights for the transients.

there is shown the static pressure behaviour at the wall during the transient II. Comparing for the same transient, the time behaviour of the cross-section averaged void fraction at different heights, one can recognize a great similarity in the time behaviour of the average void fraction and the static pressure.

Before we discuss further the transients itself, lets first observe the development of the radial void fraction profiles in flow direction. Figure 8.7 shows the radial void fraction profile at the steady-state conditions of the start and end flow-points (at $t<0$ s and at $t>15$ s) of the transients I, II and III. The profiles for S2, S3 and S4 show wall peaks. This agrees well with the information in the literature [3–5].

Figure 8.8 shows the time behaviour of the local void fractions for transient I, II and III at three radial positions. These figures allow the following conclusions:

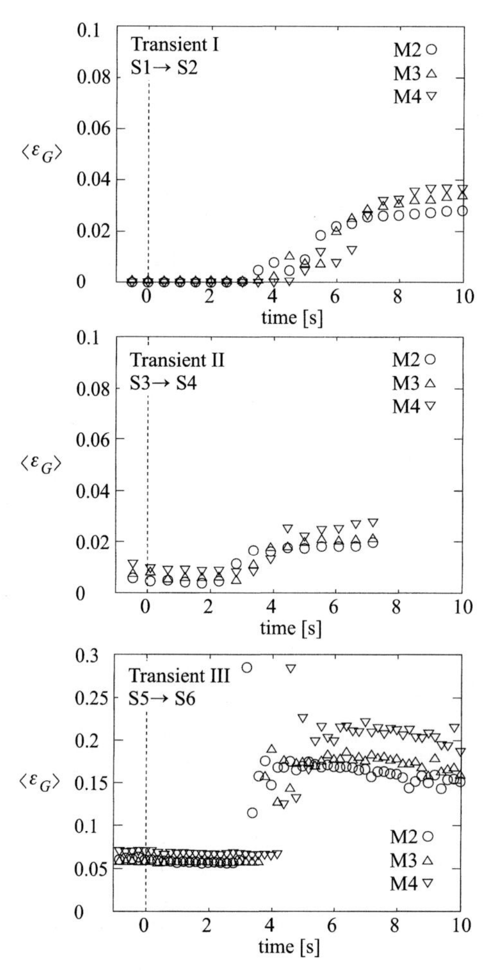

Figure 8.5: Variation of cross-section averaged void fraction with time.

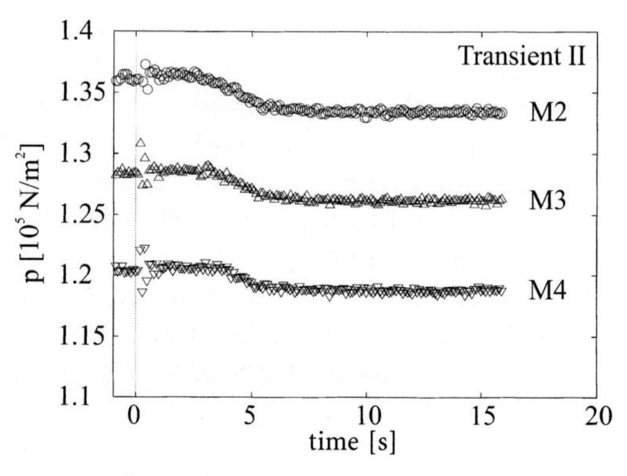

Figure 8.6: Static pressure at the wall with time for transient II.

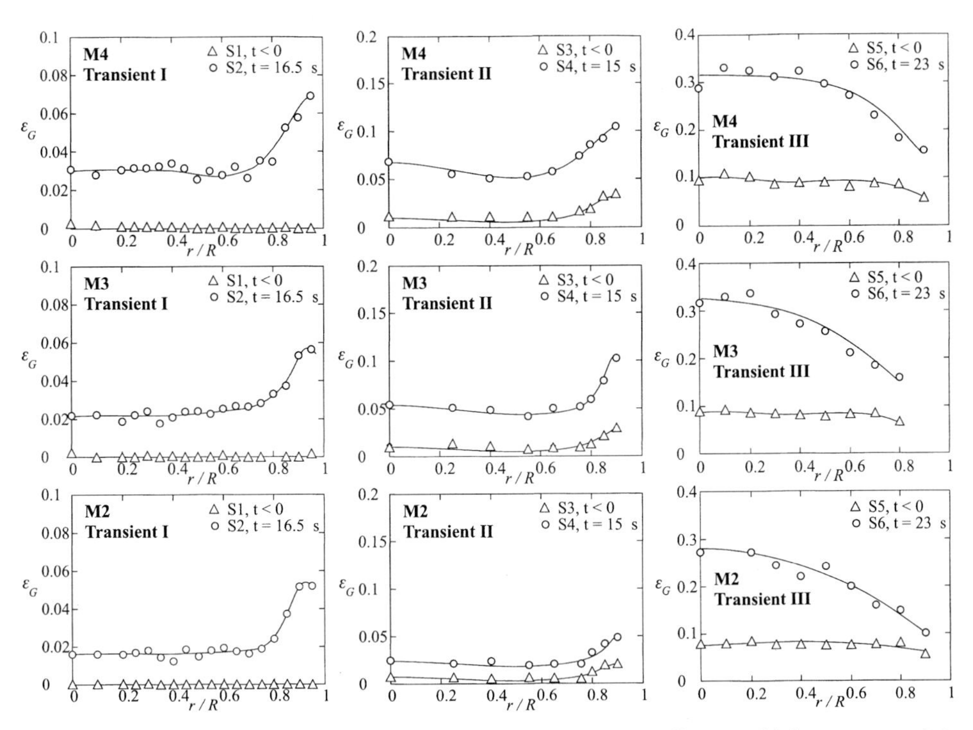

Figure 8.7: Local void fractions. Radial profiles for different flows, which correspond to the beginning and end of a transient.

For transient I: The transient starts first near the pipe wall for all heights with a higher gradient than in the center of the pipe. The development of the wall peak is obvious.

For transient II: This transient starts already with a wall peak of the void fraction. This peak decreases at the beginning of the transient with a certain time lag downstream of the flow, and increases accordingly after several seconds up to the new higher steady-state values of the end flow point. This behaviour can be explained more easily in looking also at the velocity behaviour of the phases.

For transient III: This transient starts with a flat radial void fraction profile at all heights and ends with profiles with maxima at the pipe center (see Figure 8.7). The first measured high values during the transient shown in Figure 8.8 result from few Taylor-bubbles running through the pipe. That had also been observed with a high speed video system.

Further information on the time behaviour of the local phase velocities are available for transient II. Figure 8.9 shows the time behaviour of the gas- and liquid velocities at three different radial positions ($r/R=0$, 0.4, 0.925) along the pipe. Figure 8.10 summarizes again the measured radial phase velocity profiles measured at three pipe heights at the beginning flow point S3 and the end flow point S4 of transient II. One can observe from these figures:

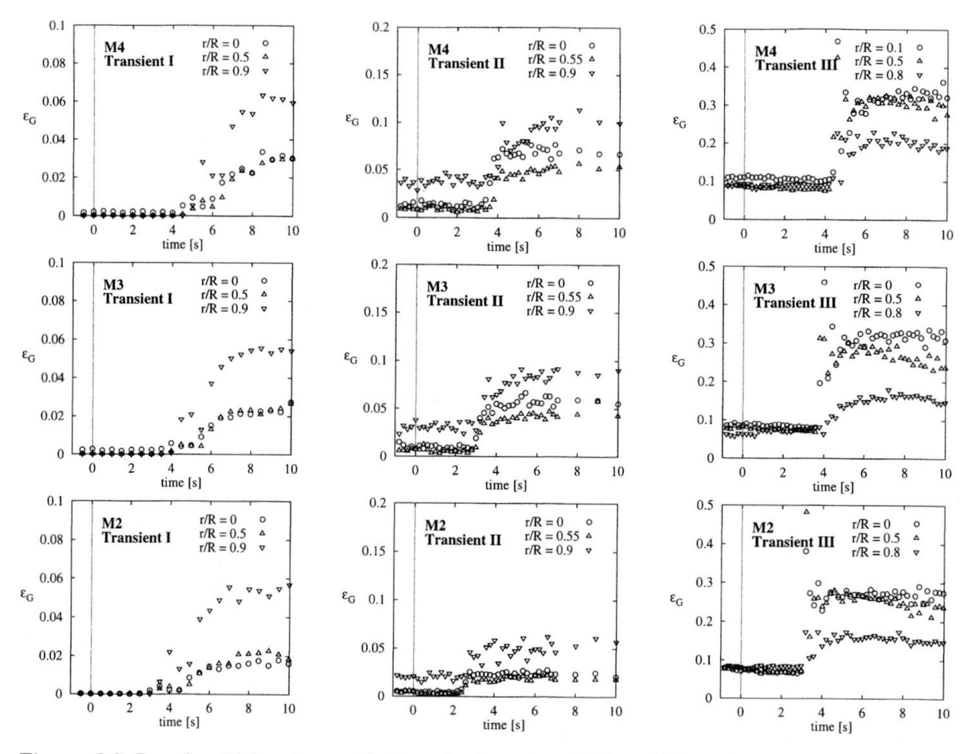

Figure 8.8: Local void fraction with time for transient I, II and III at three radial positions.

- Almost simultaneously with the start of the transient increase at all measuring levels both phase velocities. The velocity gradients near the wall positions are almost half of the size than at the pipe center.
- The radial spreading of the velocities, from the center to the wall of the pipe, stays almost the same during this transient and increases slightly downstream.
- The radial spreading of the liquid velocity is larger than that of the gas velocity at all heights.
- The absolute values of the two phase velocities are very similar, i.e. there is almost no slip for this transient II.

Very interesting are the local bubble chord length distributions. For the steady-state flow situations S3 and S4, Figure 8.11 shows distributions at position $r/R = 0.8$ along the pipe. With regard to the geometrical dimension of the rectangular fiber-optic sensor, the distribution is built for chord lengths larger than 1 mm, with 0.5 mm classes, up to 6 mm.

For S3 ($t < 0$), the maximum of the unsymmetrical distribution shifts to larger chord lengths and gets broader downstream. This means that coalescence of small bubbles occur.

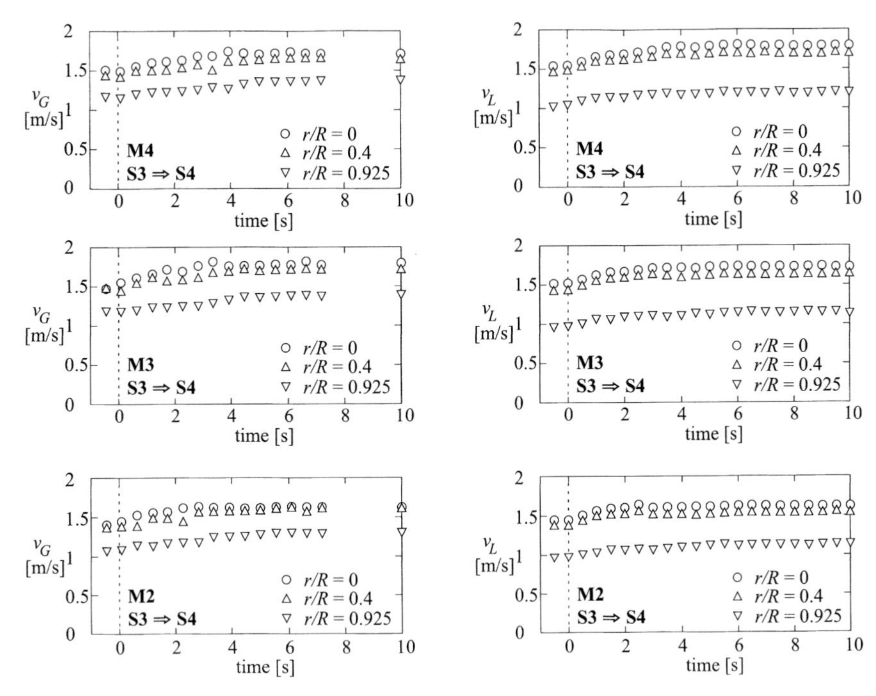

Figure 8.9: Phase velocities with time at different radial positions at the height M2, M3 and M4 for transient II.

For S4 (t=15 s), the distribution behaviour is similar but more extended because of the higher gas volume flow than at S3.

Figure 8.12 shows snapshots of flow S3 and S4 at M2. They only illustrate the bubble density in these flows.

8.5 Conclusions

Local two-phase flow parameters have been measured during several transient bubbly flows. An experimental way together with an efficient measuring technique could be successfully demonstrated while investigating some rapid two-phase flow transients in the bubbly flow regime. It could also be shown how two-phase flow parameters behave locally and with time during a transient flow. Further, it was demonstrated that for bubbly flows, with an average void fraction of some few percent, the processes of bubble coalescence and break-up play an important role in describing and simulating more correctly these

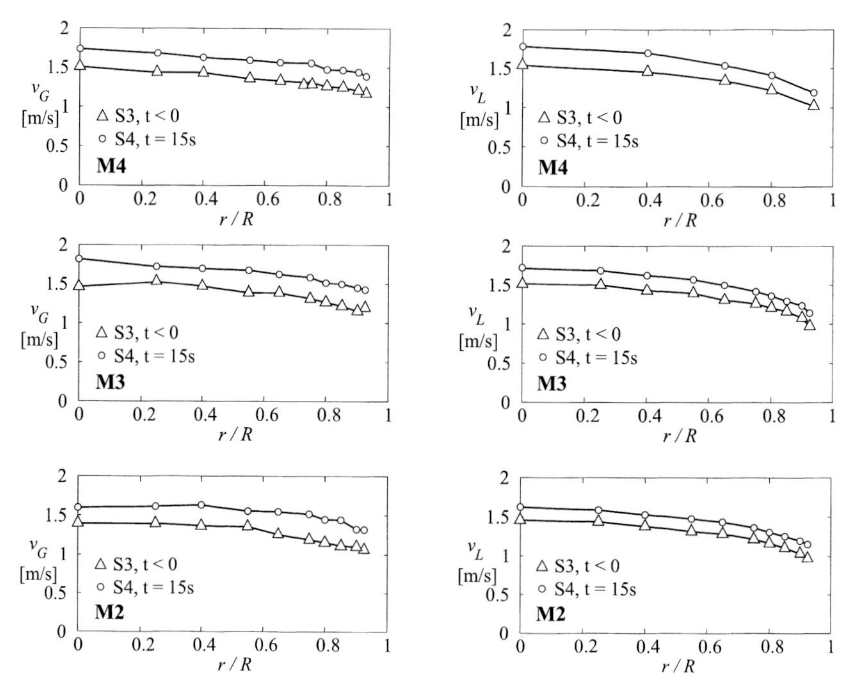

Figure 8.10: Radial phase velocity profiles at different heights for flow condition S3 (t<0 s) and S4 (t=15 s).

transients. The rate of coalescence is a function of the relative velocity between the bubbles. The turbulence in the continuous phase and the non-uniform mean velocity profiles are two important sources of the relative bubble motion, and therefore should be further investigated in more detail.

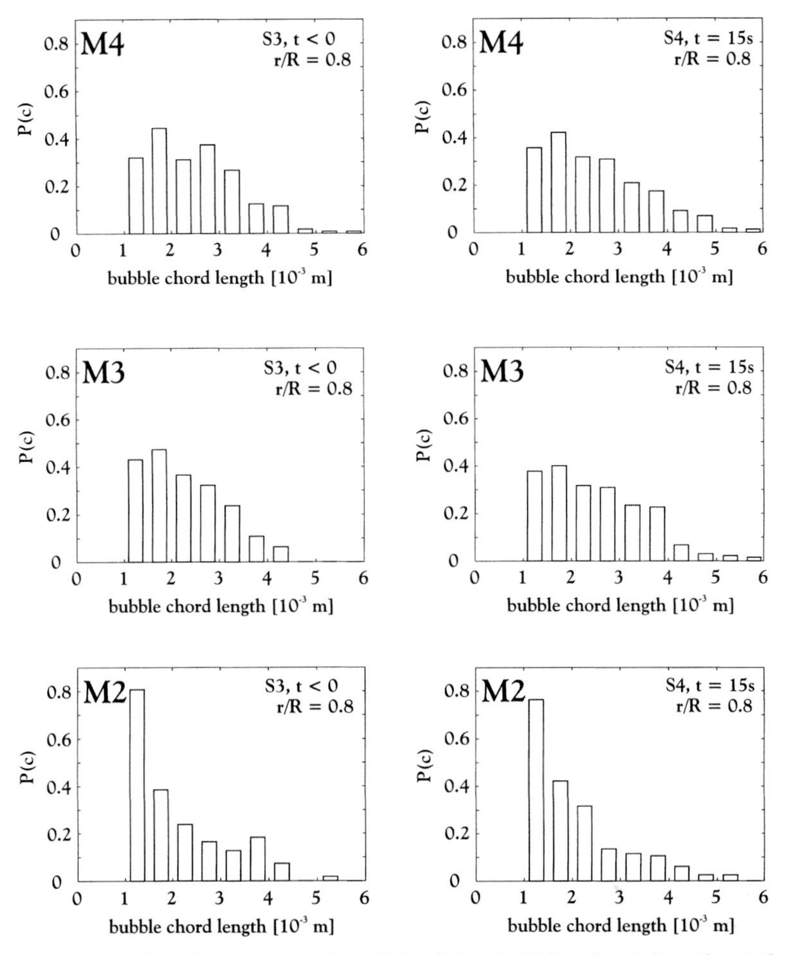

Figure 8.11: Probability density function $P(c)$ of the bubble chord length at the radial position $r/R = 0.8$ for S3 and S4.

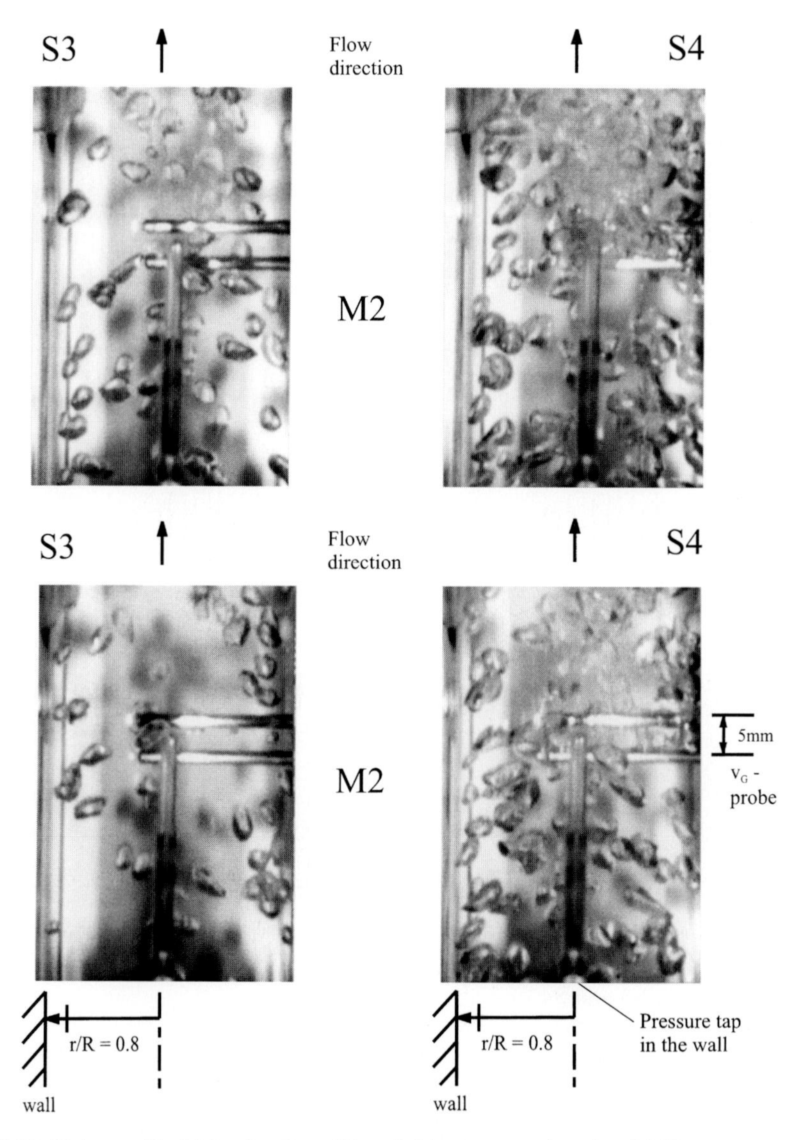

Figure 8.12: Picture of bubbles for flow S3 and S4 at measuring level M2.

List of Symbols

c bubble chord length (m)
D diameter of pipe (0.04 m)
j superficial velocity (m/s)
M measuring level (–)
p pressure (N/m^2)
P probability density function (–)
r radial distance from center line (m)
R radius of pipe (0.02 m)
t time (s)
v phase velocity (m/s)
X phase indicator function (–)
z axial position in test section (m)

Greek symbols

ε_G local void fraction (–)
$<\varepsilon_G>$ cross-section averaged void fraction (–)

Subscripts

G gas
L liquid

References

[1] A. Schmitt, K. Hoffmann, R. Loth: *A Transputer-Based Measuring System for Decentralized Signal Processing, Applied to Two-Phase Flow.* Rev. Sci. Instrum. **66** (1995) 5045–5049.
[2] Y. Taitel, D. Bornea, A.E. Dukler: *Modelling Flow Pattern Transitions for Steady Upward Gas-Liquid Flow in Vertical Tubes.* AIChE J. **26** (1980) 345–354.
[3] T.J. Liu, S.G. Bankoff: *Structure of Air-Water Bubbly Flow in a Vertical Pipe – II. Void Fraction, Bubble Velocity and Bubble Size Distribution.* Int. J. Heat Mass Transfer **36** (1993) 1061–1072.
[4] A. Rinne: *Experimentelle und numerische Bestimmung lokaler Größen adiabater vertikaler Flüssigkeits-Gas-Zweiphasenströmungen in einem Rohr mit plötzlicher Querschnittserweiterung.* Diss. Technische Hochschule Darmstadt, 1997.
[5] A. Schmitt: *Untersuchungen zur Bestimmung lokaler Parameter von Flüssigkeits-Gas-Zweiphasenströmungen mit einer neuentwickelten Meßkette.* Diss. Technische Hochschule Darmstadt, Fortschrittberichte VDI, Reihe 8 Nr. 411, 1994.

9 Identification of Fractal Structures in Two-Phase Flow

Holger Skok and Erich Hahne *

Abstract

The investigation presented here is based on the idea of applying the reconstruction techniques developed for the investigation of chaotic dynamical systems to flow regime identification in gas/liquid two-phase flows. Some results reported in the literature hint at the viability of such an approach to flow regime identification. It was our hope that by reconstructing trajectories for typical two-phase flow regimes, strong differences in the trajectory shapes would result. The differences would then serve to more clearly differentiate between the flow regimes.

We created two-phase flows of water and air in simple plexiglass tubes of circular cross section. They were mounted on a balance allowing their inclination to be changed from horizontal to vertical. For the horizontal position, the flow regimes ranged from stratified to slug flow. For the vertically oriented tubes, the flow regimes ranged from bubbly to slug flow. Besides the mass fluxes, temperatures and pressure of the two phases, two physical quantities were measured for the trajectory reconstruction: the static pressure and the void fraction. The results contradict some of the theoretical predictions of chaos theory. Conditions for a successful application of the reconstruction techniques were determined.

* Universität Stuttgart, Institut für Thermodynamik und Wärmetechnik, Pfaffenwaldring 6, D-70569 Stuttgart, Germany

9.1 Flow Regimes and Chaos Theory

Due to the complexity of the physics governing two-phase flows, the techniques of chaos theory have been applied to two-phase flows only in a few special cases. Complex behaviour usually implies correspondingly complex trajectories, which are very difficult to extract from measured data.

Despite the difficulties, some authors report having found systematic changes in quantities derived from chaos theory when treating flow regime transitions.

9.1.1 Flooding transition

Biage and Delhaye [1] investigated the flooding transition in counter-current two-phase flow. This transition occurs when a liquid film flowing down the wall of a vertically oriented channel begins to reverse its flow direction under the influence of an upward gas flow in the center of the channel.

The interaction between gas and liquid is strongly influenced by the waves forming on the film surface both in upward and downward motion of the liquid. Consequently, Biage and Delhaye [1] measured the film thickness with high temporal resolution and calculated the dimension of the resulting signal. They observed that the correlation dimension fell significantly shortly before the flooding took place. While the correlation dimension D_c remained between values of 8 and 9 for the flow further away from transition, D_c dropped to a value of 6 at the point of flooding.

According to Biage and Delhaye [1], this relatively sharp change corresponded with a similarly strong change in the pressure drop. The change of the correlation dimension indicated a corresponding qualitative change in the nature of the flow, i.e. a change of the flow regime.

9.1.2 Flow regime identification in horizontal tubes

Franca et al. [2] investigated flow regime transitions in horizontal tubes. They measured the pressure drop in fully developed flows of water and air and employed both traditional signal analysis techniques as well as some techniques derived from chaos theory. Among other quantities, they reconstructed Poincaré-sections from the pressure drop signal.

The Poincaré sections show marked differences for stratified and dispersed flows (see Figures 9.1 and 9.2). Wavy flow generated a spread-out section with an evenly covered centre, while slug flow resulted in a section with a more densely covered, cross-like area in the centre. Both shapes are surrounded by additional "clouds" of points.

The authors conclude that flow regime identification should be possible based on trajectory reconstruction and calculation of its correlation dimension. Their conclusions remain qualitative, however.

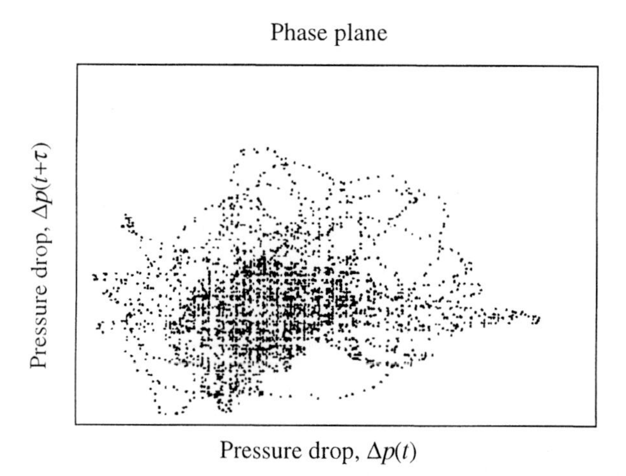

Figure 9.1: Poincaré section reconstructed from measurements of the pressure drop – wavy flow.

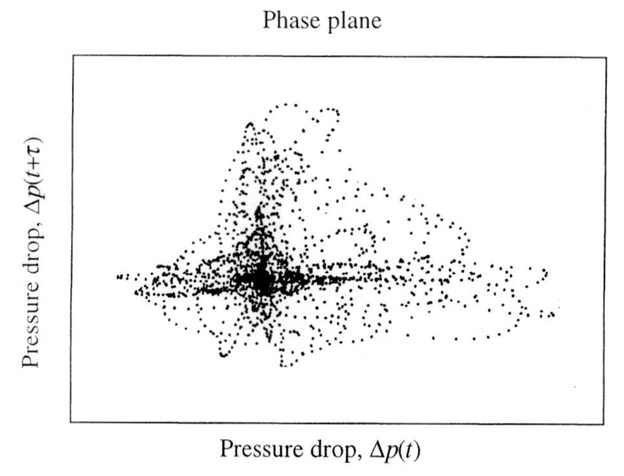

Figure 9.2: Poincaré section reconstructed from measurements of the pressure drop – slug flow.

9.1.3 Flow regime identification in a rectangular channel

Cai et al. [3] investigated flow regimes in a horizontal channel with rectangular cross section. It was much wider than high and had a hydraulic diameter of $d_h = 5.45$ mm. Air was injected into the channel from top and bottom through porous wall sections. Cai et al. [3] measured the static pressure in the flow, 79 d_h downstream from the entrance and applied both conventional and "chaotic" data analysis techniques to the signals. They found correlation dimensions around $D_c = 6$ for plug flow and $D_c = 9$ for slug and ring flow. The sampling rate used was 1 kHz and the samples contained 65 000 data points each.

9.2 Experimental Set-Up

In order to test our hypothesis – namely that the reconstruction of trajectories from data measured in two-phase flows and the determination of their dimension would allow to identify flow regimes – we conducted experiments in a simple experimental set-up. It allows us to create two-phase flows of water and air at temperatures close to room temperature and at ambient pressure. Its central part are two plexiglass tubes mounted on a balance. The inner diameter of the tubes is 22 mm and 14 mm, respectively. The balance can be tilted to any angle between the horizontal (0°) and the vertical (90°) orientation. Only one of the tubes is used for experiments at a time (Figure 9.3).

Two-phase mixtures are created by allowing demineralized water and air to enter a mixing chamber located at the entance of the tube. The water loop is closed. Water from a storage tank is pumped through a coaxial heat exchanger. After leaving the heat exchanger, the water passes through one out of a set of volume flow meters and then enters the mixing chamber. After having passed through the tube, it flows back down into the storage tank.

The air is taken from the pressurized air network in the building, which supplies oil-free air. The pressurized air passes through one of a set of volume flow meters and a throttling valve before it enters the mixing chamber. Together with the water, it flows through the tube, flows down into the storage tank and finally escapes from the tank into the surroundings.

The mixing chamber contains static stirrer elements made from stacks of corrugated plastic sheets. They create a very intimate mixture of the two

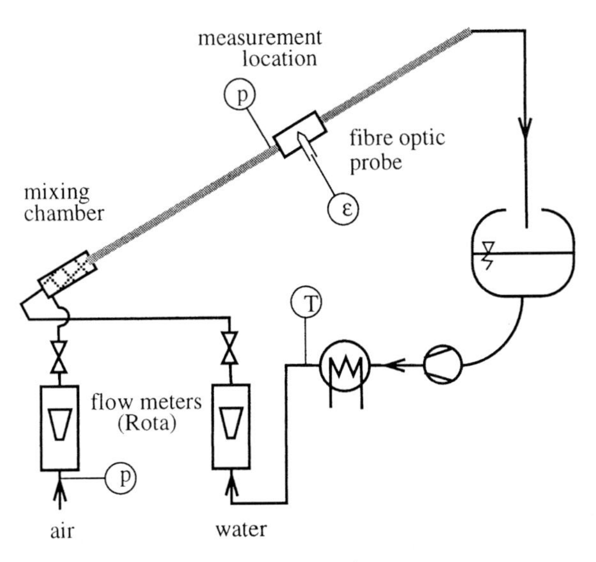

Figure 9.3: Schematic diagram of the experimental set-up.

phases such that the flow at the exit appears milky and non-transparent. Downstream from the mixing chamber, the phases separate quickly and flow structures typical for the chosen mass fluxes and tube inclination develop.

Roughly 100 inner diameters downstream from the mixing chambers, the flow passes the actual measurement location, where static pressure and local void fraction are measured.

9.2.1 Data acquisition

Volume fluxes of air and water are determined with volume flow meters. The air pressure in the flow meters and at the measurement location is determined with Bourdon-tube pressure gauges. The temperature of the water entering the flow meters is determined with a platinum resistance thermometer. The volume flow rate of air at the measurement location is calculated from the pressures and temperatures, assuming ideal gas behaviour and identical temperatures of the two phases at the measurement location. Finally, mass fluxes are calculated from the volume fluxes.

9.2.1.1 Measuring the absolute pressure

At the measurement location, a pressure tap is mounted flush with the wall. An air-water mixture enters it as long as the counter pressure is smaller than the pressure in the tube. Consequently, the two phases have to be separated in a small plexiglass separator vessel and only the gas leg of the pressure tap is connected to the actual measurement assembly. The assembly consists of a variable inductance *differential* pressure transducer (accuracy: ±0.1 mbar, 800 Hz resonance frequency) and a water-filled U-tube connected by tubing and cocks allowing two alternate modes of connection.

During a preparatory stage, the U-tube is connected to the measurement location. A diaphragm in the tubing right before the entrance into the U-tube dampens the pressure oscillations such that the U-tube is subjected to the time-averaged pressure. During the actual measuring stage, the differential pressure transducer is connected to the set-up such that the actual pressure enters one of its legs and the average pressure supplied by the U-tube enters the other. Thus, only the difference signal needs to be measured by the transducer, resulting in a much higher resolution than would have been achievable without the U-tube.

9.2.1.2 Measuring the void fraction

A fibre optic probe is located approximately 80 mm downstream from the pressure tap. Its measurement principle is based on the difference between the index of refraction of water and air. Whenever a bubble displaces the water surrounding the probe tip, a signal is recorded.

Detection of gas is impossible if the bubble passing the tip is too small to displace the water covering the tip. Since the presence of the probe alters the flow field, small bubbles in particular flow past the tip and remain undetected. Since the probe tip has a diameter of 0.5 mm, we assume the minimum bubble size still generating a signal to be of similar magnitude. Consequently, the actual void fraction is always higher than the measured one. The difference is small for flows with large bubbles or for stratified flows. The difference is larger for churn and bubbly flows with small bubble sizes. The smallest bubble sizes for typical flow patterns in our set-up were determined by analyzing photographs of the flow and were found to be on the order of 0.5 mm.

According to measurements made by Spindler [4] in air-water two-phase flows, the value for the void fraction *averaged* over the entire cross section is at worst 40% smaller than the value determined by a gamma ray densitometer for the same flow. Under favourable conditions, the difference is below 10%. Spindler [4] used a fibre optic probe with a tip diameter of 1.0 mm and integrated local measurements numerically to obtain the average void fraction in the measurement plane. Since the integration over the entire cross section introduces a sizeable error itself, the measurement error for the local void fraction with the smaller probe used here is bound to be much smaller.

9.2.1.3 Recording the data

The two quantities chosen for further analysis, namely the static pressure and the void fraction, were recorded by means of a PC-based data acquisition system. Both quantities were measured with a sampling rate of 20 Hz and 40 000 data points each were recorded for every experiment. The data were later transferred to UNIX workstations for further analysis.

The sample size was chosen to keep the run times of the analysis programs manageable. For 40 000 data points, a typical run took two hours on a SGI Indy with R 5400 CPU. The sampling rate was then chosen with the aim to record the passage of a large number of slugs, waves or other visually recognizable geometric structures in the flow. At 20 Hz, many hundred such structures passed the measurement location during each run, making sure that average rather than individual characteristics of slugs, waves or plugs were contained in the measured signals.

9.3 Experiments

For each experimental run, one of the tubes was connected to the air- and water-circuits. The inclination of the tubes and the mass flow rates of the two phases were adjusted and then kept constant throughout the run. Only after allowing initial transients to vanish and a flow structure to develop, the data acquisition was started. During the measurement, ambient conditions were monitored in order to assure that they remained constant. In case of severe variations of ambient pressure or temperature, the data were discarded and the run was repeated.

For each inclination of the tubes, one mass flow rate for the liquid phase was chosen. From run to run, the mass flow rate of the gas phase was changed from small values to the maximum possible. In order to further increase the void fraction of the flow, the water mass flow rate was then lowered, while keeping the air mass flow rate constant.

The following combinations of angle of inclination and mass flow rates were reached in our experiments (Table 9.1):

Table 9.1: Experimental conditions.

Angle [°]	$\dot{M}_l$ [kg/s]	$\dot{M}_g$ [kg/s]	X_{tt} [-]	$w_{l,0}$ [m/s]
90	$3.37 \cdot 10^{-2}$	$2.27 \cdot 10^{-5}$	0.651	0.089
	$6.32 \cdot 10^{-1}$	$2.03 \cdot 10^{-3}$	439	1.6
60	0.170	$7.30 \cdot 10^{-5}$	5.15	0.449
	0.605	$2.11 \cdot 10^{-3}$	170	1.61
45	0.132	$4.44 \cdot 10^{-5}$	2.96	0.350
	0.390	$1.39 \cdot 10^{-3}$	177	1.04
30	0.292	$3.71 \cdot 10^{-5}$	7.63	0.768
	0.454	$1.63 \cdot 10^{-3}$	224	1.20
0	0.0947	$1.23 \cdot 10^{-4}$	2.05	0.250
		$1.58 \cdot 10^{-3}$	20.5	

9.4 Reconstruction Techniques

While many of the reported incidents of chaos have been found in numerical simulations, reports on chaos in experiments are comparatively rare, particularly in the case of two-phase flow.

9.4.1 The object to identify

When attempting to identify chaotic behaviour in experimental data (or in numerically calculated data for that matter), the elusive object that needs to be found and analyzed is the *attractor* of the system being investigated. Every dissipative physical process that is in steady state on average (i.e. oscillations around a constant mean value can occur) must possess at least one such attractor (see e.g. Cvitanovic [5]). The attractor in turn is mapped out in phase space by the trajectory of the system. Only the trajectory is accessible to measurements. The properties of the attractor then have to be deduced from the properties of the trajectory.

Both, in numerical calculations and in experiments, only a number of points on that trajectory can actually be measured or calculated, preferably at equally spaced time intervals. The resulting points in phase space are located on the trajectory, which in turn delineates the geometry of the attractor. Each point is represented by its coordinate vector (e.g. $\vec{\boldsymbol{P}}_i$); these vectors can be grouped into the trajectory matrix Ξ with

$$\Xi = (\vec{\boldsymbol{P}}_1, \vec{\boldsymbol{P}}_2, \ldots, \vec{\boldsymbol{P}}_N)^{\mathrm{T}} . \quad (1)$$

It is this matrix Ξ, which we have to reconstruct from our data.

9.4.2 The method of identification

Attempts to identify chaos in experimental data are commonly based on Taken's embedding theorem [6] and its modifications by Grassberger and Procaccia [7] and by Broomhead and King [8]. Takens [6] proved that a time series of a single variable, measured (or calculated) for a given physical system in steady state, can be used to reconstruct the trajectory of the system by means of a procedure called an embedding. Grassberger and Procaccia [7] have later suggested modifications for the original embedding process taking into account the inaccuracies and noise present in experimental data. Broomhead and King [8] suggested using singular value decomposition (SVD) to separate noise from the actual signal. Pilgram et al. [9] developed a filtering method based on that approach.

The separate methods suggested by the authors mentioned above can be grouped together into the following process to reconstruct the trajectory from our data:

1. At a given steady state of the flow, the static pressure and the void fraction in the flow are measured 40 000 times with a sampling rate of 20 Hz.

2. The first local minimum of the mutual information of the time series is used to determine an optimal delay for the embedding as suggested by Fraser and Swinney [10].

3. The data are then subjected to a SVD-filtering [8, 9].

4. The actual embedding is then performed with the filtered data and the optimal embedding delay obtained in step 2.

5. The correlation dimension D_c is determined from the reconstructed trajectory. The first three Rényi-dimensions of the trajectory are calculated for some of the experiments.

6. The results are checked for systematic dependence on flow regime. Actually observed flow regimes are compared to predictions from common flow pattern maps.

9.4.3 Terms and concepts

The mutual information I is similar to the autocorrelation function. The autocorrelation function measures the linear dependence between successive points of a time series. The mutual information on the other hand measures a general dependence between successive points. The mutual information $I(p(t), p(t+\tau))$ between two hypothetical measurements $p(t)$ and $p(t+\tau)$ indicates the amount, by which the measurement of $p(t)$ reduces the uncertainty of $p(t+\tau)$. If the distance τ between the two measurements is small, I will be large since we can predict the new value by extrapolation quite accurately. Consequently, we do not gain much information by actually performing the measurement of $p(t+\tau)$.

If we increase τ to very large values, no reliable prediction of $p(t+\tau)$ is possible from the previous measurements and the mutual information $I(p(t), p(t+\tau))$ falls to zero. If the time series for which the mutual information is being calculated possesses some periodicity, I will be periodic as well and can possess local minima.

The correlation dimension D_c is a well known parameter describing the shape of trajectories or attractors. It is a generalized dimension that can take on fractional values. It is consistent with our everyday notion of dimension in that it yields values of $D_c=2$ for a plane or $D_c=3$ for a sphere for example. D_c is determined from the correlation function $C(r)$:

$$D_c = \lim_{r \to 0} \frac{\log_2 C(r)}{\log_2(r)} \ . \tag{2}$$

$C(r)$ in turn is a measure for the fraction of points on a trajectory that fits inside a sphere of radius r.

The Rényi-dimensions are based on a generalization of the box-counting algorithm. The determination of the dimension of a trajectory by means of box-counting is based on subdividing the region of phase space $\mathbb{B}$ occupied by the trajectory into equally sized boxes. One then counts the number of boxes, which contain at least one point of the trajectory. This number will be a power function of the box size r.

Since the trajectory can visit certain regions of phase space more often than others, it makes sense to weight the individual boxes with the fraction of all points contained in them. The fraction is identical to the probability P, with which a randomly chosen point from the trajectory will fall into the box under consideration:

$$P(\mathbb{B}) = \lim_{i \to \infty} \frac{1}{i+1} \sum_{j=10}^{i} \mathbf{1}_{\mathbb{B}}(\vec{\boldsymbol{P}}_j) \ . \tag{3}$$

The function $\mathbf{1}_{\mathbb{B}}(\vec{\boldsymbol{P}}_j)$ represents the counting of points. It is equal to 1 if the point $\vec{\boldsymbol{P}}_j$ is contained within the box j under consideration, and zero otherwise.

In this manner, not only the shape but also part of the dynamics of the trajectory are taken into account. Instead of simply counting the boxes containing a part of the trajectory, one sums the fractions of points contained in each box. This approach has the additional advantage of requiring less of a computational effort for obtaining an accurate estimate of the dimension than the plain box-counting approach.

The last step towards calculating the Rényi-dimensions is to switch from summing the plain probabilities for each box to summing powers of these probabilities. We obtain a generalized form of the correlation function $C(r)$ such that:

$$C_q(r) = \frac{1}{1-q} \log_2 \sum_{j=1}^{N(r)} [P(\mathbb{B}_j)]^q . \tag{4}$$

As in case of the correlation dimension, the generalized correlation function is a power function of the box size r. The exponent resulting for each value of q is then called the Rényi-dimension of the q-th order.

It can be shown that when taking the zeroth power, we obtain the box-counting algorithm (Rényi-dimension of the zeroth order). Taking the probabilities to the power of one (Rényi-dimension of the first order), the correlation dimension results. Thus, the Rényi-dimensions provide a generalized definition for the dimension of geometric objects comprising the definitions we have introduced previously.

In contrast to the calculation of the correlation dimension, we have chosen a direct approach for the Rényi-dimensions. That means that the probabilities in the definition were calculated directly and summed according to Equation (4). The correlation dimension on the other hand was calculated by considering the distances between pairs of points on the trajectory. This approach is less computationally expensive, but judging by our results, it is also less accurate and more sensitive to noise.

9.4.4 Embedding in pseudo-phase-space

In order to reconstruct a trajectory from a series of measurements, the series of scalar values has to be transformed into a series of vectors which represent successive points on the trajectory. A detailed description of the process can be found in Skok and Hahne [11]. The pressure p_o at the measurement location is measured repeatedly with a time step Δt between successive measurements (here $\Delta t = 0.05$ s). The pressures are grouped into vectors and these are grouped together into the trajectory matrix Ξ. This process is called an embedding.

The embedding as described by Grassberger and Procaccia [7] proceeds as follows. We choose a dimension n for the pseudo-phase-space. It is equal to the number of values of the pressure p_o that form one reconstructed vector. Next, we choose a time delay τ that is used to determine which of the measured pressure values will be grouped into one vector. Out of N measured values of p_o–$p_{o1}, p_{o2}, \ldots, p_{oN}$ – the first one – p_{o1} – becomes the first component of the first vector $\vec{\boldsymbol{P}}_1$. The second component is the pressure measured τ seconds later than p_{o1}. In case of a trajectory with optimal delay of $\tau = 0.5$ s, this is the pressure p_{o11} out of the original series of measurements. p_{o1} and p_{o11} are $j = 10$ time steps apart, which add up to the desired τ. The third and last component of the first vector is p_{o21} which was measured another τ seconds later. Thus:

$$\vec{\boldsymbol{P}}_1 = [p_{o1}, p_{o11}, p_{o21}] \; . \tag{5}$$

For the next vector, we repeat the same process starting with p_{o2} instead of p_{o1}. Hence:

$$\vec{\boldsymbol{P}}_2 = [p_{o2}, p_{o12}, p_{o22}] \; . \tag{6}$$

This continues until all pressure values have been used. The vectors are finally placed into the rows of Ξ. In general, the trajectory matrix Ξ is of the form

$$\Xi = \begin{bmatrix} p_{o1} & p_{o1+j} & p_{o1+2j} & \cdots & p_{o1+(n-1)j} \\ p_{o2} & p_{o2+j} & p_{o2+2j} & \cdots & p_{o2+(n-1)j} \\ \cdots & \cdots & \cdots & & \cdots \\ p_{oN-(n-1)j} & \cdots & p_{oN-2j} & p_{oN-j} & p_{oN} \end{bmatrix} \; . \tag{7}$$

Neither the time delay τ nor the embedding dimension n are known beforehand. They have to be determined by analyzing the data.

9.5 Application and Results

In contrast to our previous positive experience with the reconstruction technique outlined above [11], the data recorded here proved to be rather difficult to analyze.

9.5.1 Calculating the dimension

The first problem encountered is the lack of a pronounced first minimum of the mutual information. Often, no minimum existed. And even if a minimum existed, it was never as pronounced as in case of more clearly periodic or quasi-periodic signals. The power spectrum of the data corroborated the lack of periodicity. No peaks were to be found. Thus, even weak minima in the mutual information had to be used to determine the embedding delay τ.

The next difficulty is the behavior of the correlation function $C(r)$. In case of signals containing low-dimensional components (e.g. the Lorenz-system,

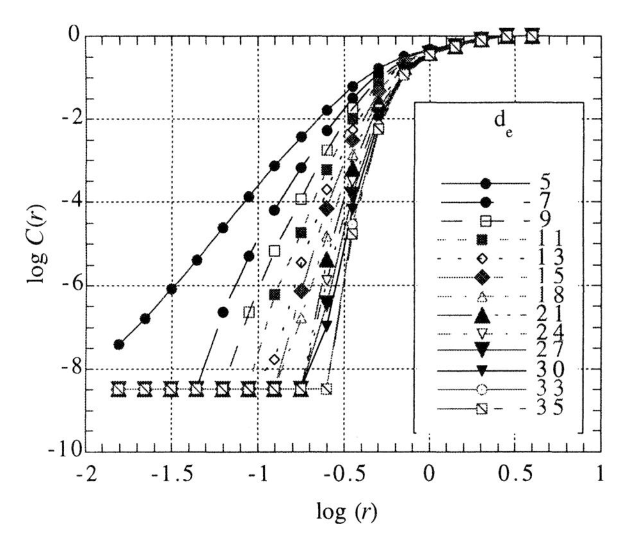

Figure 9.4: The correlation function $C(r)$ for different embedding dimensions – vertical tube, slug flow.

Lorenz [12]), the slope of $C(r)$ increases with the embedding dimension d_e only initially. For values of d_e substantially greater than the actual dimension of the signal being analyzed, the slope converges to the correlation dimension D_c and then remains constant[1] (see Equation (2)). As shown in Figure 9.4, this effect can also be observed for data from our experiments. For large values of the correlation radius r, the slope of the curves of $C(r)$ does indeed converge to a constant value. For smaller values of r, on the other hand, the slope continues to increase with the embedding dimension d_e.

This is a commonly observed phenomenon related to the presence of noise. It can be reproduced for the Lorenz-system by artificially adding low amplitude noise to the signals. The noise, being essentially infinite-dimensional, prevents the convergence of the slope to the correct value of the correlation dimension in the low range of the correlation radius r. Unfortunately, the defining equation for the correlation dimension (Equation (2)) requires taking the limit of $r \rightarrow 0$. The example of the Lorenz-system on the other hand shows that the slope of $C(r)$ converges to D_c even at large values of the correlation radius r. It is by exploiting this effect only that the correlation dimension of noisy signals can be determined. SVD-filtering improves this situation only slightly by increasing the usable range of r towards smaller values. Unfortunately, even with filtering only a few of the sets of data recorded in our experiments showed a convergence of the slope of $C(r)$ and allowed the determination of the correlation dimension. Often, the slope continued to increase instead and D_c could not be determined.

[1] In case of the Lorenz-system, this happens for $d_e \geq 5$, while the actual correlation dimension is found to be $D_c = 2.06$.

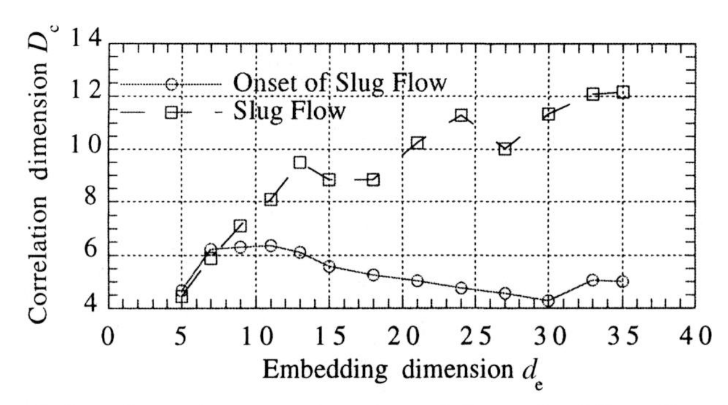

Figure 9.5: Correlation dimension as a function of the embedding dimension – vertical tube, slug flow.

By visually selecting the linear sections of the curves for $C(r)$ at large values of r and fitting power functions to them, one typically obtains results such as those presented in Figure 9.5. The lack of low dimensional components is apparent in the second curve in Figure 9.5, which does not level off with d_e. It is interesting to note that the resulting values of the correlation dimension differ strongly, despite the fact that the flow regimes are similar in the two cases. In fact, the correlation dimension shows no systematic dependence on the flow regime.

Consequently, we also employed the Rényi-dimension to characterize the shape of the trajectory. Its convergence behavior is slightly different from that of the correlation dimension and its calculation requires larger sample sizes, but the overall result did not change: convergence was found only for a few of the sets of data and no systematic relationship between Rényi-dimension and flow pattern could be found (Figure 9.6).

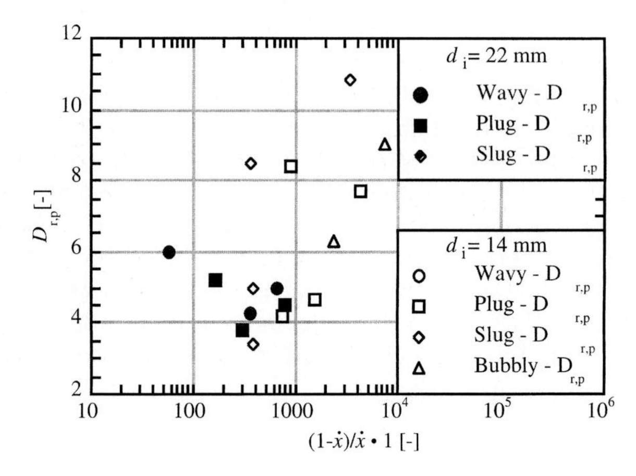

Figure 9.6: Rényi-dimension for horizontal flow, calculated from the static pressure.

9.5.2 Comparing results for pressure and void fraction

According to theory (see [6]), every physical quantity of a system should be suitable for reconstructing its trajectory. As long as the system is in a stable, steady state, it possesses a single attractor representing that state and every trajectory lies on that attractor. Consequently, the dimensions of the trajectories must be identical, no matter which physical quantity the trajectory was reconstructed from.

Our results contradict Takens' prediction. Dimensions calculated from the void fraction signal are markedly different than the dimensions calculated from the pressure signal. There are even experiments, in which the calculations converged for one signal but not for the other.

These discrepancies can probably be attributed to the fact that Takens' embedding theorem calls for infinite amounts of infinitely accurate data, which are impossible to acquire in an experiment. In practice, therefore, a judicious choice of the quantity to be analyzed is necessary. Since neither of the two quantities chosen here proved to be useful in terms of flow regime identification, recommending one over the other is impossible based on our results. Some words on the different physical nature of the two quantities are in order, nevertheless.

The void fraction is a local quantity measured at just one location in the tube cross section always near the gas-liquid interface. The pressure on the other hand is influenced by the conditions many diameters up- and downstream from the location of the pressure tap. The intermittent blocking of the tube by liquid slugs in case of slug flow are particularly obvious in the pressure signal for example. They result in large pressure oscillations, which have no equivalent in the void fraction signal.

9.5.3 Comparison to classical flow pattern maps

The difficulties described above would lead one to the conclusion that attempts at flow regime identification in the manner proposed in this article offer no improvement over other methods described in the literature. One has to keep in mind, however, that traditional methods such as flow pattern maps are not very reliable either. Comparing the flow pattern map according to Baker [13] (see Figure 9.7) with the flow patterns actually observed in our experiments, some discrepancies become apparent. Deciding if a separated flow regime or an intermittent one is to be expected is particularly uncertain. The points for wavy flow in the horizontal tube are very close to the line separating the two types of flow regimes. In addition, they are very close to the points resulting from measurements in plug flow. Using the flow pattern map according to Baker [13], it is impossible to differentiate between the two kinds of flow regimes.

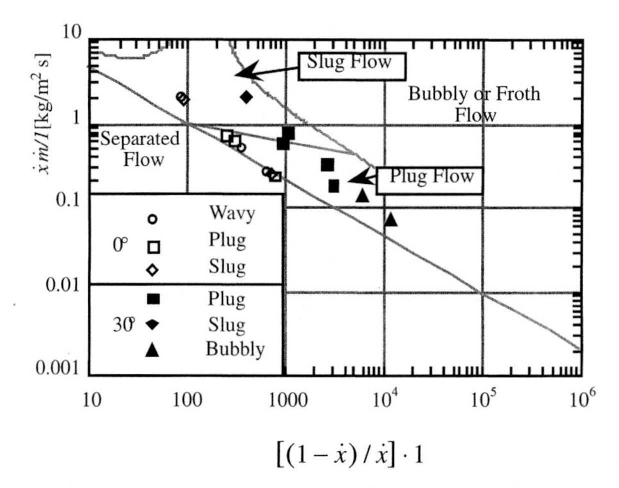

Figure 9.7: Flow pattern map for horizontal flow – comparison with experimental results (**1** – fluid property coefficients are equal to 1).

9.6 Conclusions

Our original hope to apply the reconstruction technique to any kind of gas-liquid two-phase flow in inclined tubes and obtain results allowing to clearly distinguish different flow regimes was not fulfilled. In the course of our experiments, it became clear that care had to be taken when choosing a physical quantity to subject to the reconstruction technique. Contrary to theoretical predictions, the reconstructed trajectories depend on the quantity they are reconstructed from. Dimensions calculated from them differ strongly.

The limitations of the filtering based on singular value decomposition became apparent as well. SVD-filtering works best for signals containing periodic or quasi-periodic components of sizeable amplitude. The signals obtained from steady two-phase flow do not contain such components. In addition, even with SVD-filtering the influence of noise on the reconstructed trajectory is not attenuated much.

With these conclusions in mind, the question remains whether further attempts to improve upon the work presented here are warranted. Flow patterns in two-phase flow are obvious to anyone observing such flows. There are recognizable structures to be found such as slugs, plugs, waves or bubbles and these structures, flowing past a measurement location, must leave corresponding recognizable traces in the physical quantities recorded there. With further improvements of the reconstruction process, of noise reduction techniques and of the algorithms to calculate dimensions, Lyapunov-exponents or other, yet to be defined measures of fractality, it will certainly be possible to uncover the signatures of these recognizable structures in the flow.

Acknowledgements

We would like to thank the DFG (Deutsche Forschungsgemeinschaft) for the financial support for the research presented in this article.

List of Symbols

$\mathbb{B}$	region of phase space	–
$C(r)$	correlation function	[–]
$C_q(r)$	generalized correlation function	[–]
D_c	correlation dimension	[–]
d_e	embedding dimension	[–]
d_h	hydraulic diameter	[m]
d_i	inner diameter	[m]
$D_{r,p}$	Rényi-dimension – from static pressure	[–]
I	mutual information	–
$\dot{m}$	mass flux	[kg/m^2 s]
$\dot{M}_l$	mass flow rate – liquid phase	[kg/s]
$\dot{M}_g$	mass flow rate – gas phase	[kg/s]
$N(r)$	number of boxes containing part of the trajectory (see Equation (4))	[–]
P	probability	[–]
$\vec{\boldsymbol{P}}$	point in phase space	[–]
p	pressure	[mbar]
Δp	pressure drop	[mbar]
r	radius for the correlation function	[–]
$w_{l,0}$	superficial velocity – liquid phase	[m/s]
$\dot{x}$	quality	[–]
X_{tt}	Martinelli parameter	[–]
ε	void fraction	[–]
$\boldsymbol{\Xi}$	trajectory matrix	–

References

[1] M. Biage, J.M. Delhaye: *The Flooding Transition: An Experimental Appraisal of the Chaotic Aspect of Liquid Film Flow Before the Flooding Point.* AIChE Symposium Series 269, 1989.

[2] F. Franca, M. Acikgoz, R.T. Lahey, A. Clausse: *The Use of Fractal Techniques for Flow Regime Identification.* Int. J. Multiphase Flow **17** (1991) 545–552.

[3] Y. Cai, M.W. Wambsganss, J.A. Jendrzejczyk: *Application of chaos theory in identification of two-phase flow patterns and transitions in a small, horizontal, rectangular channel.* J. Fluids Engineering, Transactions of the ASME **118** (1996) 383–390.

[4] K. Spindler: *Untersuchung zum Mischereinfluß auf die lokale Struktur von Zweiphasenströmung mit einem faseroptischen Sensor.* Ph. D. Thesis, Universität Stuttgart, Stuttgart, Germany, 1989.

[5] P. Cvitanovic: *Universality in Chaos.* Hilger, Bristol, 1986.

[6] F. Takens: *Detecting Strange Attractors in Fluid Turbulence.* In: D.A. Rand, L.S. Young (Eds.): Lecture Notes in Mathematics, Dynamical Systems and Turbulence. Springer, Heidelberg, 1981, pp. 366–381.

[7] P. Grassberger, I. Procaccia: *Measuring the Strangeness of Strange Attractors.* Physica D **9** (1983) 189–209.

[8] D.S. Broomhead, G.P. King: *Extracting Qualitative Dynamics from Experimental Data.* Physica D **20** (1986) 217–236.

[9] B. Pilgram, W. Schappacher, G. Pfurtscheller: *A Noise Reduction Method using Singular Value Decomposition.* 14. Annual Int. Conf. IEEE Engineering in Medicine and Biology Society, Paris, France, 1992.

[10] A.M. Fraser, H.L. Swinney: *Independent Coordinates for Strange Attractors from Mutual Information.* Physical Review A **33** (1986) 1134–1140.

[11] H. Skok, E. Hahne: *Identification of Chaotic Attractors in Natural Convection Boiling Flow.* Heat Transfer – Proc. 10th Int. Heat Transfer Conf., Brighton, UK, **7** (1994) 545–550.

[12] E.N. Lorenz: *Deterministic nonperiodic flow.* J. Atmos. Sci. **20** (1963) 130.

[13] O. Baker: *Simultaneous Flow of Oil and Gas.* Oil and Gas Journal **53** (1954) 184–195.

Measurement Techniques

10 PIV with Two Synchronized Video Cameras

Wolfgang Merzkirch and Thomas Wagner *

Abstract

CCD-cameras are used for easy and fast image acquisition in particle image velocimetry systems. Due to the given frame rate of 25 Hz, digital PIV with a single video camera can be applied only to fluid flows with low velocities. In order to make DPIV applicable to high speed flows, a set-up utilizing two synchronized video cameras was developed. Measurements with the two camera system were conducted with the flow in the model of a three phase reactor.

10.1 Introduction

Particle Image Velocimetry is an optical method for measuring velocities in fluid flows [1]. It relies on the use of tracer particles whose motion in a plane of the flow field is recorded with a camera. Figure 10.1 shows the basic elements of a set-up for PIV measurements. The laser beam is expanded in form of a plane light sheet to provide illumination of the particles. A photographic or video camera records the motion of the particles by viewing the light sheet under an angle of 90°.

In many applications of PIV, the chopped beam of a continuous wave laser or a pulsed laser is used to illuminate the particles and thus to produce two or more images of the particles on the film of a photographic camera. The film is then developed and digitized to provide the image data for an automatic computer-based evaluation. The evaluation is done by an autocorrelation method

* Universität GH Essen, Lehrstuhl für Strömungslehre, D-45117 Essen, Germany

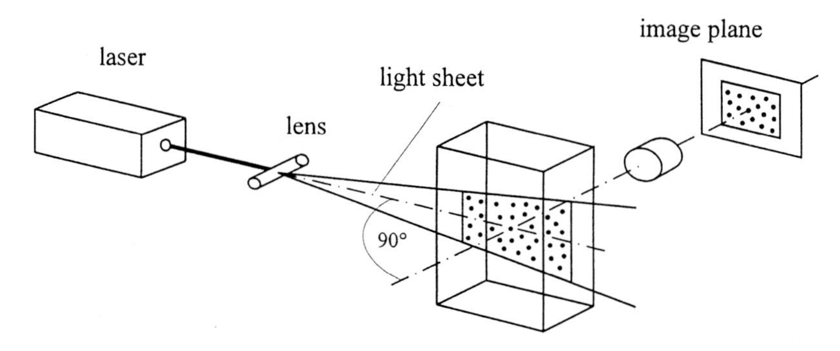

Figure 10.1: Basic elements of a set-up for PIV measurements.

as it is described in [2]. Besides the inherent directional ambiguity of the autocorrelation method, the whole procedure is time consuming and tedious.

Video cameras equipped with a CCD-chip avoid the post processing procedures necessary for photographic film and make image acquisition much faster. Since video cameras record time sequences, the use of separate images of single-exposed particle images removes the directional ambiguity in the recovered data and also allows the displacement measurement of partially overlapping particle images. According to the European video norm of the CCIR, video cameras produce 25 frames per second in the interlaced mode, which means that each frame consists of two consecutively generated and transmitted fields. With each field having an exposure time of 20 ms, measurements with digital particle image velocimetry (DPIV) utilizing a single video camera, as suggested in [3] and [4], are restricted to slow fluid flows.

10.2 The Two-Camera-System

In order to enhance the performance of video-based PIV measurements concerning application to fluid flows with higher velocities, a system with two synchronized video cameras was developed. The task was to use ordinary video cameras that are available for reasonable prices and to use them in a way that a sequence of two consecutive pictures can be taken with a time interval considerably shorter than 40 ms.

The video cameras used for this purpose are equipped with a Sony XC-77 RR-CE CCD-chip with a size of 756×581 pixels. The analogue video signal is transmitted to a frame grabber board installed into a personal computer. The frame grabber board digitizes the video signal and stores it into its own fast memory. Via this memory, access is given to the image data. The image then

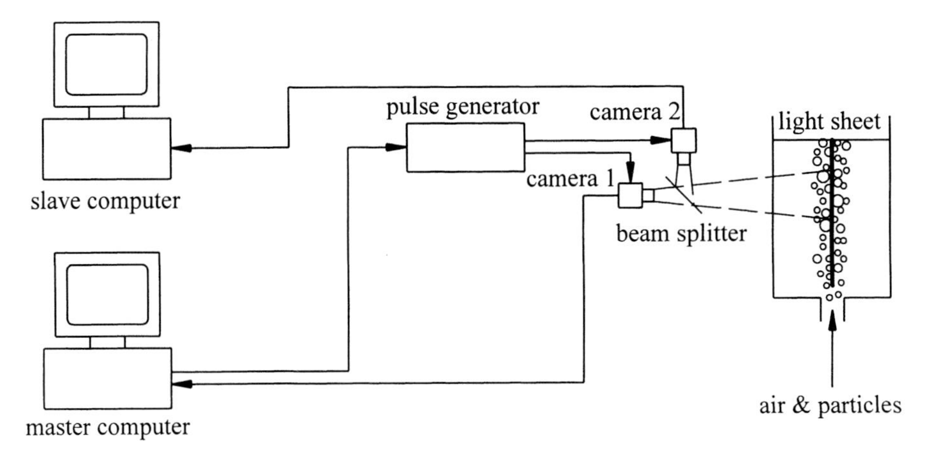

Figure 10.2: Schematic view of the two-camera-system.

has a size of 512×512 pixels with a color depth of eight bits, which means that the gray values range from 0 (black) to 255 (white).

The cameras provide two different operating modes. Besides the continuous generation of images, the cameras can produce a single picture on request. In this so called Donpisha mode, the process of recording a single picture is started by an external trigger signal, and the electronic shutter can be set from 262.8 ms for long time exposures down to 0.78 ms for short time exposures.

In Figure 10.2, the single elements of the two-camera-system and their connections are shown. Two computers are used each of which controlling one camera. Recording sequences are started by the master computer, which then delivers a trigger signal to a pulse generator. The pulse generator transmits a trigger signal to each camera to start the image generation process with the time interval between the two trigger signals being adjusted by the pulse generator.

Since the optical axis of the two cameras are perpendicular to each other, a beam splitter, a semi transparent mirror, ensures that both cameras record identical fields of view in the flow. To align the cameras, a transparent plate with a definite mesh is placed in the light sheet. A similar grid is created in the overlay plane of the frame grabber. Both, the life image of the mesh in the light sheet and the grid of the overlay of the frame grabber, are superimposed on a monitor so that the two cameras can be aligned properly.

Illumination is provided by a continuous wave Argon laser with a wave length of 514.5 nm so that the exposure time is determined by the electronic shutter of the camera.

10.3 Experimental Apparatus

Multiphase-reactors allow the contact of two or three phases for a definite period of time. The reactor serves to allow a reaction taking place between the liquid and the solid phase, while the role of the gaseous phase is to agitate and homogenize the reactants in the reactor volume.

An example for such a reactor is a gas-stirred ladle with through-flow of molten iron [5] (Figure 10.3). Calcium carbide powder is pneumatically transported and then injected into the liquid iron through nozzles on the floor of the reactor. The inert carrier gas, e. g. nitrogen, induces circulatory flows and stirs the fluid system, while the calcium carbide reacts with the sulphur included in the iron to form calcium sulphate, better known as gypsum.

For optimizing this process and for analyzing the resulting flow phenomena, it is necessary to determine the velocities and the distribution of the phases. Model experiments for providing this information can be performed with a water analogue. The three different phases in the model reactor are represented by the through-flow of the water, the bubble column produced by introducing air into the water through a plate with holes of diameter of 100 μm, and solid particles being introduced into the water by the bubble column (Figure 10.4).

In order to perform simultaneous PIV measurements of the velocity distribution of each phase, it is necessary that the signals resulting from the different phases can be distinguished. The water is seeded with neutrally buoyant particles of a diameter of 50 μm, which reflect the light of the Argon laser at the same wave length. The air bubbles also reflect the laser light unchanged,

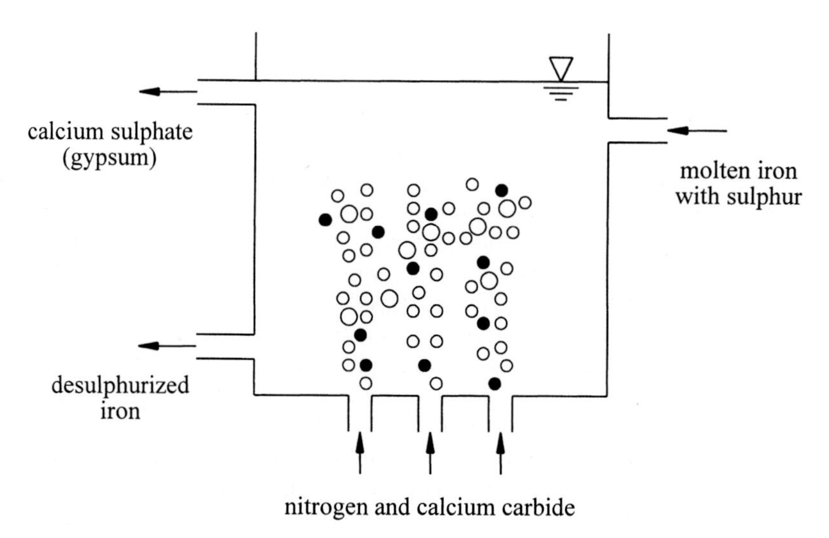

Figure 10.3: Schematic view of a gas-stirred ladle.

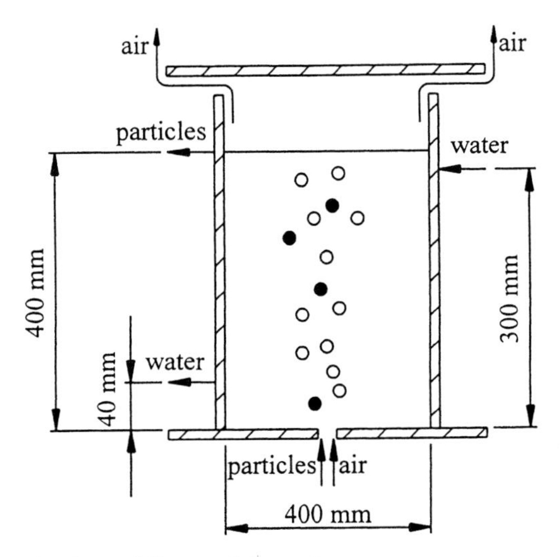

Figure 10.4: Schematic view of the model reactor.

but they are of much larger size (∼ 2 mm). To visualize the solid phase, fluorescent particles of a size of 2 mm and a density less than that of water were chosen. These particles scatter the light at a wave length shifted with respect to the laser wave length. Thus, differentiation of the three phases is achieved by different characteristic signals according to size and scattered wave length.

10.4 Measurements

In a series of experiments with the model tank, the two-camera-system was used for velocity measurements in the flow induced by the bubble column, i. e. without through-flow of the water. These measurements could be compared with PIV experiments performed with a single video camera [6]. The PIV records were evaluated with a cross-correlation algorithm applied to the sampling windows of size 32×32 pixels. The result of a single correlation is a velocity vector depicting the average velocity and direction of motion of all bubbles included in the interrogation window. In this way, the whole image is processed line by line with an overlap of the windows of 50%.

As an example of this evaluation process, a planar velocity distribution depicted as a vector plot is displayed in Figure 10.5. This vector plot is the result of the evaluation of a recording sequence taken with an exposure time of 2 ms and a time interval between the two consecutive images of 5 ms. The cross-correlation algorithm detects the correct direction of motion of the bubbles. A few vectors show a higher horizontal velocity component. This is due to bigger bub-

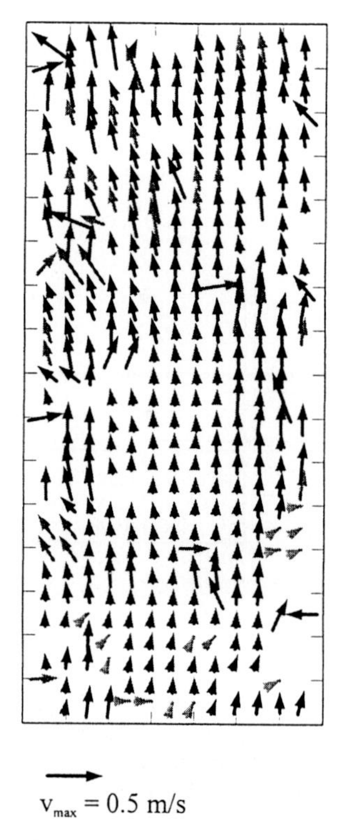

Figure 10.5: Vector plot of the velocity distribution in the bubble column.

bles, whose motion is not straight upwards but in form of a curved path. Small bubbles move with a relatively uniform velocity of approximately 0.2 m/s as expected. There is a general increase in velocity from the bottom to the top of the interrogation area indicating the acceleration the bubbles experience on their way towards the surface.

10.5 Discussion

Operating a PIV system with a video camera for the data acquisition has a number of advantages concerning data evaluation. The limitations resulting from the given video frequency can be overcome with the two-camera-system described in this paper. This allows to extend the measurements to higher velocities of the investigated flow. For the present system, this range is limited by the shutter time of the electronic cameras, $t_s = 0.78$ ms. A further increase of the maximum velocity that can be measured is possible by using a pulsed laser instead of a continuous wave laser as the light source. The pulse sequence of the laser must then be synchronized with the recording sequence of the two video cameras. The spurious vectors appearing in Figure 10.5 can be removed with one of the available correction algorithms (see, e.g., [7]).

References

[1] R.J. Adrian: *Particle-imaging techniques for experimental fluid mechanics.* Annu. Rev. Fluid Mech. **23** (1991) 261–304.

[2] W. Heckmann, S. Hilgers, W. Merzkirch, T. Schlüter: *Automatic evaluation of double-exposed PIV records by an autocorrelation method.* In: Optical Methods and Data Processing in Heat and Fluid Flow II. Inst. Mech. Eng's, London, 1994, pp. 5–8.

[3] W. Merzkirch, T. Mrosewski, H. Wintrich: *Digital particle image velocimetry applied to a natural convective flow.* Acta Mech. **4** (1994) 16–26.

[4] C.E. Willert, M. Gharib: *Digital particle image velocimetry.* Exp. Fluids **10** (1991) 181–193.

[5] S. Torii, W.-J. Yang: *Melt-particle mixing in gas-stirred ladles with throughflow.* Exp. Fluids **13** (1992) 37–42.

[6] S. Hilgers, W. Merzkirch, T. Wagner: *PIV measurements in multiphase flow using CCD- and photo-camera.* In: Flow Visualization and Image Processing in Multiphase Systems. ASME FED-Vol. **209** (1995) 151–154.

[7] J. Westerweel: *Efficient detection of spurious vectors in particle image velocimetry data.* Exp. Fluids **16** (1994) 236–247.

11 Investigation of the Two-Phase Flow in Trickle-Bed Reactors Using Capacitance Tomography

Nicolas Reinecke and Dieter Mewes *

Abstract

Investigations of fluiddynamic properties of two-phase flows in regular and irregular packings of trickle-bed reactors require a detailed knowledge of the flow phenomena inside the packing. A visual observation of the flow from the outside of the column does not yield the desired information especially for large scale pilot plant reactors. Therefore, new measurement techniques are required to obtain the necessary information for the physical description of the flow, especially for scale-up calculations. In this paper, the possibilities of tomographic imaging of the two-phase flow in trickle-bed reactors using capacitance tomography are discussed. The principle of the measurement technique is explained and the possibility of obtaining new information pointed out. Finally, results of this new imaging technique are presented.

11.1 Introduction

Tomographic measurement techniques have been applied to medical imaging for over 20 years, and they are a standard diagnostics tool for many purposes by now. The use of tomographic imaging techniques in process engineering applications has not been considered as long, but is becoming increasingly popular. These measurement techniques allow for two- or three-dimensional imaging of various processes in chemical and mechanical engineering industries. Using several possible physical principles to measure the integral value used for the reconstruction, the phase-, the temperature- and the concentration-distribution, as well as velocity distributions can be calculated. In addition, some techniques allow for the property sensitive investigation of labelled compo-

* Universität Hannover, Institut für Verfahrenstechnik, Callinstr. 36, D-30167 Hannover, Germany

nents or phases. The advantage of all these techniques is the possible space-imaging without intrusion into the process.

For the tomographic reconstruction, integral measurements have to be made. These can be obtained in numerous ways (see Reinecke et al. [1]). There are for example positron or photon emission, γ- and x-ray transmission and scattering, nuclear magnetic resonance, microwave reflection and diffraction, ultrasonic/acoustic techniques, interferometry and holographic tomography, and electrical and electromagnetical field interaction. Depending on the specific technique used, different advantages and disadvantages regarding the accuracy, frequency and resolution of the reconstructed images and thus applications can be identified. Furthermore, for industrial applications, the cost of the system as well as the fault-tolerance are decisive.

Most applications of process tomography in the chemical industry require the imaging of fast changing, so called transient processes. This is the case for most multiphase flows of interest. It requires the tomography system to have the capability of very fast data acquisition and, for control purposes, often on-line image reconstruction. Furthermore, the measurement volume is often not a simple circular pipe but a rather complex geometry, such as process equipment containing impellers or catalysts [2, 3]. In addition to that, more than two different phases can be present, making the imaging tasks even more difficult.

One tomographic measurement technique that is well suited to both research and industrial applications is capacitance tomography. The sensors required for the imaging are robust and relatively cheap and the actual data generation and acquisition well reported (see for example [3]). An on-line reconstruction of the measured phase distribution is possible, using simple reconstruction algorithms like backprojection running on parallel computing architectures. A more sophisticated iterative reconstruction like ART yields better results, but needs more computing power (see for example [4, 5]). Altogether, capacitance tomography offers a very useful tool for a number of process engineering applications. In the following, the application of capacitance tomography for the imaging of two-phase flow in trickle-bed reactors is reported. Particular interest will be paid to the highly instationary flow regimes encountered in these types of reactors.

11.2 Measurement Principle

In capacitance tomography, the interaction of a permittivity distribution with an electrical field applied to the measurement volume is used for the generation of the integral measurements required for the reconstruction. The capacitance of two electrodes is a function of the permittivity of the components in the measurement volume as well as their distribution. This is schematically shown in Figure 11.1. The relationship between the permittivity, the spatial distribution

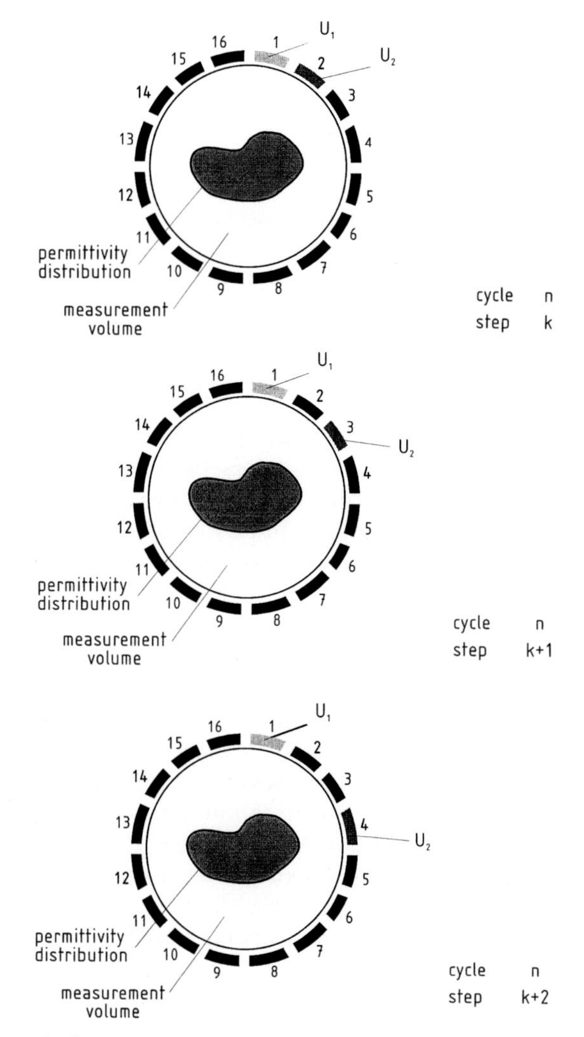

Figure 11.1: Schematical representation of the capacitance as a function of the size and distribution of the permittivity.

of the permittivity and the resulting capacitance can be derived from Maxwell's equations [6]. When the dielectric displacement $\boldsymbol{D}$ and the electrical field density $\boldsymbol{E}$ are replaced by the spatial distribution of the electrical potential $\varphi(x, y)$ and permittivity $\varepsilon(x, y)$, Poisson's equation can be derived:

$$\Delta\varphi(x, y) + \frac{1}{\varepsilon(x, y)} \operatorname{grad}(\varphi(x, y)) \operatorname{grad}(\varepsilon(x, y)) = 0 . \tag{1}$$

For a homogeneous permittivity distribution, Equation (1) can be simplified to yield the well known Laplace equation:

$$\operatorname{grad}(\varepsilon(x, y)) = 0 \Rightarrow \Delta\varphi(x, y) = 0 . \tag{2}$$

The integral capacitance C, which is measured by the peripheral electrodes 1 and 2, can then be obtained through:

$$C = \frac{Q}{\varphi_2 - \varphi_1} = \frac{\oiint_A \varepsilon(x,y)\boldsymbol{E}\mathrm{d}\boldsymbol{A}}{\varphi_2 - \varphi_1} = \frac{\oiint_A \varepsilon(x,y)\mathrm{grad}(\varphi(x,y))\mathrm{d}\boldsymbol{A}}{\varphi_2 - \varphi_1}, \tag{3}$$

where $\varphi_2 - \varphi_1$ is the potential difference between the driving and measuring electrode and $\boldsymbol{A}$ the directed normal area of integration on the electrodes surface.

From Equation (1), it is evident that Poisson's equation is linear in reference to the electrical potential distribution but non-linear in reference to the permittivity distribution. Therefore, the integral capacitance calculated from Equation (3) is a non-linear function of the permittivity distribution. In contrast to any ray transmission techniques, capacitance tomography is thus a non-linear technique yielding either very computationally intensive iterative forward solutions or linearization procedures with inherent errors in the reconstruction. In addition to that, it is also evident that an inversion of Equations (1) and (3) with reference to the permittivity distribution is not possible, which is common to most tomographic techniques mainly because of the severely underdetermined system of equations. Therefore, it is only possible to calculate the capacitance from a known permittivity distribution and not vice versa.

In order to obtain a two-dimensional tomographic image of a phase distribution inside the measurement volume, several linearly independent measurements of the capacitance between electrodes have to be made. The electrodes used for the measurement are placed peripherally around the measurement volume. Rather than having one sender and one receiver electrode as would be used for linear transmission techniques like CAT, both electrodes are involved in generating an electrical field, which is distorted by the permittivity distribution within the measurement volume. Due to the electrical field interactions between fields generated by different electrodes, the measurement of the different linearly independent capacitances is conducted by sequentially sampling different pairs of electrodes. This can be explained using Figure 11.1. In the first cycle, the capacitance is measured between electrodes 1 and 2, in the second cycle between 1 and 3, in the third cycle between 1 and 4, and so on. Thus, a set of linearly independent measurements is obtained.

In Figure 11.2, a schematic representation of the measurement chain for capacitance tomography is shown. It consists of the primary sensor system, the sensor electronics and capacitance measurement instrument and the computer for the reconstruction. This type of generalisation allows for a better exchange of the different components and a better assessment of the measurement errors associated with the individual part. The sensor generally consists of peripherally mounted electrodes manufactured from conducting foil and a casing giving it mechanical stability and electromagnetic stray immunity. The sensor electronics are used to do the sequential sampling of the electrodes, while the mea-

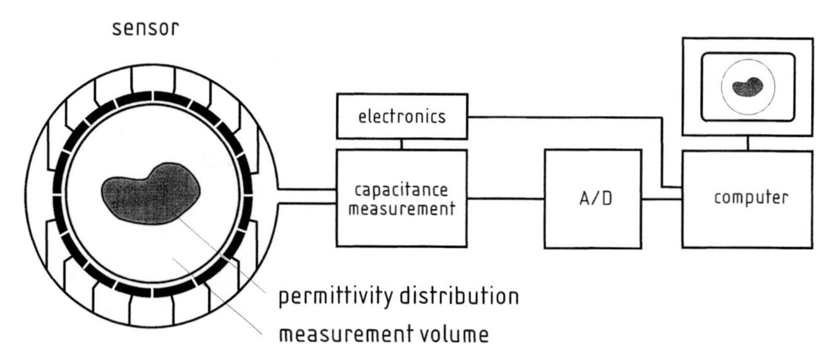

Figure 11.2: Schematical representation of the measurement chain.

surement instrument is used for the measurement of the capacitance between the electrodes. The set of measurements is transferred to a computer, where the reconstruction is done.

11.3 Capacitance Tomography System

11.3.1 Sensor

In Figure 11.3, the primary capacitance sensor used for the present study is shown schematically. It consists of three planes of identically segmented electrodes. A grounded shield is placed around them to ensure electromagnetic stray immunity. Two sets of driven shield electrodes above and below the measurement electrodes are used to focus the electrical field to a known volume. This was implemented to allow for a two-dimensional representation of the electrical field within the measurement plane as well as a defined measurement volume. In Figure 11.4, the resulting schematical representation of the axial distribution of the electrical field is shown. The electrical field across the measurement volume is very homogeneous. Therefore, change of permittivity approaching the measurement volume only results in a change of the measured capacitance when the tip of the measurement electrodes is reached. Due to the increase in homogeneity of the electrical field and the lack of effects, it is possible to reduce the dimension of the measurement plane to 30 mm. This reduction yields a more exact two-dimensional representation of the measurement and eliminates the axial integration effects.

When operating the 16-electrode sensor, a versatile programming was developed. This allows for several of the electrode segments adjacent to another to be used as one electrode. This procedure does not immediately increase the

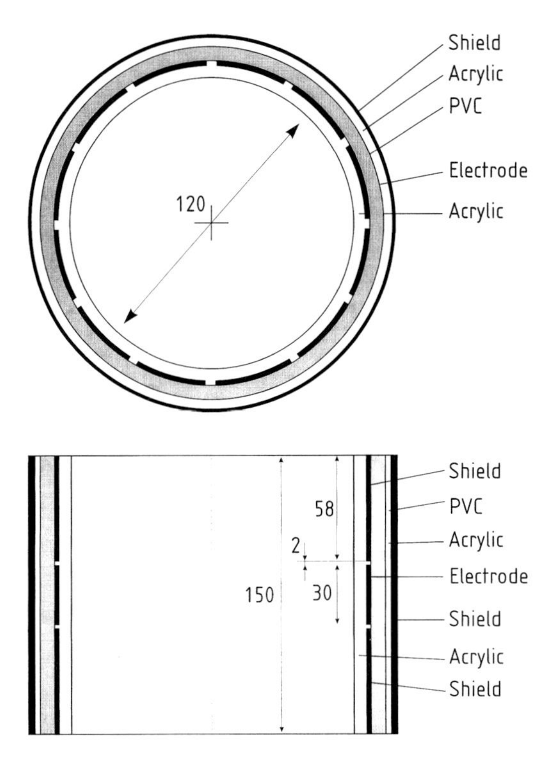

Figure 11.3: Schematical representation of the sensor.

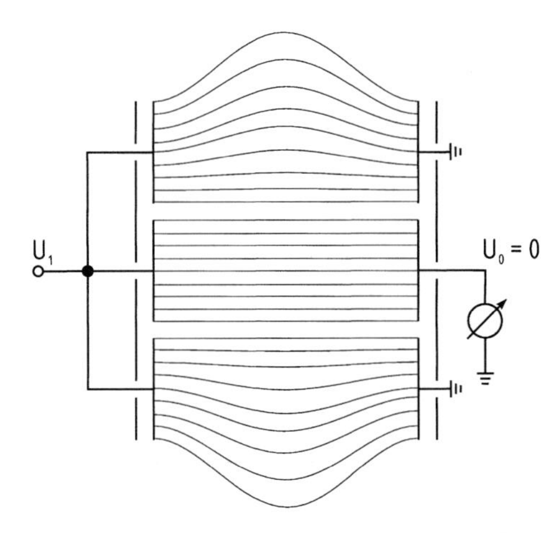

Figure 11.4: Plot of the electrical field for driven guard electrodes.

number of linearly independent measurements since they are determined by the smallest segmentation of the electrode plane, i.e. the number of single electrode segments. With an increase of the number of electrode segments per electrode, though, the signal level and thus, the signal-to-noise ration (SNR), is increased, while the spatial resolution is decreased. Considering two sensors with equal electrode area (and thus signal level) but different numbers of segments per electrode, the sensor with the higher number of segments per electrode will yield an increase in the total number of linearly independent measurements since a smaller rotational step-size is possible. The total number of measurements per image determines the imaging speed and the image resolution. The electrodes not involved in a measurement are left floating, which results in virtual source electrodes around the measurement volume. This increases the signal level, the SNR, the linearity, the spatial amplification and the resolution in the center of the sensor [4].

11.3.2 Sensor electronics and capacitance measurement

In capacitance tomography, the encountered capacitances are usually high (several pF) with the variations to be detected being very small (typically few fF) at very high sampling rates (typically 5 to 15 kHz). The electrical current then flowing onto the electrode does so in only a finite time setting a physical

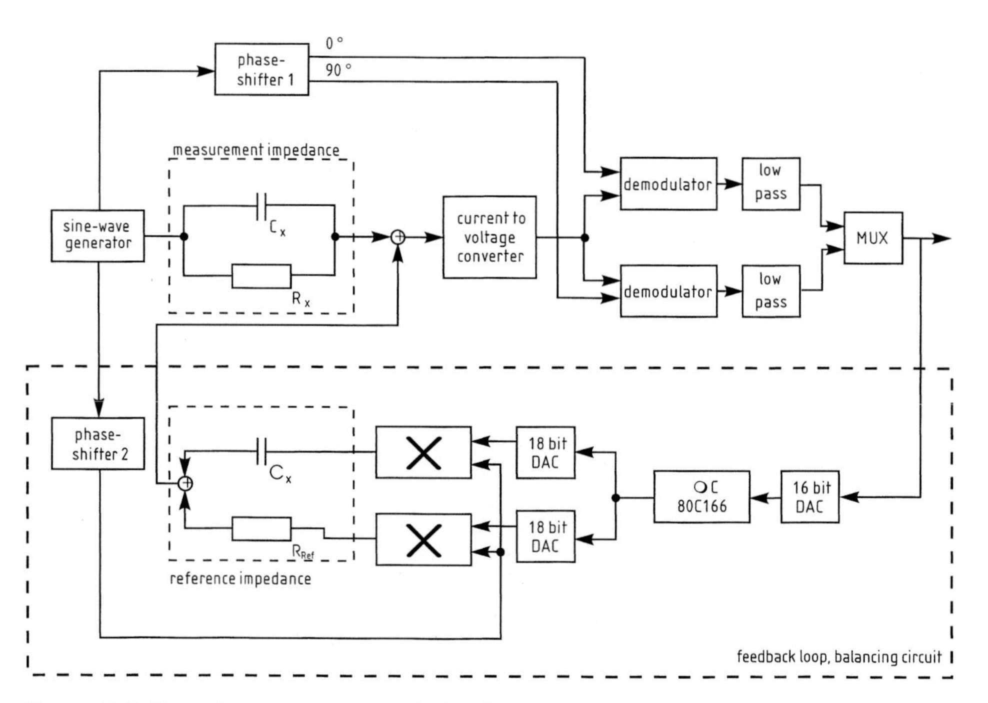

Figure 11.5: Capacitance measurement circuit.

limit to the sampling frequencies. In addition to this, the stray immunity of the circuit is very important to avoid extensive noise on the very small signal level.

The measurement technique used in the present study consists of a commercial capacitance meter based on an AC-bridge capacitance measurement circuit. This is schematically shown in Figure 11.5. A signal generator is used to drive one electrode of the unknown capacitance. The resulting current on the other electrode is converted to a voltage and then amplified. By using a reference voltage 90° phase shifted, a demodulation is possible and an analogue signal proportional to the capacitance is available. Using this technique, accurate and stable capacitance measurements are possible due to a low baseline drift and high SNR. The same technique is also employed by other investigators including Yang and Scott [7], and Klug and Mayinger [8]. Commercial instruments can typically be operated at measurement frequencies of 14 kHz with accuracies of 0.5 fF to 0.1 fF. The accuracy of the measurement increases with decreasing measurement frequency.

11.3.3 Reconstruction

A simple and very fast reconstruction algorithm is the backprojection algorithm. It is schematically shown in Figure 11.6a. The algorithm corresponds to the filtered backprojection algorithm popular for linear tomography. The integral measurement value of every measurement is distributed across the mea-

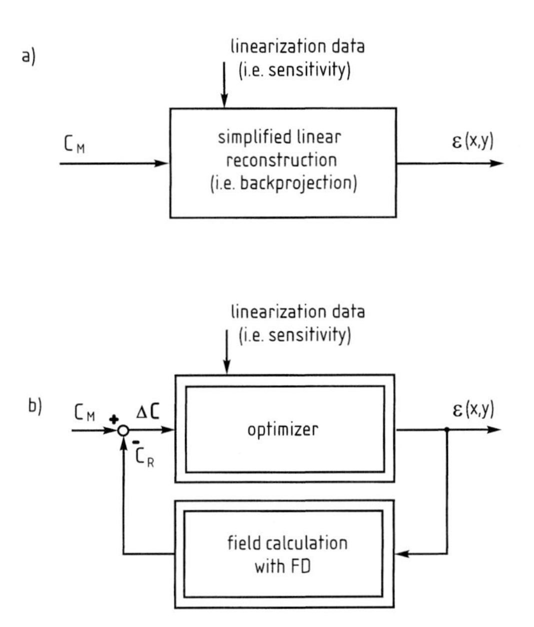

Figure 11.6: Schematical representation of the reconstruction algorithms.

surement plane according to a weighting or sensitivity matrix $S_i(x,y)$, which is unique for an electrode combination i and obtained a priori by:

$$S_i(x,y) = \left.\frac{\Delta C_i}{\varepsilon A}\right|_{A\to 0} . \tag{4}$$

It represents a spatial amplification of the sensor system (ΔC) at volume fractions of zero ($\varepsilon A \to 0$). Due to the non-linearity of this imaging technique, this linear approach results in relatively large reconstruction errors.

For the present investigation, an iterative reconstruction technique was used. This is schematically shown in Figure 11.6 b. In the iteration, the forward calculation is done using a finite difference calculation of the electrical field and thus the capacitance. Due to the computationally intensive FD calculation, the reconstruction time is high (several seconds per image), and an on-line reconstruction is not possible. The optimization of the assumed permittivity and phase distribution is done, using a set of sensitivities defined in Equation (4). This calculation procedure is similar to the algebraic reconstruction technique (ART) used for linear tomography [9]. For the electrode combination i and the iteration step k, the calculated dimensionless capacitance C^*_R and the measured dimensionless capacitance C^*_M yield the error ΔC^*_i

$$\Delta C_i^{*k} = C_{R,i}^{*k} - C_{M,i}^{*} \quad \text{and} \quad C_{R,i}^{*k} = f(a(x,y)) \quad \text{and} \quad C^* = \frac{C - C_0}{C_1 - C_0} . \tag{5}$$

The correction of the assumed phase distribution $a(x,y)$ is done using:

$$a^{k+1} = a^k + h((x,y),i,k) , \tag{6}$$

where h is an additive function of the sort

$$h = h^* \left(\frac{S_i(x,y)}{\sum\limits_{(x,y)} S_i(x,y)} \right)^n . \tag{7}$$

With the exponent $n=1$, the constant h^* can be calculated by

$$h^* = \frac{-\Delta C_i^{*k}}{\sum\limits_{(x,y)} (A_i(x,y)^2)} \quad \text{and} \quad A_i(x,y) = \frac{S_i(x,y)}{\sum\limits_{(x,y)} S_i(x,y)} . \tag{8}$$

The iteration is continued until the error ΔC^*_i is below a given value. The reconstruction is done in a Cartesian co-ordinate system with 32×32 pixels for the entire cross-section of the imaging plane [10].

11.3.4 Additional data-processing

11.3.4.1 Time-correction

In most applications of capacitance tomography, the sequentially obtained measurement values for the different combinations of the electrodes are used for the reconstruction of one image. This is no problem if the imaging time is much lower than the ratio of the length of the electrodes to the structural velocity of the flow. If the structural velocity of the flow is high, the time lag between the sequential measurements used for one frame is too large. This can be the case for instationary flows in trickle-bed reactors. Therefore, the velocity of the measured flow field is limited not only by the frequency of the individual measurements but also by the frequency of the reconstructed frames.

Using a calculation method described by Reinecke et al. [11], a time correction between the sequential measurements can be made. Therefore, the integral measurements for one frame are all at the same virtual time. Using an interpolation technique, it is possible to increase the allowable transients of the measured phases and still obtain images, which are less blurred. The general shape of the interpolation function to be fitted into the data can be determined beforehand by continuously measuring every individual electrode configuration.

11.3.4.2 Velocity calculation

When the tomographic sensor used for the measurement consists of only one measurement plane without any axial extension, the correlation of signals from two sensors is the only possible way to achieve a measurement of the velocity. In the case of capacitance tomography, though, the measurement plane of the sensor is extended in the axial direction. For the measurement of the velocity distribution, this axial extension of the measurement plane can be used for the calculation of the velocity of the interphase within the measurement volume from the measured capacitances. This is basically achieved by autocorrelating the sequentially measured capacitances. Details of the calculation are described by Reinecke and Mewes [5].

11.4 Experimental Apparatus

In order to assess the application of capacitance tomography for the imaging of two-phase flows in trickle-bed reactors, an experimental facility was built. In Figure 11.7, a schematic representation of the experimental set-up is shown. The liquid and the gaseous phase (water and air) are fed into the column at the top and flow through the packing cocurrently. They are separated at the bottom in a gas/liquid separator and the liquid phase is recirculated. The peripher-

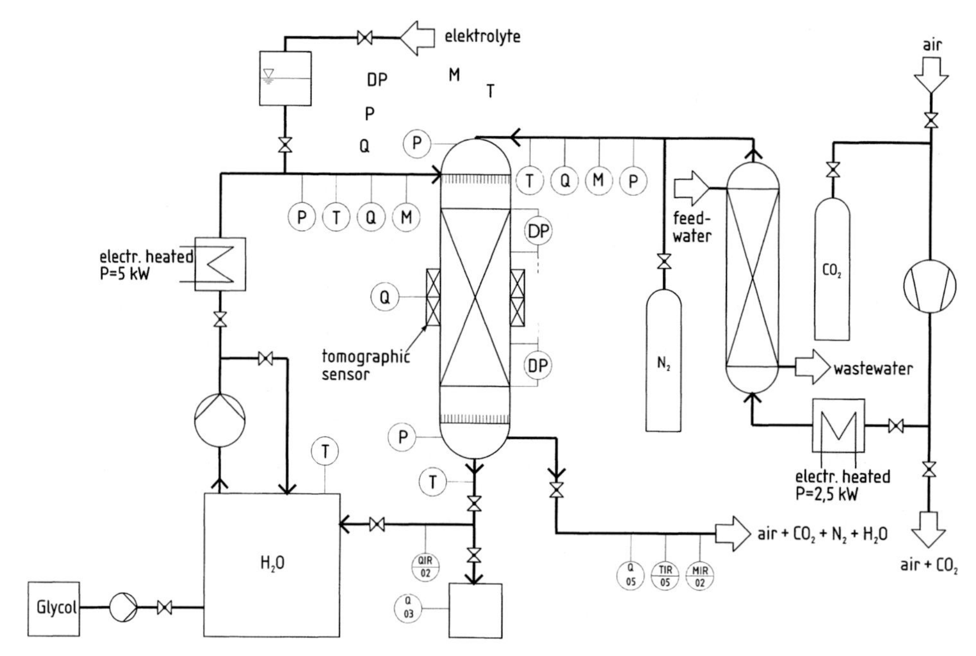

Figure 11.7: Experimental set-up.

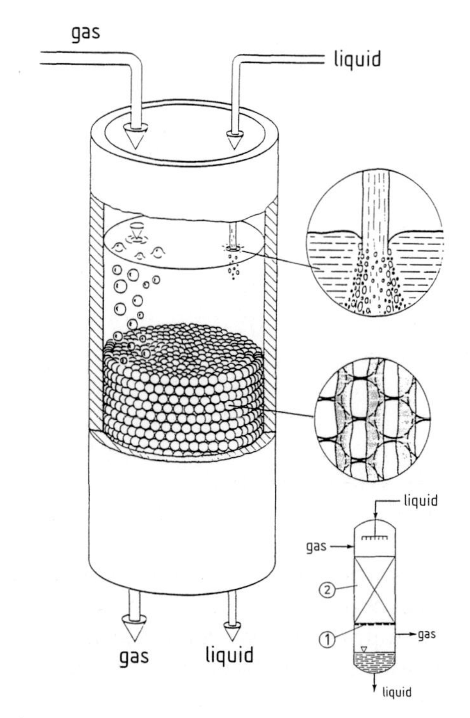

Figure 11.8: Schematical representation of a trickle-bed reactor.

al measurement techniques for the liquid and the gas feed include temperature, flow rate, pressure as well as the pressure drop across the column. The temperature of both phases can be set as well as the condition and concentration of other media.

In Figure 11.8, a trickle-bed reactor is shown schematically. The feed of both phases into the column is designed so that both gas and liquid enter the head of the column through a large number of cocurrent injection nozzles. Therefore, maldistributions at the head of the column are avoided. The inner diameter of the column is 120 mm with a total length of 2000 mm of packing. The packings used are standard spheres of 10, 5 and 3 mm hydraulic diameter made of celcore, a ceramic material. The ceramic itself is porous so that the liquid phase wets the packing thoroughly.

In order to avoid errors from a high conducting liquid phase, deionized water was used. The conductivity of the water was thus kept below 1 mS/m. The measurement plane was located at the bottom of the column, well upstream of the support tray in order to avoid interference. The calibration of the capacitance tomography was done with a fully wetted and a drained column. The full calibration was conducted with the packing being wetted for 30 min in order to assure fully soaked ceramic particles.

11.5 Experimental Results

In general, four different flow regimes encountered in trickle-bed reactors are characterised. These are plotted schematically in Figure 11.9. For small flow rates of both phases, the trickle flow regime is encountered (Figure 11.9 a), where the catalysts are completely wetted and the gaseous as well as the liquid phase is continuous. An operation below this point will yield a brake-up of the liquid film into rivulets. The resulting incomplete wetting of the packing makes the operation of most reactors in this flow regime not sensible. When the flow rates of both phases are increased, the pulse flow regime is encountered (Figure 11.9 b). Here, the liquid is bridging the gaps between the catalysts and accumulates to form a plug, which is then accelerated through the column by the expansion of the initially blocked gaseous phase. When the flow rate of the gaseous or the liquid phase is increased further, the spray flow regime (Figure 11.9 c) and the bubble flow regime (Figure 11.9 d) are encountered, respectively.

In order to fully assess the possibilities for the application of capacitance tomography for the tomographic imaging of trickle-bed reactors, highly transient flow regimes are desirable. In these flow regimes, the necessary time constants for the measurement are small and up to date, no information on the gas and liquid distribution are available. Therefore, most of the tomograms were measured for the pulse flow regime or for flow regimes close to the transition into pulse flow.

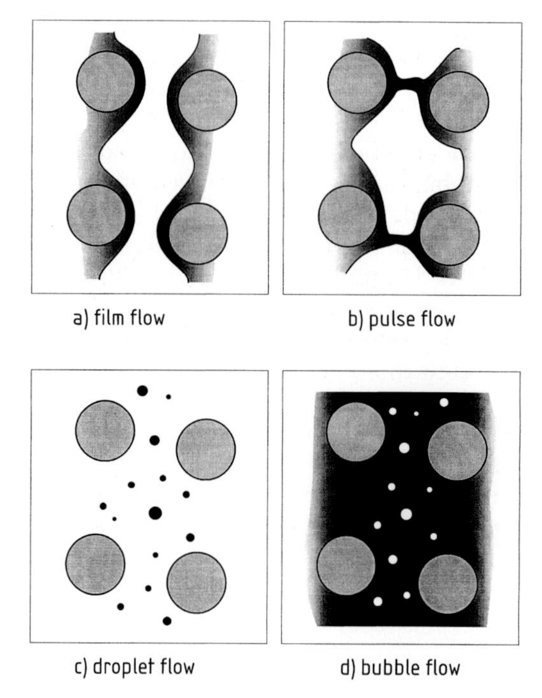

Figure 11.9: Schematical representation of the flow regimes in trickle-bed reactors.

11.5.1 Tomographic images

In Figure 11.10, exemplary results from the tomographic imaging of the pulse flow in trickle-bed reactors are shown. Plotted are individually imaged cross-sections of the pipe during the flow. The corresponding time goes from left to right and from top to bottom. The time scale between the images is 15 ms. The dark shadings represent the liquid phase, while the whiter ones represent an increase of the gas phase. From these images it can be seen that the pulse flow itself is established in a strongly three-dimensional manner. In the radial direction, even during the passage of a pulse, the distribution of the void fraction changes. Especially for the liquid rich zones (pulse), the void fraction is never 0 so that a bubble flow within these zones can be assumed. This is in accordance with other authors, who have studied pulse flow in two-dimensional flat-bed reactors [12].

In order to better understand the axial/time development of the flow field, a different representation of the images has to be chosen. In Figure 11.11, the data from Figure 11.10 and additional other data points are plotted differently. From the individual images in Figure 11.10, the central radial lines were taken and plotted one after the other. Therefore, if these lines are plotted from bottom to top, the direction of flow would be from top to bottom as is the case

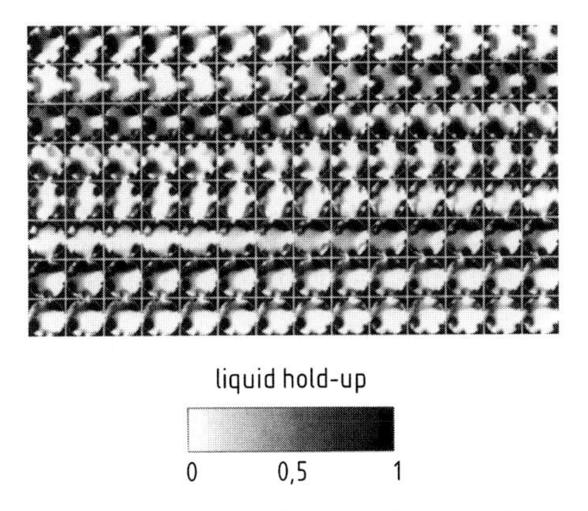

Figure 11.10: Tomograms of the two-phase flow in trickle-bed reactors, time plots.

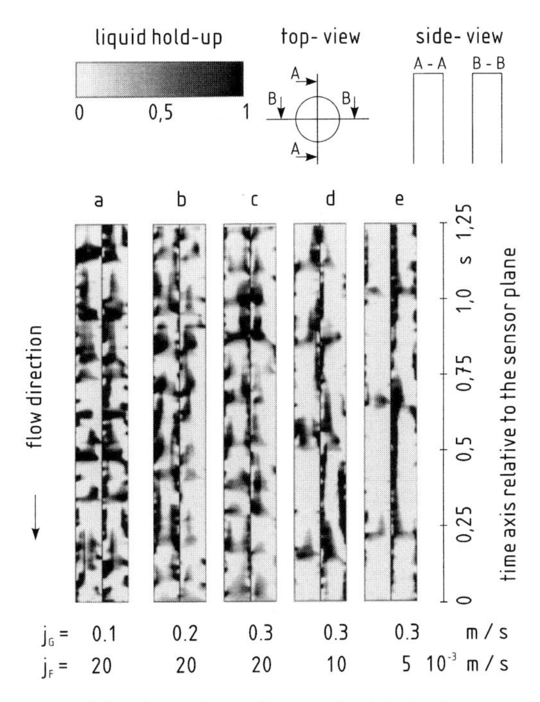

Figure 11.11: Tomograms of the two-phase flow in trickle-bed reactors, axial plots.

here. This type of representation has already been chosen for other tomographic investigations and is usually referred to as Eulerian slices. Given a constant and homogeneous velocity distribution, the images are identical to three-dimensional representations of the flow field. The colour representation is identical to Figure 11.10. In Figure 11.11, the volumetric liquid flow rate and the gas flow rate is increased from plot a to plot e.

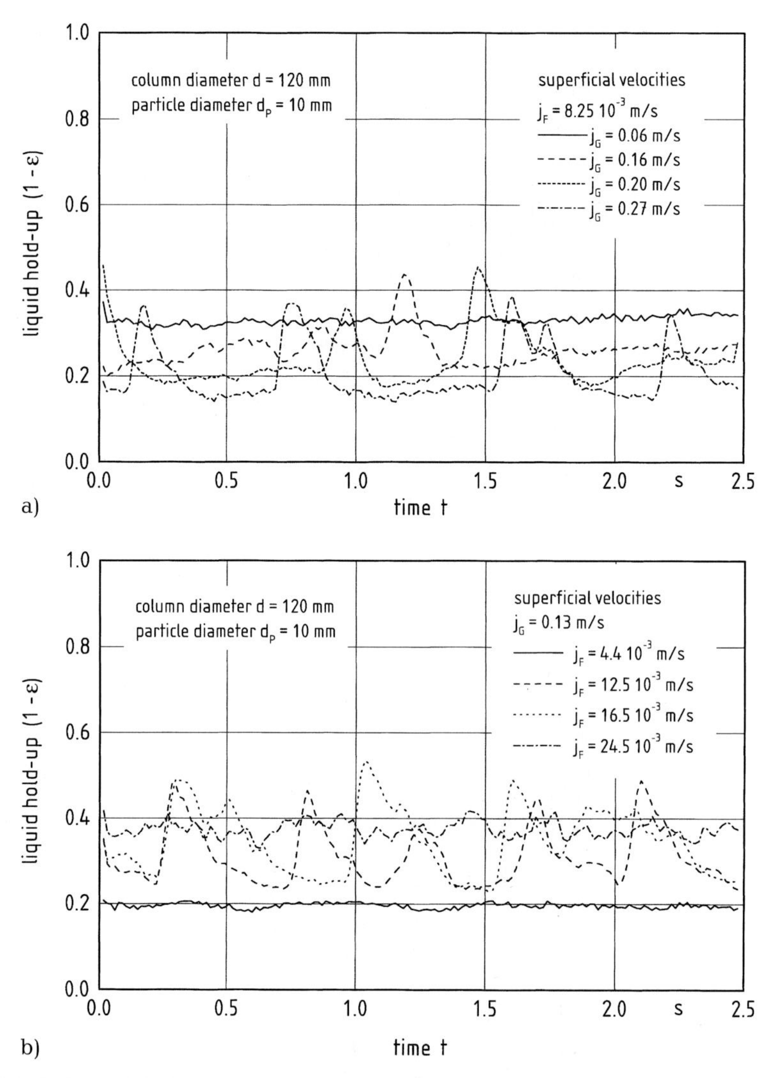

Figure 11.12: Liquid hold-up as a function of the gas and liquid superficial velocity and as a function of time.

As it is evident from the individual frame plots in Figure 11.10, the pulses passing through the column are aerated with bubbles and not of constant thickness. The three-dimensional shape and extension of the pulse is clearly evident and different for all the pulses. This observation cannot be made from the outside of the column. From visual observations from the outside, the pulses appear to be one-dimensional. This can be verified by plotting just the void fraction values from the wall region, where a constant pattern will appear. Focusing on the shape of the gas rich zones inside the column, it can be no-

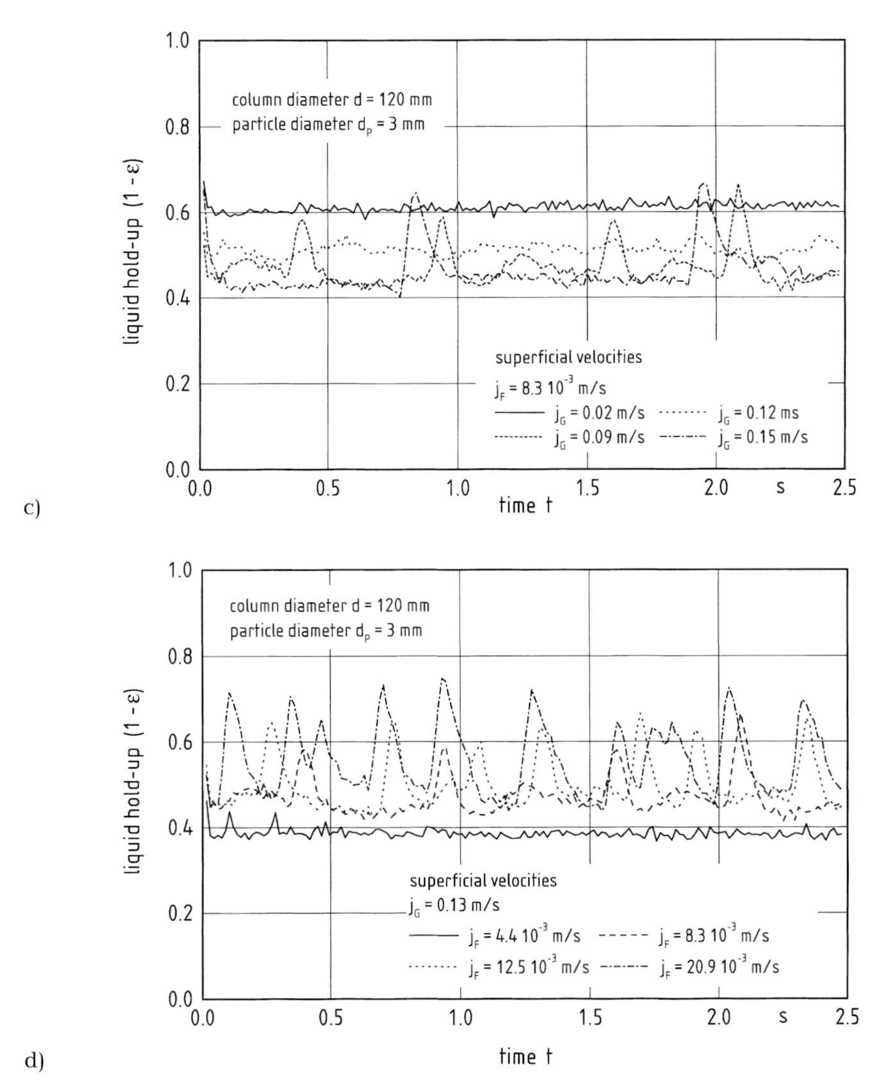

Figure 11.12: Liquid hold-up as a function of the gas and liquid superficial velocity and as a function of time.

ticed that for low gas flow rates, the leading and trailing edge of this zone is identical. For an increase of the gas flow rate, the front of the gaseous zone looses its plane shape, while the trailing end remains almost the same. The general shape of the gas rich zone can best be described by the shape of Taylor bubbles in the vertical two-phase flow in simple pipes.

As the gas flow rate is increased, the pulse frequency and thus the thickness of the liquid zones is decreased. With the decrease of the thickness of the liquid zone, a decrease of the strong three-dimensional effect is observed. This can be attributed to an increase in the momentum forces inside the column, which make a radial dispersion more difficult. An increase of the frequency can also be observed when the liquid flow rate is increased. The shape of the gaseous zones becomes more pronounced, while the radial distribution also becomes more homogeneous.

11.5.2 Integral void fraction

Using the tomographic images obtained, it is also possible to calculate an integral void or liquid fraction as a function of time. This is done not from the actual capacitance values measured, but rather from the reconstructed images. This way, the non-linear relationship between the liquid fraction and the capacitance values of the individual measurements does not yield any error for the calculation.

In Figures 11.12, four plots of the liquid hold-up as a function of time are shown. In the Figures 11.12 a and 11.12 b, the particle size is $d_P = 10$ mm, while in the Figures 11.12 c and 11.12 d, the particle size is $d_P = 3$ mm. In Figure 11.12 a, the liquid superficial velocity is kept constant at $j_F = 8{,}25\ 10^{-3}$ m/s, and the superficial gas velocity is varied. For an increase of the gas velocity, the liquid hold-up decreases. When the pulse flow regime is reached, the liquid hold-up in the liquid rich zones increases strongly, while the liquid hold-up in the gas rich zones decreases further. This effect is due to the depletion of liquid from the gas rich zone through pulses travelling over the liquid film and accumulating liquid. In Figure 11.12 b, the gas superficial velocity is kept constant at $j_G = 0{,}13$ m/s, and the superficial liquid velocity is varied. For an increase in the liquid velocity, the liquid hold-up in the column also increases. When the pulse flow regime is reached, the liquid hold-up in the gas rich zones is almost constant. Therefore, additional liquid fed into the column must be inside the liquid rich zones either by increase of size or frequency of the pulses. The same can be observed in the Figures 11.12 c and 11.12 d, where the same plots are given for the smaller particle size. Even though the overall value of the liquid hold-up is increased due to a decrease in the particle size, the general behaviour of the system is identical above.

In Figure 11.13, the average liquid hold-up is plotted as a function of the superficial gas velocity (Figure 11.13 a) and the superficial liquid velocity (Figure 11.13 b). In these plots, the effect discussed above is clearly visible, with the liquid hold-up decreasing for an increase in the superficial gas velocity and the liquid hold-up increasing for an increase in the superficial liquid velocity. When the liquid hold-up as a function of time is integrated, the increase respectively decrease of the hold-up seems to be independent of the flow regime. Even though the pulse flow regime is established, the monotony of the curve remains the same.

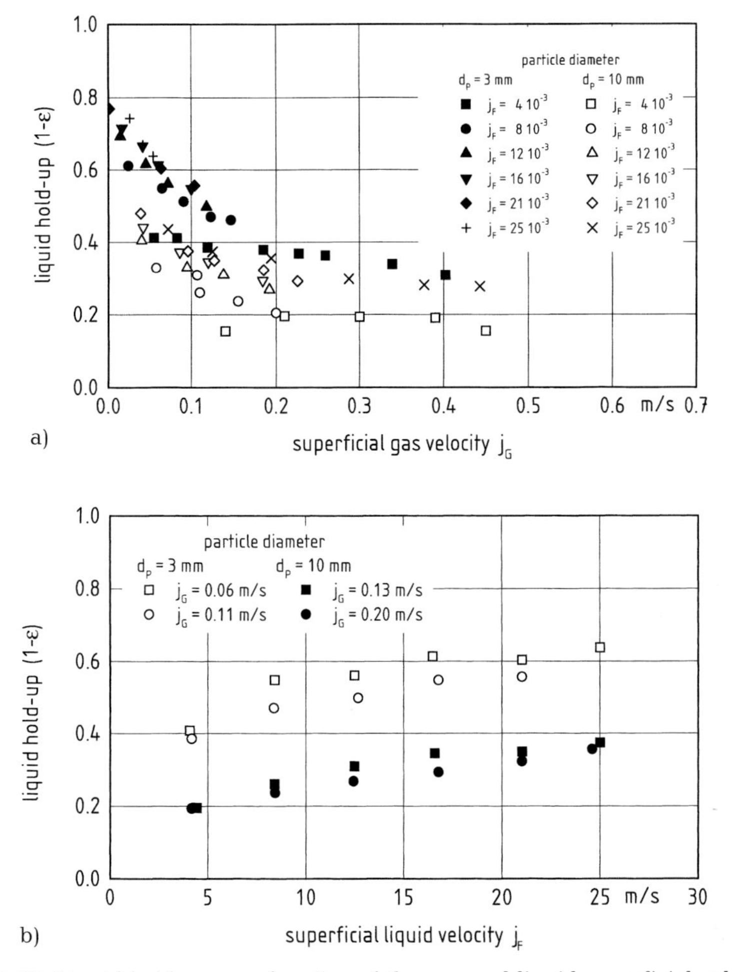

Figure 11.13: Liquid hold-up as a function of the gas and liquid superficial velocity.

11.5.3 Velocities

Using the data shown in Figure 11.12 and the equation derived in (12), it is possible to calculate the velocity of the pulses travelling through the column. In Figure 11.14 a, the liquid hold-up as a function of time is plotted. The pulses can clearly be distinguished as well as the gas rich zones. In Figure 11.14 b, the autocorrelation function

$$\Phi_{\dot{a}_M,-\dot{a}_M}(\tau) = -\Phi_{\dot{a}_M,\dot{a}_M}(\tau) = \left.\frac{da_M}{dt}\right|_t \left.\left(-\frac{da_M}{dt}\right)\right|_{t+\tau} \stackrel{!}{=} \Phi_{AKF} \tag{9}$$

proposed by Reinecke and Mewes [5] is plotted as a function of the time step τ. A clear maximum can be observed for about 0.1 s. This data was taken with a

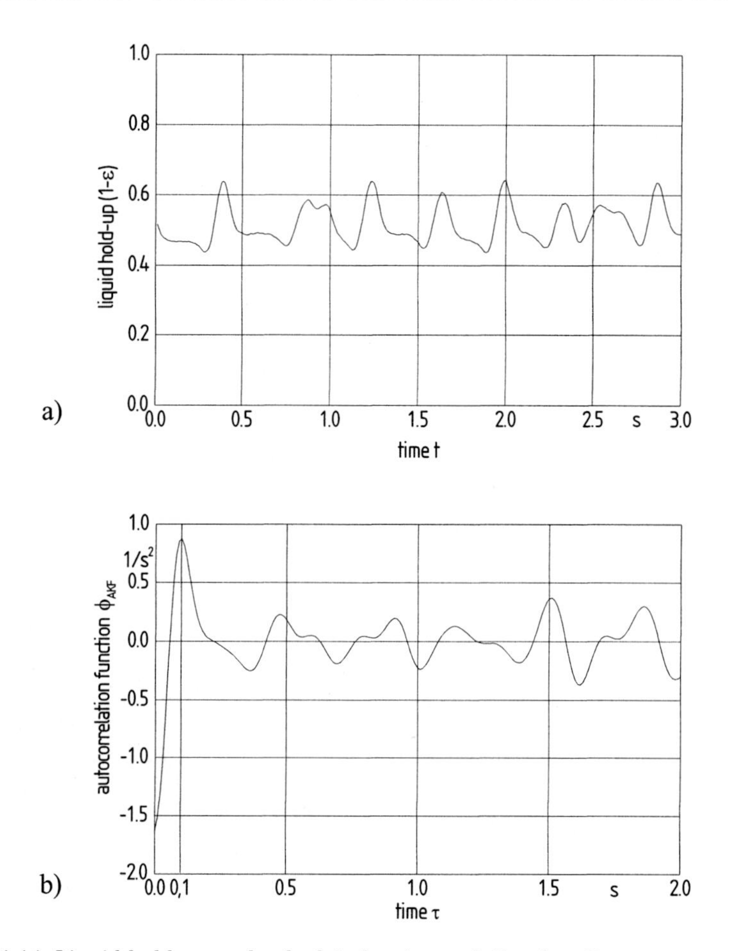

Figure 11.14: Liquid hold-up and calculated autocorrelation function.

sensor with a slightly longer axial extension of 75 mm. The corresponding velocity of the pulses is therefore about $j_{\text{Pulse}} = 0.75$ m/s. This is also in accordance with the data that can be obtained from the literature.

11.6 Conclusions

In this article, the possibilities of tomographic imaging of the two-phase flow in trickle-bed reactors using capacitance tomography are discussed. A newly developed sensor system is shown. The sensor consists of 16 peripherally mounted electrodes, which are sampled sequentially. The measurement of the capacitance is done using an AC-bridge and the reconstruction is done using an iterative ART-algorithm with a full electrical field calculation. It is shown, how the data-processing can be improved for this application, and how an additional velocity measurement can be made.

The results of the measured tomograms show the possibilities of the technique presented. The three-dimensional shape of the gas and the liquid rich zones inside the column for the pulse flow are visualised. It can be seen that this quality of data cannot be determined from the outside of the column. When the resulting integral values for the liquid hold-up are plotted, it is evident that in the pulse flow regime, the liquid is depleted from the gas rich zones to accumulate inside the liquid rich zones. The overall liquid hold-up remains unaffected by the flow regime. Using the data-processing described, it is possible to calculate a velocity of the pulses.

References

[1] N. Reinecke, G. Petritsch, D. Schmitz, D. Mewes: *Tomographic Measurement Techniques-Visualization of Multiphase Flows.* Chemical Engineering and Technology **21** (1998) 7–18.

[2] S.L. McKee, R.A. Williams, A. Boxman: *Development of solid-liquid mixing models using tomographic techniques.* Proc. ECAPT '94, Oporto, Portugal, Process Tomography – A Strategy for Industrial Exploitation – 1994, Eds.: M.S. Beck et al., 1994, pp. 342–353.

[3] C.G. Xie, N. Reinecke, M.S. Beck, D. Mewes, R.A. Williams: *Electrical tomography techniques for process engineering applications.* Chem. Eng. J. **56**(2) (1995) 127–133.

[4] N. Reinecke, D. Mewes: *Improvement of linearity and resolution of multielectrode capacitance sensors for the tomographic visualization of transient two-phase flows.* Proc. Symp. on Flow Visualization and Image Processing of Multiphase Systems, ASME, Fluids Engineering Devision, Hilton Head, South Carolina, USA, Eds.: W.-J. Yang et al., 1995, pp. 259–266.

[5] N. Reinecke, D. Mewes: *Calculation of interface velocities from single plane tomographic data.* Proc. ECAPT '94, Oporto, Portugal, Process Tomography – A Strategy for Industrial Exploitation – 1994, Eds.: M.S. Beck et al., 1995, pp. 41–50.

[6] J.C. Maxwell: *A treatise on electricity and magnetism.* Clarendon Press, Oxford, 1892.

[7] W.Q. Yang, A.L. Scott: *Tomographic techniques for process design and operation.* CEC Brite Euram Computational mechanics publications, 1993.

[8] F. Klug, F. Mayinger: *Novel Impedance Measuring Technique for Flow Composition in Multi-Phase Flows.* Proc. ECAPT '93, Karlsruhe, Germany, Process Tomography – A Strategy for Industrial Exploitation – 1993, Eds.: M.S. Beck et al., 1993, pp. 152–155.

[9] D.W. Sweeney: *Interferometric measurement of three-dimensional temperature fields.* Dissertation Univ. of Michigan, 1972.

[10] N. Reinecke, M. Buchmann, G. Petritsch, D. Mewes: *Enhancement of resolution for reconstruction algorithms for limited view absorption spectroscopy.* Int. Sem. on Optical Methods and Data Processing in Heat and Fluid Flow, IMECHE, London, United Kingdom, 1996.

[11] N. Reinecke, M. Boddem, D. Mewes: *Improvement of tomographic reconstructions by time-correction of sequential measurements*. Proc. ECAPT '94, Oporto, Portugal, Process Tomography – A Strategy for Industrial Exploitation – 1994, Eds.: M. S. Beck et al., 1995, pp. 381–392.

[12] T.R. Melli, L.E. Scriven: *Theory of Two-Phase Cocurrent Downflow in Networks of Passages*. Ind. Eng. Chem. Res. **30**(5) (1991) 951–969.

Drop Impact on Hot Walls

12 Transient Phenomena during Drop Impact on Heated Walls

Humberto Chaves, Artur M. Kubitzek and Frank Obermeier *

Abstract

The impact of single drops of ethanol upon temperature controlled horizontal surfaces has been investigated. A new classification scheme of the dominating phenomena in a Weber number/initial wall temperature plane is presented for Weber numbers between 12 and 3864 and wall temperatures between 20 °C and 480 °C. Furthermore, newly discovered effects are described. The results were obtained by the analysis of flash images. In a leading order analysis, the main effect corresponding to an increasing of the Weber number is to fragment the incoming drop into smaller droplets or films and thus to increase the surface area effective for the heat transfer between the wall and the liquid.

12.1 Introduction

The impact of liquid sprays on heated walls is a very important process for many technical applications, e.g. spray cooling [1] (steel production/nuclear power stations), fire fighting [2], and combustion in direct injection engines [3]. In order to cope with and to optimize such applications, an improved understanding of the underlying phenomena of single drop impact on heated walls is needed as a first step. Despite a great number of such investigations in the past, many fundamental questions still remain to be answered, and we are far from having a physical model, which includes all the manifold effects and their complex dependences on specific boundary conditions. Most of the publications cover single effects in rather limited ranges within a multidimensional pa-

* TU-Bergakademie Freiberg, Institut für Fluidmechanik und Fluidenergiemaschinen, Lampadiusstr. 2, D-09696 Freiberg, Germany

rameter space relevant to interactions between a wall and impinging drops. Basic to these interactions is the Leidenfrost phenomenon that results in a considerably reduced evaporation rate of drops deposited on heated walls above the so-called Leidenfrost temperature. The phenomenon was observed independently by Boerhaave [4] and Leidenfrost [5] already back in the 18th century. A survey on studies concerning drop impact and/or evaporation can be found in review articles by Rein [6], Bell [7] and Obermeier [8].

The aims of the present paper are

- to introduce a new wide-range classification scheme of dominating phenomena for vertical impact processes on polished horizontal surfaces,
- to present novel effects,
- to suggest how the understanding of any of the phenomena observed can be improved.

12.2 Experimental Set-Up

The experimental apparatus consists of three basic components

- a drop generating system,
- a temperature regulated wall,
- visualisation techniques.

In order to generate single drops of defined size and velocity, two different methods were applied. On the one hand, a fall apparatus, consisting of a well prepared hypodermic needle connected to a fluid reservoir, covered the range of drop diameters between 1.8 mm and 3.2 mm and of drop velocities between 0.08 m/s and 3.1 m/s, respectively. These values correspond to Weber numbers between 10 and 1000. On the other hand, a liquid jet was disintegrated into single drops by means of superimposed periodic pressure perturbations from a piezo-ceramic transducer inside the nozzle taking advantage of convective instabilities (Grabitz and Meier [9]) and/or Rayleigh instabilities (Weber [10]). Here, single drops can be separated from the resulting chain by charging and deflecting all other drops in an electric field. This apparaturs was exploited for drop diameters between 0.2 mm and 1 mm in combination with drop velocities between 4 m/s and 30 m/s. In this way, Weber numbers as high as 5000 could be realized. To meet the demand of high accuracy in the drop's velocity, its dimensions and its impact position, two major technical difficulties had to be surmounted [11]. On the one hand, an extremely undisturbed laminar liquid jet at first place turned out to be the requisite needed to achieve a highly reproducible modulated jet. Self made smooth quartz glass nozzles give very good re-

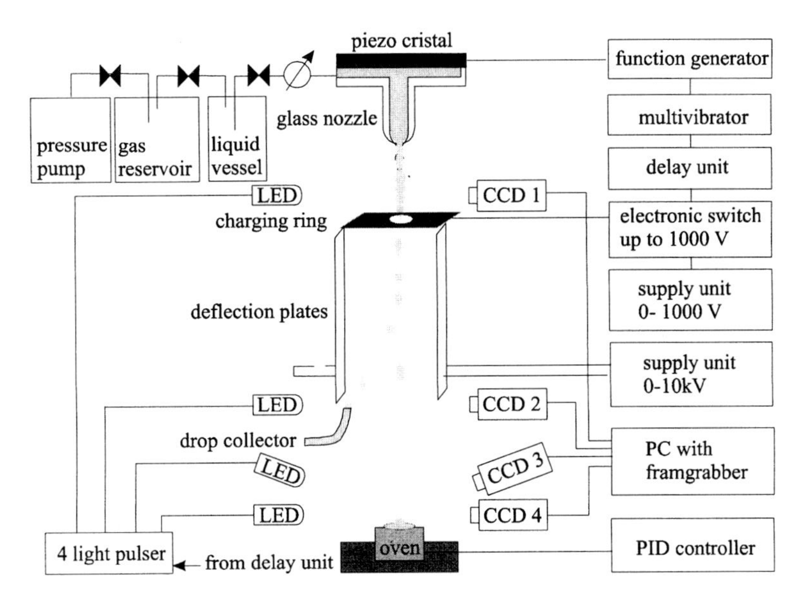

Figure 12.1: Experimental set-up.

sults. On the other hand, a very fast charging technique had to be developed, which allows charging each drop of the chain – independently of neighbouring drops – up to repetition frequencies of 60 kHz. A sketch of the deflection apparatus is given in Figure 12.1. On the right side, four CCD cameras are numbered 1, 2, 3, and 4. The cameras 1 and 2 help to control the deflection process; the cameras 3 and 4 capture the impact process from different angles. For technical details we refer to [11].

Inlays of materials of different heat conductivities, which could be inserted into a temperature regulated oven, served as heated walls. The PID-controlled wall temperature was adjustable between ambient temperature and 500 °C. Most results were obtained by analysing flow visualisation images obtained by CCD cameras. Advanced TTL synchronizing techniques led to film-like sequences – put together from series of single but reproducible events –, which resolve the dynamics of the interactions between the wall and an impinging drop.

12.3 Results

12.3.1 Regimes of drop fragmentation

On Figure 12.2, the regimes valid for different phenomena occurring during drop impact on heated walls are shown in a Weber number versus wall temperature diagram. Pictures representing the phenomena are shown in Figure 12.3 and Figure 12.4 (see also [12]).

Regime I is limited by the boiling temperature of the liquid. In this regime, only the effects of impact kinetic energy are important, i.e. spreading of a liquid film on the wetted wall and splashing. These purely hydrodynamic effects have already been documented elsewhere. With increasing initial wall temperature above the boiling temperature, nucleate boiling in the liquid film is enhanced and the collapse of bubbles occurring in the film leads to "jetting" of liquid from the wall and its subsequent miniaturisation (regime II). The upper temperature limit for this regime is determined by the onset of lift-off of the film (due to a steam layer between the film and the wall). It first occurs at the outer parts of the spreading film; with increasing temperature, it reaches also the inner parts of the film flow. This phenomenon is strongly dependent on the heat conductivity of the wall material, because the heat flux occurring after the impact reduces the wall temperature. This local transition from nucleate

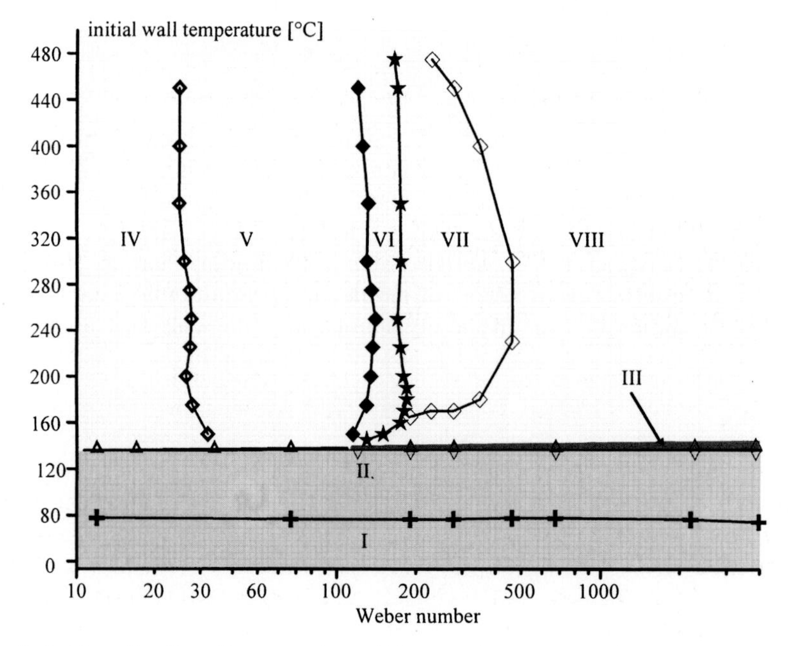

Figure 12.2: Classification diagram of the different drop impact phenomena.

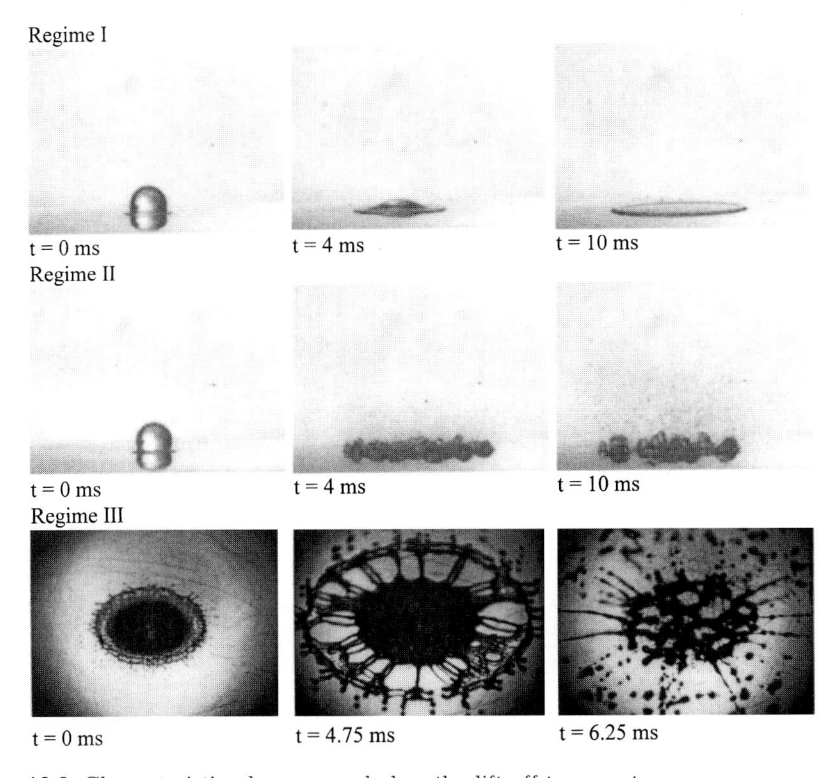

Figure 12.3: Characteristic phenomena below the lift-off temperature.

boiling (regime II) to film boiling (regimes IV–VIII) happens monotonically (regime III). The lower limit of regime III corresponds to the Leidenfrost temperature for sessile drops. However, the temperature for complete lift-off of the film (equal in the limit for $We = 0$ to the Leidenfrost temperature for sessile drops) is strongly dependent on the impact Weber number and the wall material. Therefore, the temperature for the onset of film boiling in the case of impacting drops is called lift-off temperature in this paper to avoid confusion with the Leidenfrost temperature valid for sessile drops.

With respect to the Weber number, the region above the lift-off temperature is divided into five areas, surprisingly the dividing lines between these areas do not depend strongly on the initial wall temperature. In regime IV, the drop oscillates at its eigen-frequencies and bounces over the hot wall during evaporation. With increasing Weber number in regime V, satellite droplets separate from the bouncing and deforming drop. Here, two new categories of bouncing are observed. First, in the area VI, secondary droplets break off due to rim perturbations when the liquid starts to contract. Then a new outer bulge arises and confines the rest of the liquid, which now bounces as in regime V. Second, in the area VII, a so-called coalescence bouncing is observed. The phenomenon in VI entails also the disruption of the spreading (nonwetting) liq-

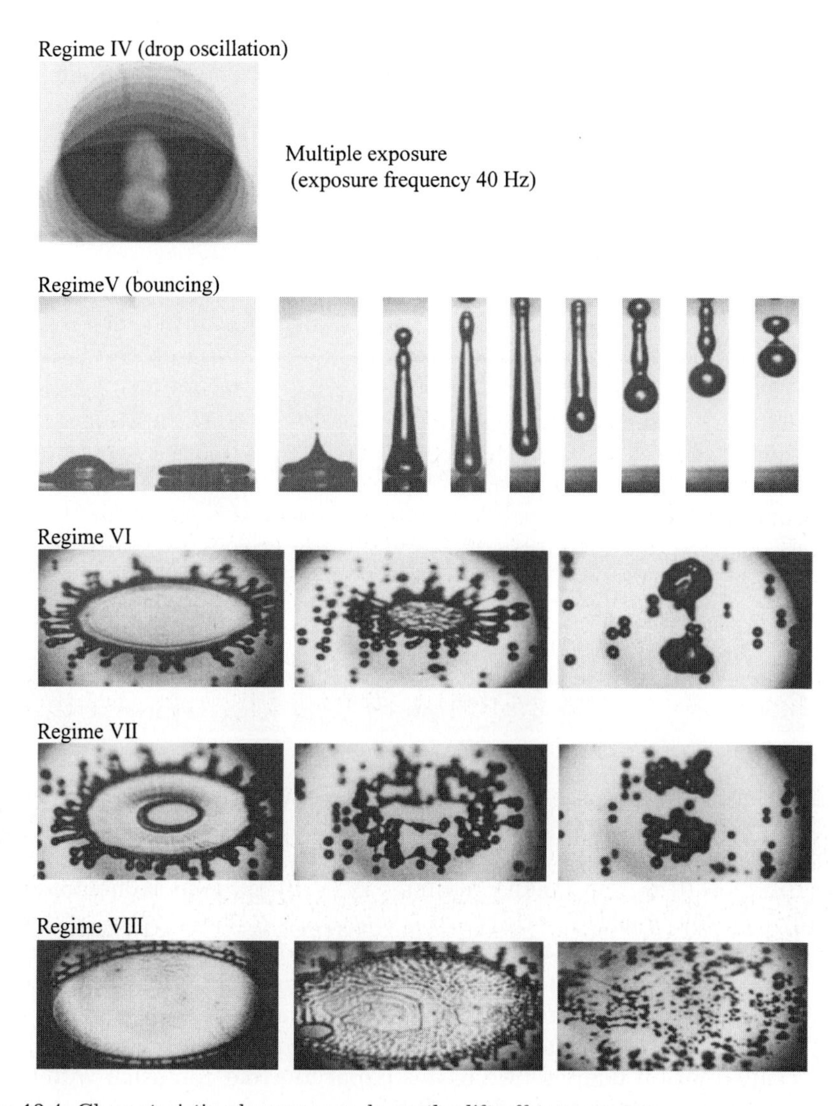

Figure 12.4: Characteristic phenomena above the lift-off temperature.

uid in its centre. When the liquid contracts to a ring-shaped bulge, the remaining horizontal component of the momentum is crucial for the further dynamics. In VII, the remaining momentum is aligned inwards, hence, the secondary fragments coalesce in the centre and bounce. At higher Weber numbers in region VIII, the time, at which the extending liquid starts to disintegrate (measured in the relative time scale of the spreading processes), becomes the dominant factor for the further dynamics. It is decisive for the remaining momentum of the secondary fragments. Here, usually many mesh-like structures can be observed indicating that the film evaporates rather homogeneously. In certain areas, these meshes seem to organize to a symmetrically distributed structure. This surprising effect suggests that Bénard convection cells occur in the spreaded liquid film.

12.3.2 Nucleate boiling in the liquid film

In the temperature regime II, i.e. at a temperature above the boiling temperature, a single bubble is observed at the impact centre, which is due to air entrainment; in addition, a regularly distributed circle of bubbles appears (Figure 12.5 a). The latter are probably caused by a rarefaction wave within the drop, which itself results from the reflection of the primary compression wave at the free surface of the impinging drop. With increasing wall temperature in this regime, additional bubbles are produced spaced irregularly (Figure 12.5 b). It is well known [13] that in this temperature range, the roughness of the surface has a strong effect on nucleate boiling. At even higher wall temperatures, the bubbles coalesce and produce a steam layer between the drop and the wall (Figure 12.5 c), and finally the cavities display a ring structure around the impact centre (Figure 12.5 d). It seems reasonable that they result from coalescing bubbles of a primary bubble ring created in the early stage of impact.

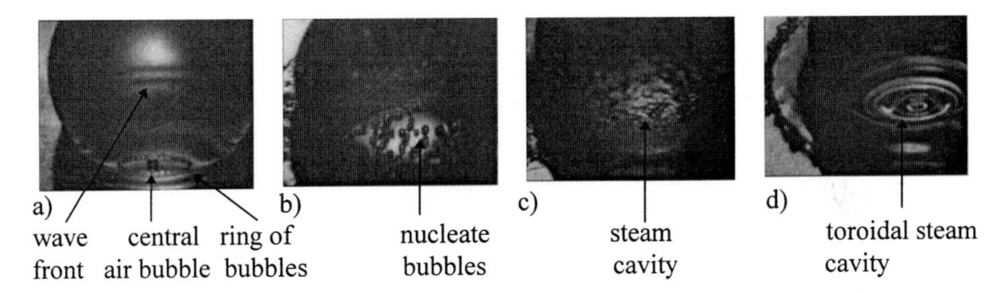

Figure 12.5: Steam cavities and bubbles close to the liquid-wall interface at the beginning of the spreading of drops, $t<1$ ms, $We=300$; a) $T=105\,^{\circ}$C, b) $T = 135\,^{\circ}$C, c) $T=200\,^{\circ}$C, d) $T=400\,^{\circ}$C.

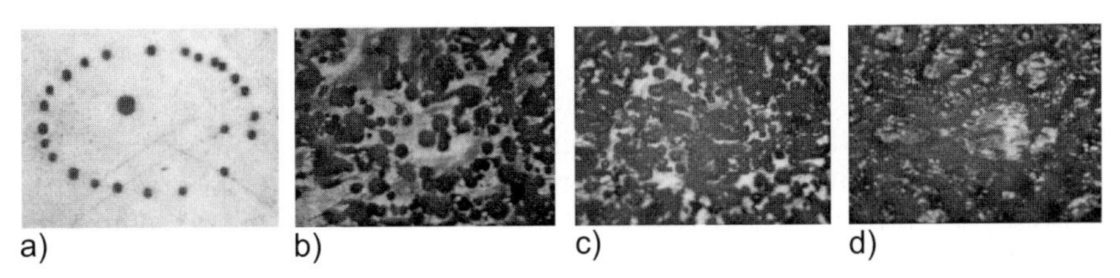

Figure 12.6: Bubble arrays and steam cavities in the developed liquid film, $t=$ 3 ms, We = 300; a) $T=105\,^{\circ}$C, b) $T=135\,^{\circ}$C, c) $T=$ 145 $^{\circ}$C, d) $T=155\,^{\circ}$C.

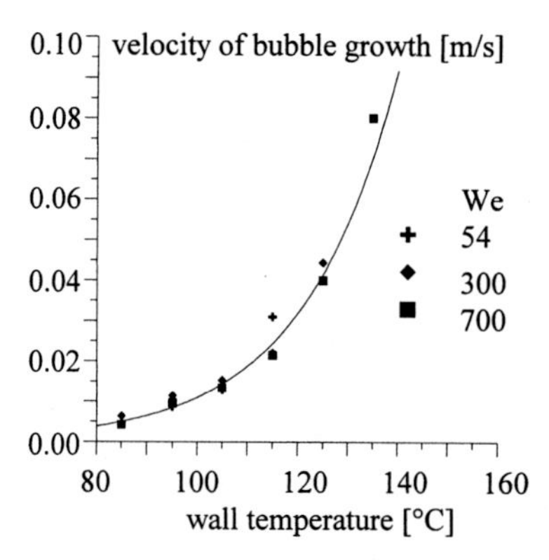

Figure 12.7: Velocity of the bubble growth until collapsing as a function of initial wall temperature.

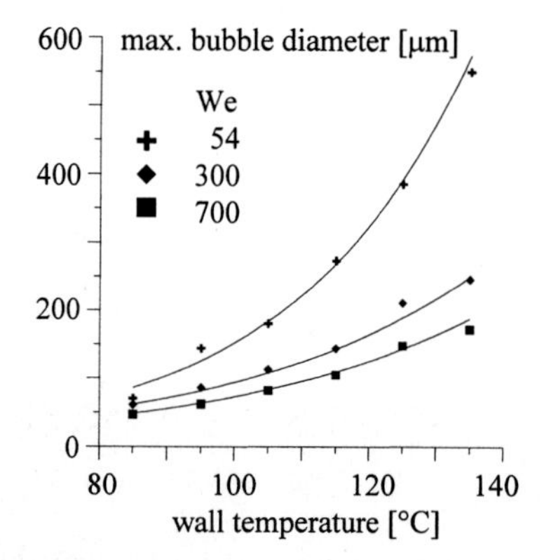

Figure 12.8: Maximum bubble diameter.

Figure 12.6 a–d displays bubbles in the spreading liquid film for different wall temperatures 3 ms after the impact.

The growth of the bubbles of the ring was measured, small bubbles grow approximately linearly with time. From the increase of the bubble diameter with time, the velocity of the bubble growth could be calculated for different Weber numbers and different wall temperatures, typical values are shown in Figure 12.7 and Figure 12.8. Later on, the bubbles either collapse or coalesce with each other.

The maximum diameter, which the bubbles can attain, is limited by the film thickness. In addition, it depends on the Weber number and on the wall temperature. This outcome may be explained in the following way: With increasing Weber number, the liquid film, which is produced during spreading of the primary drop, gets thinner and larger (Figure 12.9).

Furthermore, the bubbles rise to the free surface due to buoyancy. Because of the high temperature gradient in the film, also thermocapillarity ef-

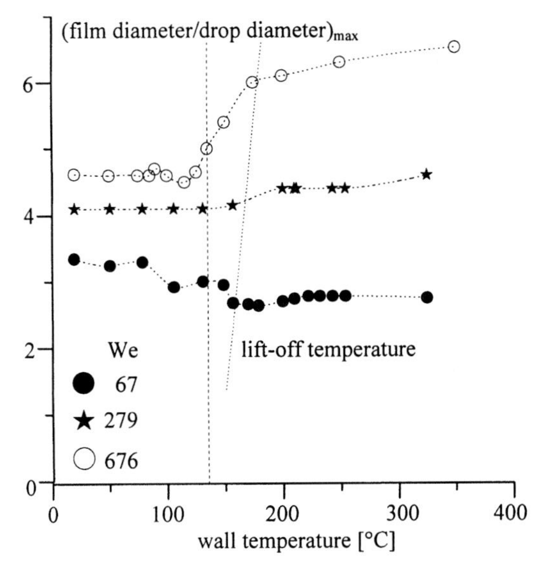

Figure 12.9: Maximum diameter of the liquid film.

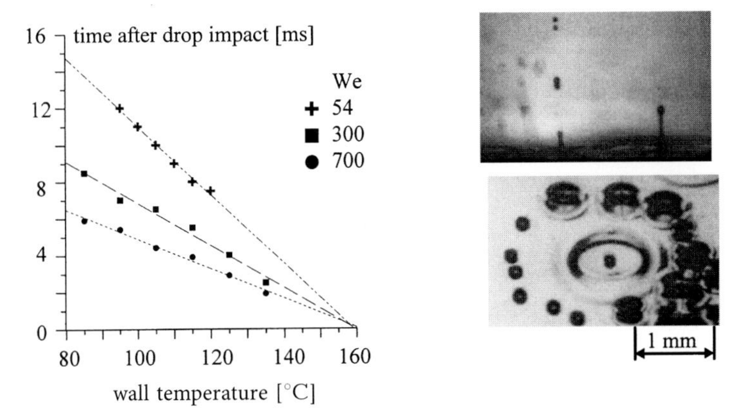

Figure 12.10: The limitation of the life time of the bubbles and typical exposures of collapsing bubbles taken at two different viewing angles relative to the impact.

fects contribute to this motion. The bubbles life time after drop impact depends again on the Weber number and wall temperature (Figure 12.10). For times above the dashed lines the bubbles disappear. Collapsing bubbles are displayed in detail in Figure 12.10.

Details of the collapse of the bubbles are illustrated in Figure 12.11. As soon as the bubble reaches the free liquid surface, the thin liquid layer separating the steam within the bubble and the surrounding air ruptures. Next, the adjacent liquid fills the cavity immediately and steam is discharged. Along the bubble axis perpendicular to the surface, the liquid rises and forms a jet from which droplets

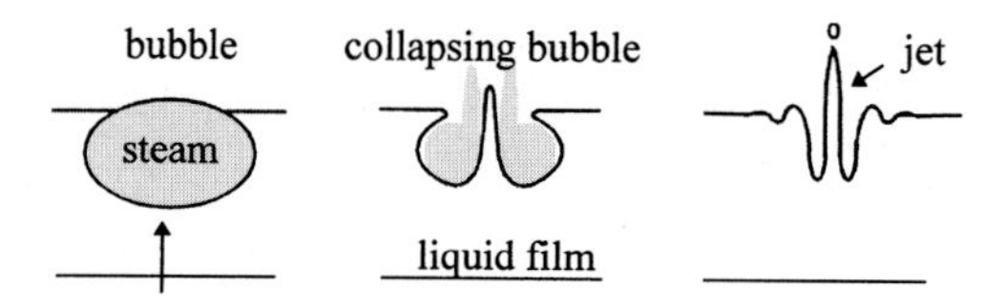

Figure 12.11: Illustration of bubble collapse.

can separate (Rayleigh jet). With increasing wall temperature, also the length of the jet and the number of droplets increase. Additionally, circular waves propagate outwards from the centre of collapse and interfere. Moreover, in the temperature range II above 135 °C, the number of bubbles increases, the spreading fluid atomizes and produces a lot of tiny droplets. This powerful process named miniaturisation corresponds to an immense increase of the free surface of the liquid and, consequently, to a decrease of the evaporation time.

Further details of those phenomena are described in [14, 15].

12.3.3 Convection in the liquid film

In regime VIII of Figure 12.2, polygonal structures appear on the film. They are the result of Rayleigh-Bénard convection in the fluid. An estimate of the Rayleigh number, i.e. for ethanol films with a defined thickness shown on Figure 12.12, corresponds roughly to the critical value [16]. The time evolution of these cells is presented in Figure 12.13. With increasing wall temperatures, the

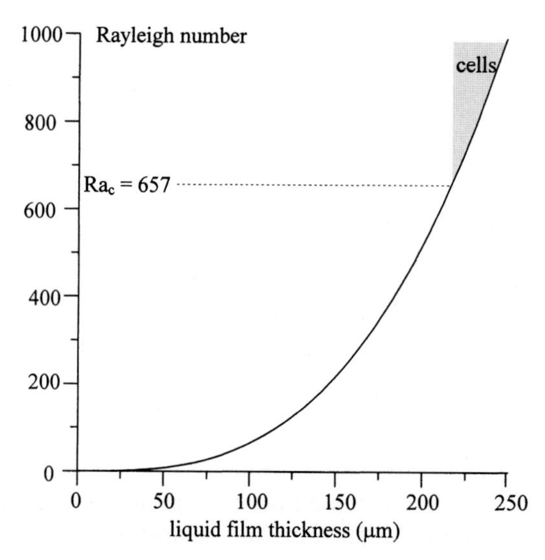

Figure 12.12: Appearance of convection cells in the liquid film. Critical Rayleigh number from theoretical calculation for films of ethanol with two free surfaces.

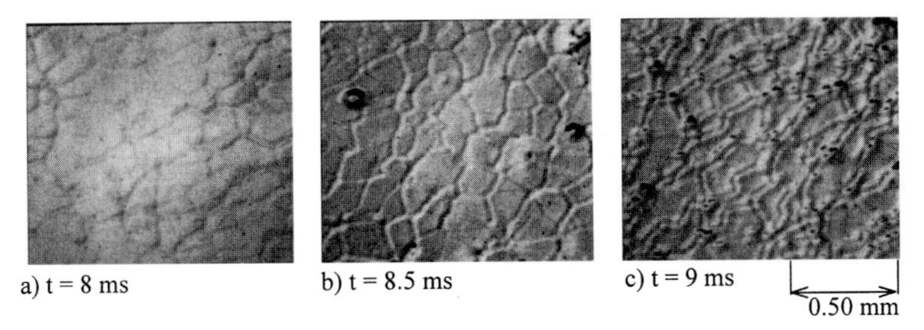

Figure 12.13: Evolution of convection cells in a spreading liquid film after drop impact; $We=300$, $T=200\,^{\circ}\mathrm{C}$.

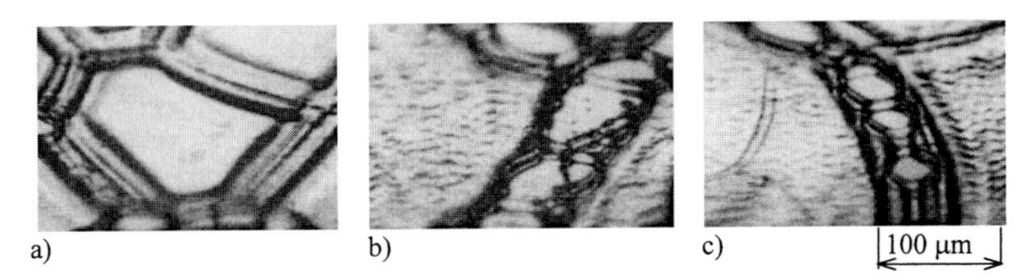

Figure 12.14: a) Convection cells, b) and c) break-up process of cells edges; $We=300$, $T=200\,^{\circ}\mathrm{C}$.

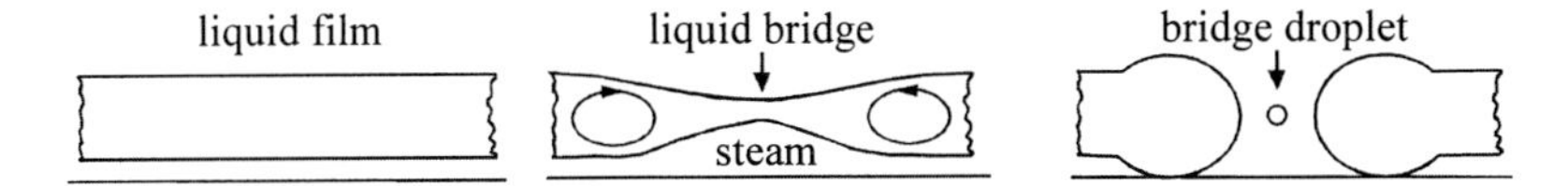

Figure 12.15: Illustration of the liquid film disruption.

cell size decreases and the structure becomes more irregular and chaotic (transition to turbulent convection).

The cells display a typical length scale between 0.1 mm and 0.2 mm (Figure 12.13 a); with progressing time, these cells become more pronounced (Figure 12.13 b). Next, the edges of the cells grow in such a way that the cell space shrinks and the cells start to break-up within less than 0.2 ms (Figure 12.13 c). This fast process of the disruption of a cell is documented in a series of pictures (Figure 12.14).

The mechanism of disruption is illustrated schematically in a cross-sectional view of the liquid film (Figure 12.15). The convection rolls in the liquid film lead to the formation of thin liquid bridges around the cells. When the film reaches a characteristic minimum thickness, these liquid bridges break leaving tiny bridge droplets behind. At the locations, where the edges of neighbouring cells meet the liquid, bridges are more stable than elsewhere. This observation reflects the creation of mesh like pattern shown in Figures 12.16.

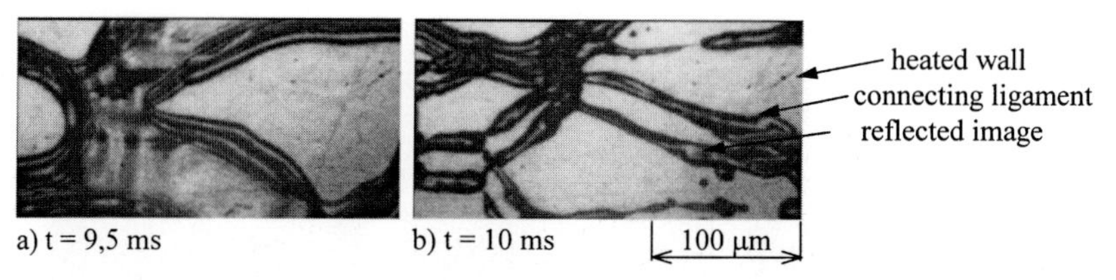

Figure 12.16: a) Shrinking of the liquid film, b) formation of the mesh structures; $We=300$, $T=400\,^{\circ}$C.

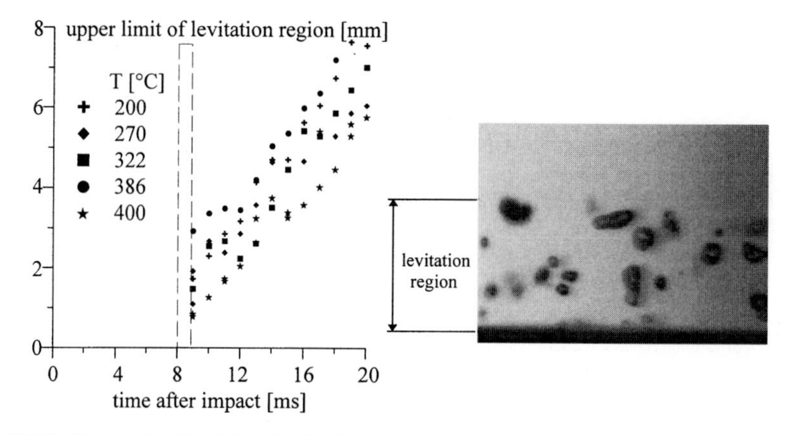

Figure 12.17: Upper limit of the levitation region as a function of time ($We=300$) and a typical detail of an ensemble of levitated droplets.

This intermediate state disappears very quickly. The connecting ligaments tear on their knots and contract to droplets. Then, these droplets rise from the hot surface and collect in a well defined region above the wall. A most likely reason for this motion is a collective "Leidenfrost" effect. For a given Weber number, a typical detail and the upper limit of levitation as a function of time are shown in Figure 12.17. A dependence of this limit upon the wall temperature could not be detected. The dashed lines characterize the transition time between the spreading and the fragmentation of the drop and the subsequent motion of droplets and their evaporation. The width of the dashed region is due to trigger uncertainties. Finally, it was observed that with increasing Weber number, the transition occurs at earlier times, and that the diameters of the droplets decrease due to smaller cells at the beginning. Consequently, the evaporation rate is enhanced with increasing Weber number.

12.4 Conclusions

Apart from the number of new phenomena, which were observed, it is important to emphasize that the influence of the Weber number and the wall temperature decouple in some of the regimes described. The main effect of the Weber number, i.e. the ratio of kinetic energy of the impact relative to the surface energy, is to modify the topology, the structure, and the dynamics of the flow in comparison to the almost spherical case of sessile drops. The spreading of the fluid into a film due to the stagnation flow at impact is the first important difference. Secondly, fragmentation of the initial drop either due to splashing or to drop break-up increases the surface area of the liquid. Here, the wall temperature has a two-fold effect. It causes heat flux from the wall and evaporation, but it also modifies the boundary conditions for the fluid flow from a no-slip condition at low wall temperatures to a slipstream condition after lift-off of the liquid. This implies some consequences on the limit of the onset of splashing for example. The limiting Weber number for splashing is lowered for the lift-off case because now less energy is dissipated than within the boundary layer of a wetting film. Therefore, the remaining kinetic energy is sufficient to overcome surface tension and it causes the generation of small droplets at the rim of the film. Also the diameter of the film is increased after lift-off implying a smaller film thickness. The last quantity is important for many of the subsequent phenomena like bubble collapse or the Rayleigh-Bénard convection in the liquid film.

Acknowledgement

We acknowledge gratefully experimental contributions by T. Jonas.

References

[1] E.A. Mizikar: *Spray Cooling Investigation for Continuous Casting of Billets and Blooms.* Iron and Steel Engineer (1970) 53–60.
[2] Y.S. Ko, S.H. Chung: *An Experiment on the Breakup of Impinging Droplets on a Hot Surface.* Experiments in Fluids **21** (1996) 118-123.
[3] C. Arcoumanis, J.-C. Chang: *Heat Transfer Between a Heated Plate and an Impinging Transient Diesel Spray.* Experiments in Fluids **16** (1993) 105–119.
[4] H. Boerhaave: *Elementa Chemiae* Vol. 1 Experiment XIX. Lugduni Batavorum, 1732, p. 257.

[5] J.G. Leidenfrost: *On the Fixation of Water in Diverse Fire*. Int. J. Heat Mass Transfer **9** (1966) 1153–1166.
[6] M. Rein: *Phenomena of Liquid Drop Impact on Solid and Liquid Surfaces*. Fluid Dynamics Research **12** (1993) 61–93.
[7] K.J. Bell: *The Leidenfrost Phenomenon: A Survey*. Chemical Engineering Progress Symposium **22** (1967) 73–82.
[8] F. Obermeier: *Experimental Aspects of Spray Impact on Walls and Films*. Invited Lecture, ILASS, Florence, 1997.
[9] G. Grabitz, G.E.A. Meier. *Laufzeiteffekte beim geschwindigkeitsgestörten Flüssigkeitsstrahl*. ZAMM **71** (1991) 471–474.
[10] C. Weber: *Zum Zerfall eines Flüssigkeitsstrahles*. ZAMM **11** (1931) 136–154.
[11] T. Jonas: *Transiente Phänomene beim Tropfenaufprall auf temperierte Wände*. Diplomarbeit, Georg-August-Universität, Göttingen, 1996.
[12] T. Jonas, A. Kubitzek, F. Obermeier. *Transient Heat Transfer and Break-Up Mechanisms of Drops Impinging on Heated Walls*. Experimental Heat Transfer, Fluid Mechanics and Thermodynamics 1997, Proc. of the 4th World Conference on Experimental Heat Transfer, Fluid Mechanics and Thermodynamics Brussels, 1997, pp. 1263–1270.
[13] H.D. Baehr, K. Stephan: *Wärme- und Stoffübertragung*. Springer-Verlag, 1994, pp. 463–466.
[14] A.M. Kubitzek: *Experimentelle Untersuchungen des Phasenübergangs beim Tropfenaufprall auf heiße Wände*. Forschungsbericht 97-21 der Deutschen Forschungsanstalt für Luft- und Raumfahrt e.V., 1997.
[15] H. Chaves, A.M. Kubitzek, F. Obermeier: *Dynamical Processes Occurring during the Spreading of Thin Liquid Films Produced by Drop Impact on Hot Walls*. ILASS Manchester, 1998, pp. 135–140.
[16] G.P. Merker: *Konvektive Wärmeübertragung*. Springer-Verlag, 1987, pp. 360–369.

13 Investigation of Droplets Impacting on Hot Walls

Arnold Frohn, Alexander Karl and Martin Rieber *

13.1 Introduction

Two-phase flows containing droplets carried by a gas are quite common in nature and engineering. Examples are cloud droplets, fuel droplets in combustion processes or medical sprays. For better understanding of such systems, the investigation of the involved elementary processes is important. The behavior of individual droplets has been studied since a long time. In recent years, interactions between droplets and solid walls have been investigated in detail by many researchers as these processes have considerable influence on two-phase flows [1–6]. The impact process between droplets and walls depends essentially on the wall temperature. At wall temperatures above the Leidenfrost temperature, a vapor cushion forms between droplet and wall and impinging droplets are reflected [7, 8]. For small impact energies, the droplets remain intact during the reflection by the wall, and the size distribution does not change. There is however a considerable loss of normal momentum during the impact resulting in a change of the velocity distribution of the droplets [9, 10]. For higher impact energies, the droplets are deformed strongly by the impact and disintegrate during the interaction process, which results in considerable influence on the size distribution of the droplet spray [11]. The size distribution on the other hand influences vaporization and combustion processes. For even higher impact energies, the droplets shatter during the interaction process. This behavior is well known as splashing [12]. In this paper, a detailed investigation of the interaction between liquid droplets with diameters ranging from 70 µm to 260 µm and hot walls above the Leidenfrost temperature has been performed. Detailed measurements of important parameters like loss of momentum and droplet deformation during the impact process are carried out, using monodisperse droplet streams and a special illumination technique. Different phenomena, ranging from nearly perfect reflection of the droplets by the hot wall to disintegration of

* Universität Stuttgart, ITLR Institut für Thermodynamik der Luft und Raumfahrt, Pfaffenwaldring 31, D-70550 Stuttgart, Germany

the droplets during the interaction process, are observed with a video imaging system. The experimental results are compared with numerical calculations obtained with a new code solving the 3D-Navier-Stokes equations for incompressible fluids [13, 14].

13.2 Experimental Set-Up and Measurement Techniques

A direct observation of small droplets in the diameter range from 70 μm to 260 μm impinging on solid walls is difficult due to the short interaction times, which are typically in the range of some hundred microseconds. The visualization of such dynamic processes requires usually high speed cameras, which are able to take several ten thousand exposures per second. In the presented work, a new observation technique has been developed, which enables the observation of highly dynamic impact phenomena of droplets, using monodisperse droplet streams in combination with a normal video camera. The monodisperse droplet streams consist of equally sized and equally spaced droplets. These monodisperse droplet streams are generated with a vibrating orifice droplet generator [15–18]. This generator, which utilizes the well known phenomenon of laminar disintegration of liquid jets according to the theory of Rayleigh, will not be described in more detail [19, 20].

Figure 13.1 shows a schematic representation and a photo of the experimental set-up. The incident and the reflected droplet streams appear as thin bright lines on the photo. Individual droplets cannot be detected because of the long exposure time. The incident droplets move upwards and impinge on the hot wall. The impinging angle of the droplets with respect to the heated wall can be varied continuously over a wide range. The wall temperature is measured with a thermocouple a short distance below the surface of the hot wall. The material of the walls was chromium-plated copper or steel. Due to the high heat conduction of the wall material, the measured temperature is assumed to be equal to the true surface temperature. Neglecting the small temperature difference between the location of the thermocouple and the surface is justified as all experiments are performed well above the Leidenfrost temperature. The desired wall temperature is maintained constant within some degrees by means of a temperature control unit.

The impact parameters are measured directly at the impact point by means of a digital image processing technique described later. For observation of the impact phenomena, the impact region is illuminated with a flash light, which generates strong flashes with a duration of approximately 200 ns at a maximal repetition rate of 50 Hz. These short flashes are necessary to get sharp images of the fast moving droplets. The repetition rate of the flashes is exactly equal to the number of half-frames of the European video norm, so the flashes illuminate each half-frame. Two lenses are used to adjust the light intensity

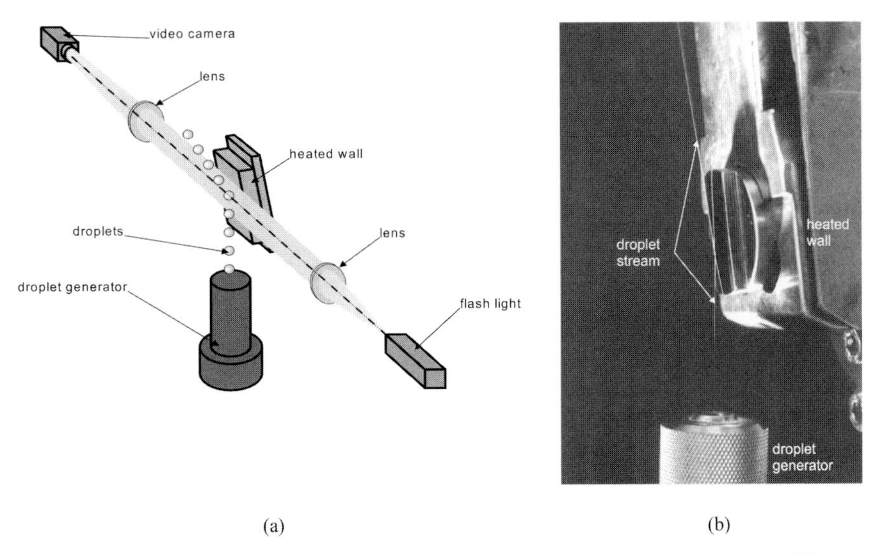

Figure 13.1: Schematic diagram (a) and photo (b) of the experimental set-up. The impinging and reflected droplet streams are visible as thin bright lines on the photo. The reflected images of the droplet streams are visible on the chromium-plated wall. The optical set-up is shown in the schematic diagram.

and the magnification of the observation system. With a small magnification, an overview of the phenomena is possible, whereas a large magnification allows detailed observations of the impact phenomena. The observation system consists of an ordinary CCD-video camera with 756×581 pixels. Shadow images are generated with the camera positioned exactly opposite to the flash light. This means that the droplets appear as black shadows on a bright background. The video images, which are evaluated directly by a PC-based image processing unit using a frame grabber, can be recorded with a video recorder for later post-processing. During recording of the images, a VITC [1] is written on the video tape. This code labels each image with a unique time label, so the individual images can be grabbed exactly and reproducibly during later post-processing.

The illumination and observation equipment is synchronized by a custom-made digital frequency generator, which enables two different imaging modes: still video images and slow motion recordings of the droplet wall interaction. For both observation methods, a stable monodisperse droplet stream is essential. All droplets of such a droplet stream impinging on the hot wall show exactly the same deformation and interaction behavior. The excitation frequency f_G of the droplet stream is usually in the range from 10 kHz to 100 kHz. This means that 10^4 to 10^5 droplets impinge per second on the hot wall. As mentioned above, only 50 half-frames are recorded per second, which results in the observation of only a few of the impinging droplets. The digital frequency gen-

[1] Vertical Interval Time Code, professional time coding of video images.

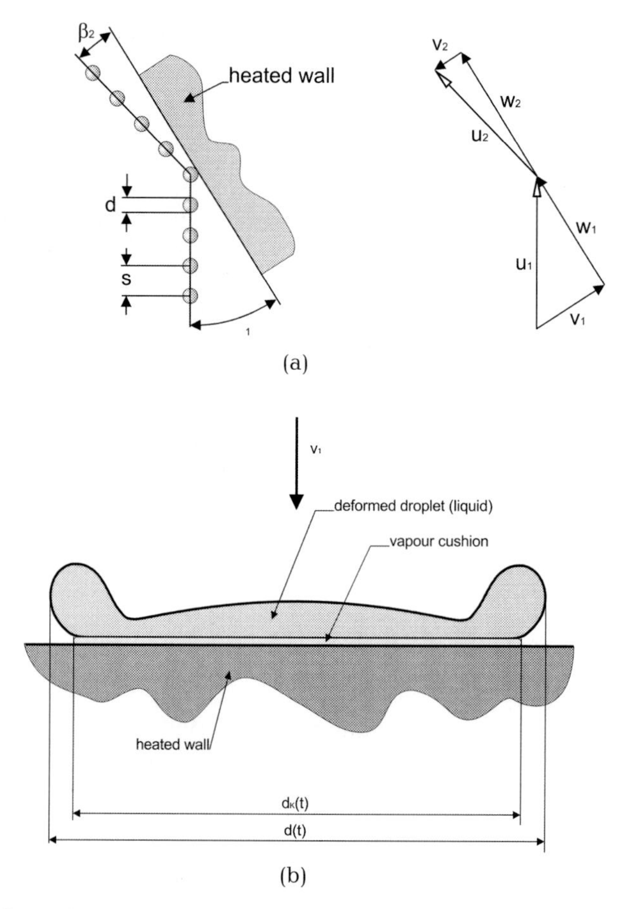

Figure 13.2: Schematic representation (a) of the impinging droplet stream with definitions of droplet diameter d, impinging angle β_1, reflection angle β_2 and all velocity components, and (b) deformed droplet with definitions of important time-dependent diameters.

erator uses the synchronization signal of the video camera in order to make sure that a trigger for the flash light is only generated when the video camera is ready to take an image. Two trigger modes are possible with this digital frequency generator. In the first mode, the trigger for the flash light is generated always at the same phase of the excitation signal of the droplet generator. Using a stable monodisperse droplet stream, always the same deformation and interaction state is observed but with different droplets. This results in still video images as all droplets are in the same deformation state. In the second mode, the trigger for the flash light is shifted each time by a small phase angle $\Delta\phi$ with respect to the excitation signal. In this mode, each image records a slightly different interaction state and slow motion recordings are obtained through the stroboscopic effect. Using this slow motion images, time measurements with a time resolution below 1 μs are possible by simply counting half frames. A schematic representation of the droplet stream and of a single deformed droplet with important characteristic parameters is shown in Figure 13.2.

For the determination of the collision parameters, a digital image processing unit with the commercial software OPTIMAS[2] is used. Custom-made macro programs are used to measure the impinging angle as well as the reflection angle. After calibration of the digital image processing system, also the droplet diameter d and the droplet spacing s can be measured. With the droplet spacing and the excitation frequency, the velocity of the droplets is known. With this velocity and the known impinging angle, all velocity components can be calculated. The droplet temperature is measured with a thermocouple within the droplet generator directly at the exit. For the investigation of the droplet deformation, the time-dependent diameter $d(t)$ is determined using the above described slow motion recordings. For describing the heat transfer, the diameter $d_K(t)$ is important. However, this diameter cannot be determined precisely during the measurements due to the limited resolution of the video camera and the small droplet diameters. But in a first step, this diameter can be set equal to the diameter $d(t)$. A more detailed description of the experimental set-up, measurement techniques and results for the droplet deformation has been given in [21].

13.3 Results

13.3.1 Loss of momentum

An important parameter for the interaction of liquid droplets with hot walls above the Leidenfrost temperature is the loss of momentum. This parameter influences the behavior of droplets in two-phase flows. The loss of momentum can only be determined for hot walls, which lead to regular reflection of the impinging droplets. This parameter is not meaningful for cold walls when the droplets form a liquid film and all momentum is transferred to this liquid film. The loss of mass of the droplet due to boiling processes during the interaction with the hot wall can be neglected [22–25], so the loss of momentum can be calculated with the measured velocities before and after the interaction process. The experiments show that the loss of tangential momentum can be neglected in most cases; its maximum values reach only 5%. In contrary, the loss of normal momentum can be quite considerable. For a dimensionless representation of the loss of normal momentum, the parameter

$$L = \frac{mv_1 - mv_2}{mv_1} = 1 - \frac{v_2}{v_1} = 1 - r_n \qquad (1)$$

[2] OPTIMAS 5.2, BISC.

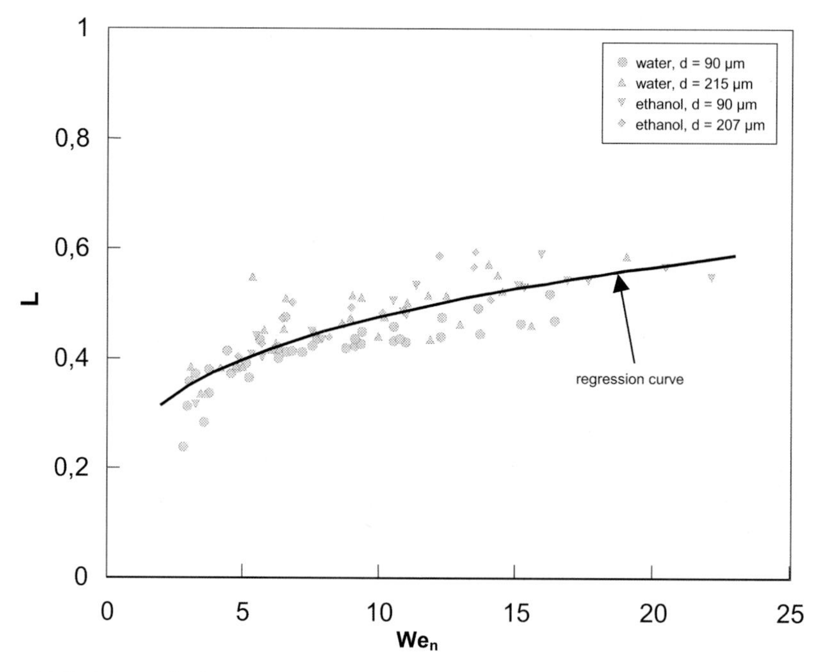

Figure 13.3: Dimensionless representation of loss of normal momentum L for different droplet liquids and different droplet diameters in dependence on the Weber number We_n. The Weber number is calculated with the normal velocity v_1. The wall temperature is T_W=548 K for ethanol and T_W=598 K for water. The regression curve $L=0.263\ We_n^{0.257}$ has been obtained by non-linear curve fitting.

is introduced. The velocity ratio $r_n=v_2/v_1$ is determined by measuring the normal velocities v_1 and v_2 before and after the interaction of the droplet with the hot wall. All experiments are performed for impinging angles $\beta_1<50°$. This range of impinging angles has not been investigated yet in detail. Most results reported in the literature are for impinging angles $\beta_1\sim 90°$. Results for the loss of momentum normal to the heated wall are given in Figure 13.3 for different droplet diameters and liquids as a function of the Weber number

$$We_n = \frac{\rho d v_1^2}{\sigma}\ , \qquad (2)$$

which is associated with the impact velocity v_1 normal to the heated wall.

It is clearly visible that this parameter is suited to describe the loss of normal momentum for different droplet diameters and droplet liquids as all measurements are in the same regime of the diagram. The loss of momentum shows a considerable increase with increasing Weber number or impact energy normal to the wall. There seems to be a level off for higher Weber numbers. The loss of normal momentum is considerable and may have a large influence on the velocity distribution of sprays. For the measured data, a regression curve is determined by means of non-linear curve fitting. A function of the form

$$L = a \cdot We_n^b \tag{3}$$

is chosen, which incorporates the condition that the loss of momentum vanishes when the Weber number approaches zero. Small Weber numbers mean very stiff droplets exhibiting nearly elastic droplet wall interactions with little loss of momentum. For the investigated parameters, the function

$$L = 0.263 \cdot We_n^{0.257} \quad \text{for} \quad 0 \leq We_n \leq 25 \tag{4}$$

has been found to fit the experimental data best. An extrapolation of this function to higher impact Weber numbers is not possible as the reflection behavior of the droplets changes considerably for higher Weber numbers as will be shown in the next section.

13.3.2 Secondary droplet formation

In the following, the behavior of the droplet during the second part of the interaction process is described in more detail. As already stated, wetting of the wall does not occur when the wall temperature is above the Leidenfrost temperature of the droplet liquid and the droplets are regularly reflected. This section describes in detail the formation of secondary droplets during such reflection processes. Different modes of secondary droplet formation are shown in Figure 13.4.

These prints are obtained by digitizing video recordings and transforming them subsequently to black and white pictures with an image processing program. For all three cases, the droplet streams come from the left-hand side and impinge on the hot wall at a rather shallow angle. In Figure 13.4 a, the beginning of secondary droplet formation is shown. In this case, the secondary droplets and the primary droplets coalesce after travelling a short distance. During the interaction with the wall, the droplet is strongly deformed, which results in large internal velocities after the reflection by the wall. The surface tension is not able to maintain a closed surface, and a small secondary droplet is formed. This disintegration process is related to the Rayleigh instability of liquid jets [19, 20]. For the case in Figure 13.4 a, the disintegration takes place when the droplet mass is already in a receding motion. As a consequence, the formed droplets have a relative velocity towards each other and coalesce after a short distance. For a slightly larger impact energy, the secondary droplets remain permanently separated from the primary droplets, and two reflected droplet streams form as shown in Figure 13.4 b. Here, the disintegration takes place when the droplet is still in a stretching motion, so the formed droplets do not coalesce. Usually, the primary droplet is much larger than the secondary drop-

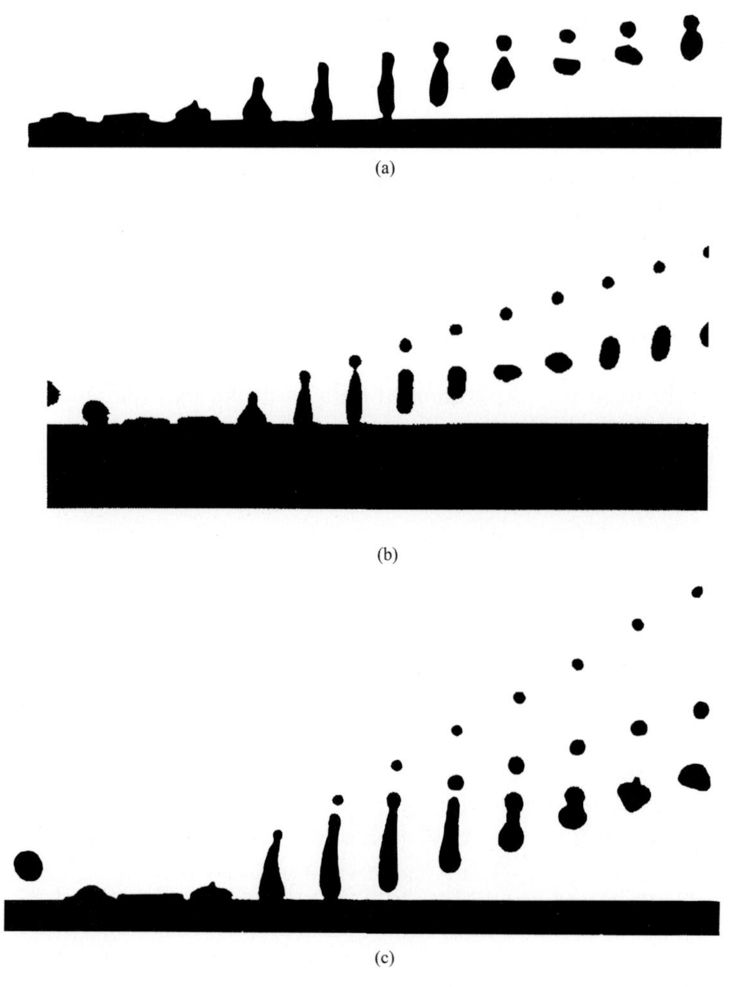

Figure 13.4: Different modes of the formation of secondary droplets at wall temperatures well above the Leidenfrost temperature. The impact energy increases from (a) to (c).

let and exhibits strong oscillations. For even larger impact energies, a second disintegration occurs during the stretching motion. In the case shown in Figure 13.4 c, three stable droplet streams are formed in the reflection process. Even more secondary droplets can be observed when the impact energy is increased until the energy for splashing is reached. The droplet size increases from the first secondary droplet to the primary droplet. In the following, it is assumed that the case shown in Figure 13.4 a represents the transition to formation of secondary droplets. For the experimental determination of this limit, the impinging angle is increased in small steps at constant droplet velocity u_1 until secondary droplets are observed. This procedure is repeated for different droplet diameters and droplet velocities in order to determine the transition to formation of secondary droplets.

For a first analysis of these measurements, the Weber number is introduced. If different fluids are compared, an additional parameter has to be used, which includes the viscosity of the droplet liquid. For a dimensionless description of the phenomena, the Reynolds number is used to introduce the viscosity. According to the Ohnesorg number for atomization processes, a combination of the Weber number and the Reynolds number is searched, which describes secondary droplet formation. For the available results, the empirical parameter

$$We_n \cdot Re_n^{0.8} \tag{5}$$

formulated with the velocity v_1 of the droplets normal to the wall seems to give the best fit to the experiments. Results for $We_n Re_n^{0.8}$ for secondary droplet formation for the droplet liquid ethanol are shown in Figure 13.5 in dependence on the impinging angle.

Two different regions can be identified clearly. For larger impinging angles, the parameter $We_n Re_n^{0.8}$ seems to have a constant value of approximately 2000 for the transition between regular reflection and secondary droplet formation. This corresponds to a minimum impact energy needed for the formation of secondary droplets independently of the impinging angle. This behavior is as expected. However, for smaller impinging angles, a steep increase of the parameter $We_n Re_n^{0.8}$ is observed, which corresponds to a minimum angle for the

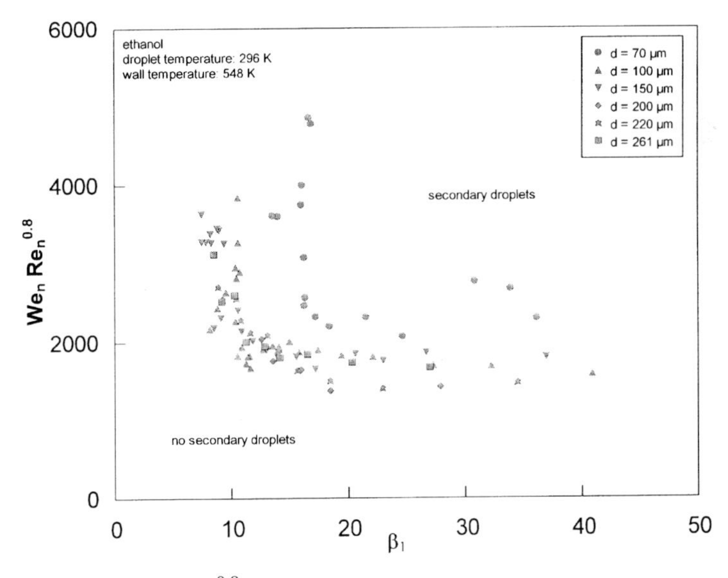

Figure 13.5: Parameter $We_n Re_n^{0.8}$ of the impinging droplets in dependence on the impinging angle β_1 for the formation of secondary droplets. The wall temperature is $T_W = 548$ K for ethanol.

formation of secondary droplets. This minimum angle depends on the droplet diameter. This new phenomenon is only observed for impinging angles below 20° and will be described in more detail in the following section.

13.3.3 Influence of surface roughness

It can be shown that the minimum impinging angle for the formation of secondary droplets described in the last section is directly related to the friction between the droplets and the wall, which depends on the surface roughness of the wall. For technical applications, the surface roughness is usually described with the average depth of roughness R_Z. This value is directly related to the maximum depth of the surface roughness over a defined length. For the presented investigation, the depth of roughness ranges from 0.41 μm for chromium-plated copper walls to 2.4 μm for ground steel walls. The smaller value represents a very smooth surface, which can be obtained by polishing. The larger value is usually obtained by grinding the surface. With some effort, this value can also be obtained directly by machining with a lathe or a milling machine. The experimental results show that the general behavior of the impinging droplets is not altered. But a considerable influence on the transition from regular reflection to secondary droplet formation is observed, especially for small impinging angles, where a minimum impinging angle for the formation of secondary droplets seems to exist. This behavior can be explained as follows: When the impinging angle decreases along the transition limit for secondary droplet formation, a strong increase of the velocity $w_1 = v_1/\tan(\beta_1)$ in the direction tangential to the wall occurs at constant normal impinging velocity. With increasing tangential velocity, the velocity gradients normal to the wall within the vapor cushion increase also. As shown in [26], the interaction time is independent of the normal impinging velocity in the investigated parameter range. For higher tangential velocities, this results in greater distances travelled by the droplet during the contact with the heated wall. Both effects cause an increase of energy dissipation and loss of momentum. Thus, for smaller impinging angles, there is not sufficient kinetic impact energy for forming a secondary droplet. As the dissipation at the hot wall depends on the surface roughness, this minimum impinging angle for the formation of secondary droplets should also depend on the surface roughness. This behavior has been investigated for different droplet diameters. For all investigated droplet diameters, the behavior is similar; for constant surface roughness and increasing droplet diameter, the minimum impinging angle decreases. So, the minimum impinging angle depends not only on the surface roughness but also on the droplet diameter. Therefore, all results for the minimum impinging angle are given in Figure 13.6 in dependence on the ratio of the depth of roughness and the droplet diameter R_Z/d. This parameter seems to be the controlling factor at least for a given droplet liquid.

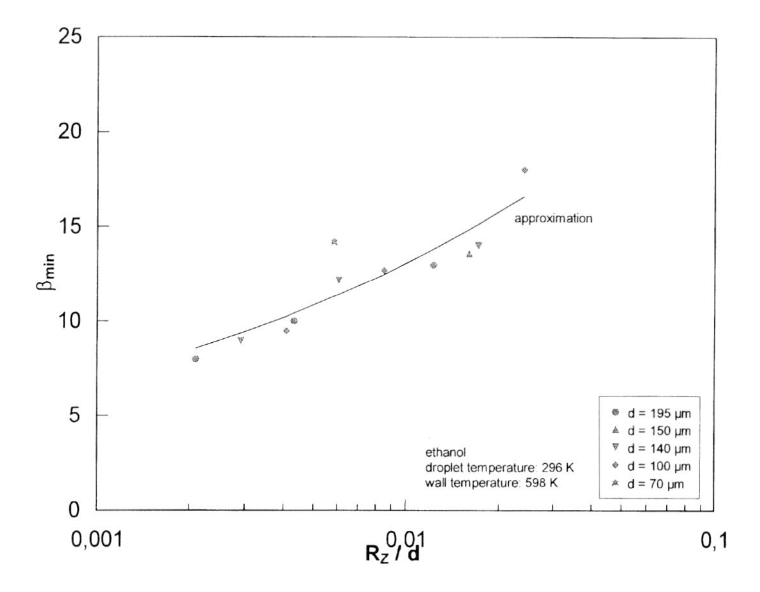

Figure 13.6: Experimental results for the minimum impinging angle β_{min} for the formation of secondary droplets in dependence on the parameter R_Z/d. The solid line represents the approximation $\beta_{min} = 45.7\,(R_Z/d)^{0.272}$.

It can be seen that for most of the experimental data points, a reasonable correlation exists. So, the minimum impinging angle can be described by the empirical relation:

$$\beta_{min} = 45.7 \cdot \left(\frac{R_Z}{d}\right)^{0.272}, \tag{6}$$

which is shown as solid line in Figure 13.6.

13.3.4 Numerical results

Due to the limited space, only a brief comparison of numerical results with experimental results will be presented. A new numerical code has been developed at ITLR for solving the 3D-Navier-Stokes equation for incompressible fluids [27]. For this code, the volume-of-fluid (VOF) method with a continuous surface force (CSF) model is used to simulate the droplet liquid including surface tension effects. The energy equation is not solved in this numerical method. Instead, a special boundary condition simulating the vapor cushion between the droplet and the hot wall has been developed.

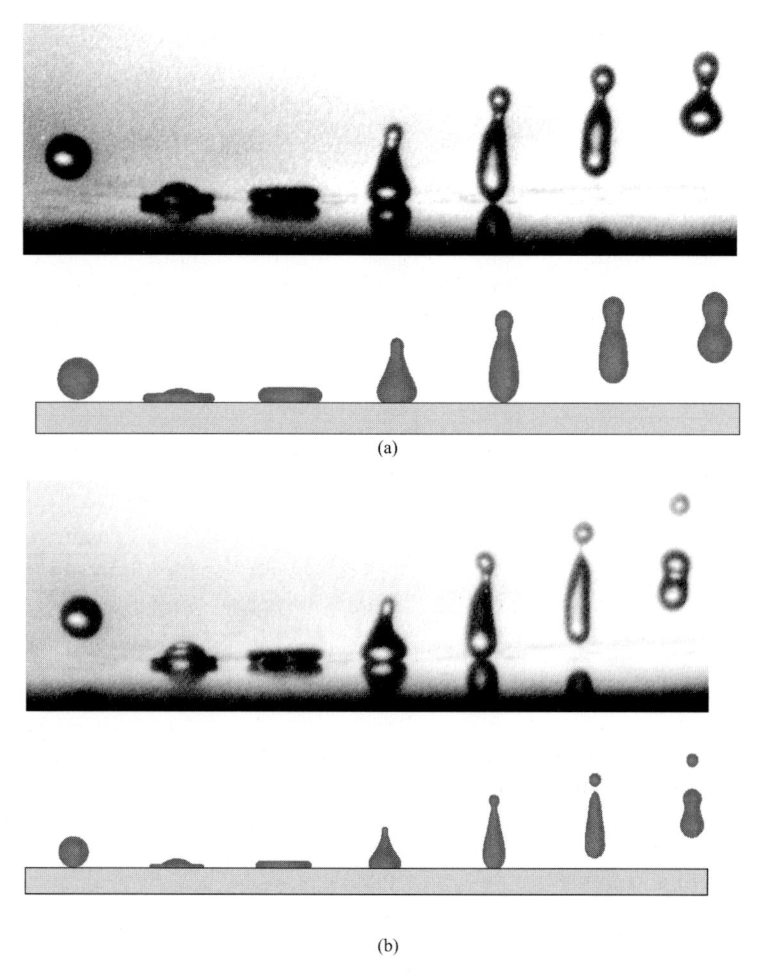

Figure 13.7: Comparison of experimental and numerical results for the formation of secondary droplets: (a) a case without secondary droplet formation and (b) a case with secondary droplets.

In Figure 13.7, the calculated droplet shapes and the observed droplet shapes are compared for a case without (a) and a case with (b) secondary droplet formation.

The upper panels show the experimental results, whereas the lower panels depict the numerically obtained droplet shapes arranged in the form of a droplet stream. For both cases, the numerical results agree very well with the experimental results. Even the formation of the small secondary droplet and the subsequent movement of the droplets is predicted very well by the numerical simulation.

13.4 Conclusion

Detailed investigations of interaction processes of small liquid droplets with hot walls well above the Leidenfrost temperature have been described. Wetting does not occur at these temperatures, and the droplets are reflected regularly by the hot wall. For the observation of the phenomena, a new experimental technique using monodisperse droplet streams in combination with a standard video camera is described. This technique has a very high time resolution, and video images of the highly dynamic interaction processes are obtained. The loss of momentum has been studied. Detailed investigations of a special mode for the formation of secondary droplets during the interaction process have been performed. Correlations describing the different phenomena have been developed for all experimental results, which can be used to improve the numerical modelling of two-phase flows.

The loss of tangential momentum can be neglected for the experiments of the presented paper, whereas the loss of normal momentum reaches considerable amounts of approximately 60%. For the loss of normal momentum, a correlation with the Weber number We_n based on the normal component of the impinging velocity of the droplet has been developed. Finally, secondary droplet formation well below the splashing threshold has been investigated in detail, especially the transition between regular droplet reflection and secondary droplet formation. The parameter $We_n Re_n^{0.8}$ has been identified as governing parameter for this type of droplet disintegration. A new phenomenon, the minimum impinging angle for the formation of secondary droplets, has been discovered for small impinging angles. This phenomenon, which is the direct result of the energy dissipation at the hot wall, depends on the ratio of the surface roughness and droplet diameter. For the investigated ethanol droplets, a correlation for this minimum impinging angle could be deduced.

With the presented work, the understanding of basic interaction processes between droplets and hot walls has been improved. The developed correlations may be used in numerical models to improve two-phase flow calculations. With the correlation for the loss of momentum, the influence of the wall on the velocity distribution within spray systems can be considered as well as the influence on the size distribution with the correlation for secondary droplet formation. All correlations and theoretical approximations are important for practical applications concerning fuel preparation, combustion processes and spray cooling processes.

References

[1] J. Fukai, Z. Zhao, D. Poulikakos, C.M. Megaridis, O. Miyatake: *Modeling of the droplet deformation of a liquid droplet impinging upon a flat surface*. Phys. Fluids A **5** (1993) 2588–2599.

[2] J. Fukai, Y. Shiiba, T. Yamamoto, O. Miyatake, D. Poulikakos, C.M. Megaridis, Z. Zhao: *Wetting effects on the spreading of a liquid droplet colliding with a flat surface: Experiment and modeling*. Phys. Fluids A **7** (1995) 236–247.

[3] N. Hatta, H. Fujimoto, H. Takuda: *Deformation process of a water droplet impinging on a solid surface*. Transactions of the ASME Journal of Fluids Engineering **117** (1995) 394–401.

[4] H. Fujimoto, N. Hatta: *Deformation and rebounding processes of a water droplet impinging on a flat surface above Leidenfrost temperature*. Transactions of the ASME Journal of Fluids Engineering **118** (1996) 142–149.

[5] D.A. Weiss: *Periodischer Aufprall monodisperser Tropfen gleicher Geschwindigkeit auf feste Oberflächen*. Mitteilungen aus dem Max-Planck-Institut für Strömungsforschung Nr. 112, Ed.: E.-A. Müller, Göttingen, 1993.

[6] A.L. Yarin, D.A. Weiss: *Impact of drops on solid surfaces: self-similar capillary waves and splashing as a new type of kinematic discontinuity*. J. Fluid Mech. **283** (1995) 141–171.

[7] B.S. Gottfried, C.L. Lee, K.J. Bell: *The Leidenfrost phenomenon: Film boiling of liquid droplets on a flat plate*. Int. J. Heat Mass Transfer **9** (1966) 1167–1187.

[8] K.J. Baumeister, F.F. Simon: *Leidenfrost temperature – Its Correlation for Liquid Metals, Cryogens, Hydrocarbons, and Water*. Transactions of the ASME Journal of Heat Transfer, 1993, pp. 166–173.

[9] K. Anders, N. Roth, A. Frohn: *The velocity change of ethanol droplets during collision with a wall analyzed by image processing*. Experiments in Fluids **15** (1993) 91–96.

[10] A. Karl, K. Anders, A. Frohn: *Experimental investigation of the droplet deformation during wall collisions by image analysis*. ASME Experimental and Numerical Flow Visualization Symposium, FED-Vol. **172** (1993) 135–141.

[11] A. Karl, K. Anders, A. Frohn: *Disintegration of droplets colliding with hot walls*. In: A. Serizawa, T. Fukano, J. Bataille (Eds.): Advances in Multiphase Flow 1995. Elsevier Science B.V., Amsterdam, 1995, pp. 313–319.

[12] M. Marengo: *Analisi dell'Impatto di Gocce su Film Liquido sottile*. Dissertation, Politecnico di Milano, Facolta di Ingeneria, 1996.

[13] A. Karl, K. Anders, M. Rieber, A. Frohn: *Deformation of liquid droplets during collisions with hot walls: Experimental and numerical results*. Part. Part. Syst. Charact. **13** (1996) 186–191.

[14] A. Karl, M. Rieber, M. Schelkle, K. Anders, A. Frohn: *Comparison of new results for droplet wall interactions with experimental results*. ASME Symposium on Numerical Methods in Multiphase Flows, FED-Vol. **236** (1996) 201–206.

[15] R.N. Berglund, B.Y.H. Liu: *Generation of monodisperse aerosol standards*. Env. Sci. Tech. **7** (1973) 147–153.

[16] K. Anders, N. Roth, A. Frohn: *Operation characteristics of orifice generators: the coherence length*. Part. Part. Syst. Charact. **9** (1992) 40–43.

[17] M. Orme, E.P. Muntz: *The manipulation of capillary stream break-up using amplitude-modulated disturbances: A pictorial and quantitative representation*. Phys. Fluids A **2** (1990) 1124–1140.

[18] M. Orme, K. Willis, T. Nguyen: *Droplet patterns from capillary stream break-up*. Phys. Fluids A **5** (1993) 80–90.

[19] Lord Rayleigh: *On the instability of liquid jets*. Proc. Math. Soc. London **10** (1878) 4–13.

[20] Lord Rayleigh: *On the capillary phenomena of jets*. Proc. Roy. Soc. London **29** (1879) 71–97.

[21] A. Karl: *Untersuchung der Wechselwirkung von Tropfen mit Wänden oberhalb der Leidenfrost-Temperatur*. Dissertation Universität Stuttgart, 1997.

[22] L. H. J. Wachters, H. Bonne, H. J. van Nouhuis: *The heat transfer from a hot horizontal plate to sessile water drops in the spheroidal state*. Chem. Eng. Sci. **21** (1966) 923–936.

[23] L. H. J. Wachters, N. A. J. Westerling: *The heat transfer from a hot wall to impinging water drops in the spheroidal state*. Chem. Eng. Sci. **21** (1966) 1047–1056.

[24] L. H. J. Wachters, L. Smulders, J. R. Vermeulen, H. C. Kleiweg: *The heat transfer from a hot wall to impinging mist droplets in the spheroidal state*. Chem. Eng. Sci. **21** (1966) 1231–1238.

[25] L. Bolle, J. C. Moureau: *Spray cooling of Hot Surfaces*. In: G. F. Hewitt, J. M. Delhaye, N. Zuber (Eds.): Multiphase Science and Technology, Vol. 1. Hemisphere Publishing Corporation, Washington, New York, London, 1982, pp. 1–98.

[26] A. Karl, K. Anders, A. Frohn: *Measurements of interaction times during droplet wall collisions*. Proceedings of the Fourth Triennial International Symposium on Fluid Control, Fluid Measurement, Fluid Mechanics, Visualization and Fluidics (FLUCOME) Vol. **2** (1994) 731–735.

[27] M. Rieber, A. Frohn: *Navier-Stokes Simulation of Droplet Collision Dynamics*. Proc. 7th Int. Symp. on Computational Fluid Dynamics, Beijing, China, 1997, pp. 520–525.

14 Transient Phase-Change of Droplets Impacting on a Hot Wall

Norbert M. Wruck and Ulrich Renz *

Abstract

To improve understanding of the heat and momentum transfer phenomena between a droplet and a hot wall, two kinds of detailed information are needed: (a) about the hydrodynamics of impinging droplets, i.e. spreading, fluid contraction and momentum loss, and (b) about the temperature in the contact zone. In this experimental study, a high resolution thermocouple probe is placed in the wall surface to provide information during droplet contact within this nearly inaccessible area.

The transient surface temperature is directly related to the heat transfer rate and subsequently to the rate of vapour production. The transfer rate reaches a maximum in the transition boiling regime, it then decreases with increasing wall temperature. The maximum transfer rate in the considered case is indicated by a local wall temperature drop by more than 60 Kelvin for big isopropanol droplets, while in the Leidenfrost regime a typical wall temperature drop is one order of magnitude lower. In the Leidenfrost regime, a stable vapour layer in the contact zone is established due to high wall temperature. However, even at that high temperature, there is a very short period of liquid-solid contact detected in the initial moment of impingement. The surface temperature during that period of boiling delay, lasting only some ten microseconds, can be interpreted very well with a "Contact Temperature Model". The effects of very rapid heating on the thermal properties of the fluid are discussed.

* RWTH Aachen, Lehrstuhl für Wärmeübertragung und Klimatechnik, Eilfschornsteinstr. 18, D-52056 Aachen, Germany

14.1 Introduction

With continuing progress in modern microcomputers, a simulation of complex multiphase flow phenomena becomes more and more possible. For a better understanding of the transient phenomena during droplet wall interaction, models describing the transfer of momentum, mass and energy when a droplet impacts on the wall need further improvements.

In most of the commercial computer programs, the droplet wall interaction is either considered as ideal reflection or ideal sticking [1]. Andreani and Yadigaroglu [2] point out that important details of the break-up process are not known. In fact, the physics are quite complicated. The heat transfer and momentum loss of a droplet that collides with a wall depend on a number of factors.

While several recent studies (e.g. Jonas et al. [3], Karl et al. [4]) mainly deal with the hydrodynamics of droplets in the Leidenfrost regime (i.e. the wall is hotter than the Leidenfrost temperature), the present work considers a wider range of wall temperatures. The investigation deals with three different boiling regimes as shown in Figure 14.1, which is similar to the well-known Nukiyama curve.

It has to be noted that no dynamic effects are included. A chart including dynamic conditions will need a number of parameters, which are only known partially at present [5, 6].

The hydrodynamics of impacting droplets are dominated by the wall temperature and the impact Weber number (Equation (1)):

$$We = \frac{\rho_d u^2 d_d}{\sigma_d} \,. \tag{1}$$

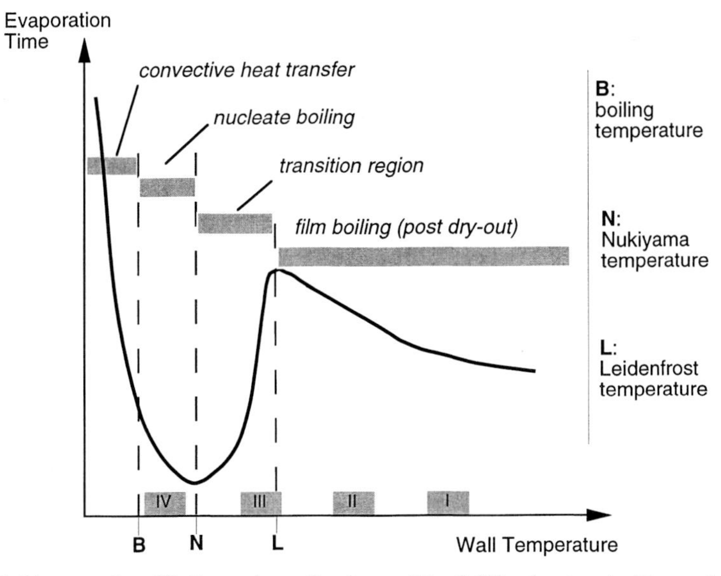

Figure 14.1: Evaporation lifetime of sessile drops (No. I–IV referring to Figure 14.5).

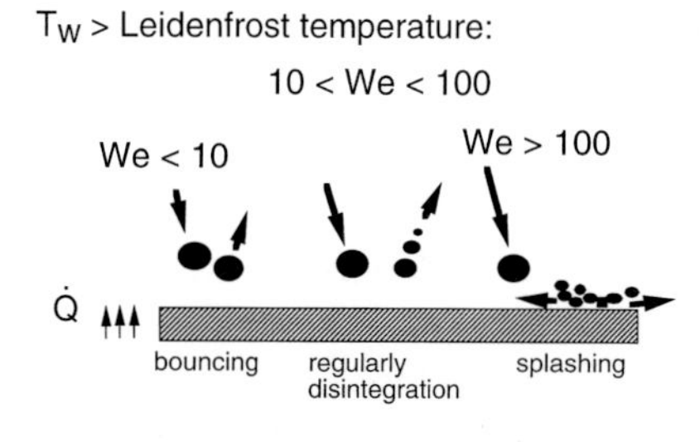

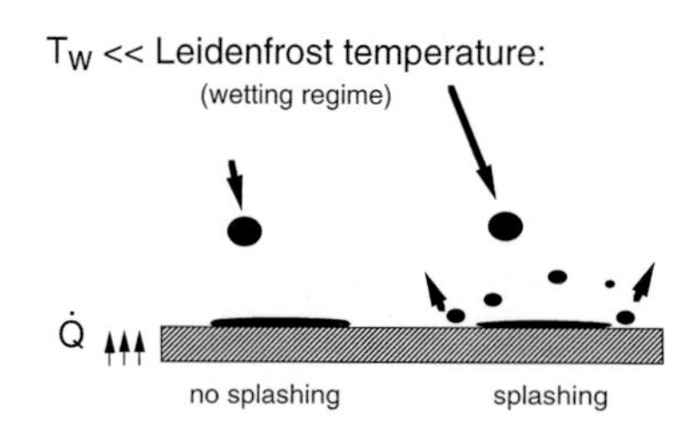

Figure 14.2: Disintegration of droplets impacting on a wall.

The main types of interaction in the Leidenfrost regime are bouncing, regularly disintegration and splashing (Figure 14.2). The "wetting regime" in this chart represents wall temperatures around the boiling temperature and does not cover the transition regime, where additional forces are induced by rapidly growing vapour bubbles.

The different phenomena involved in the droplet-wall collision process show transient behaviour. In the present study, the term "transient" means a process, in which the predominating mechanisms undergo a very rapid change. For example, the period of boiling delay, that will be discussed later, lasts only a few microseconds. To analyse these phenomena, a new surface temperature measurement technique has been developed.

14.2 Experimental Set-Up

An atmospheric free falling droplet experiment, combines a video technique with a special surface temperature probe that is characterised by high spatial and time resolution [7] (Figure 14.3).

Since the droplet-sizing device provides different apertures, droplets between 0.6 and 2.6 mm diameter can be generated. The range of droplet Weber number covers values from lower than 10 to higher than 100.

The surface thermocouple is embedded in the wall (Figures 14.3 and 14.4).

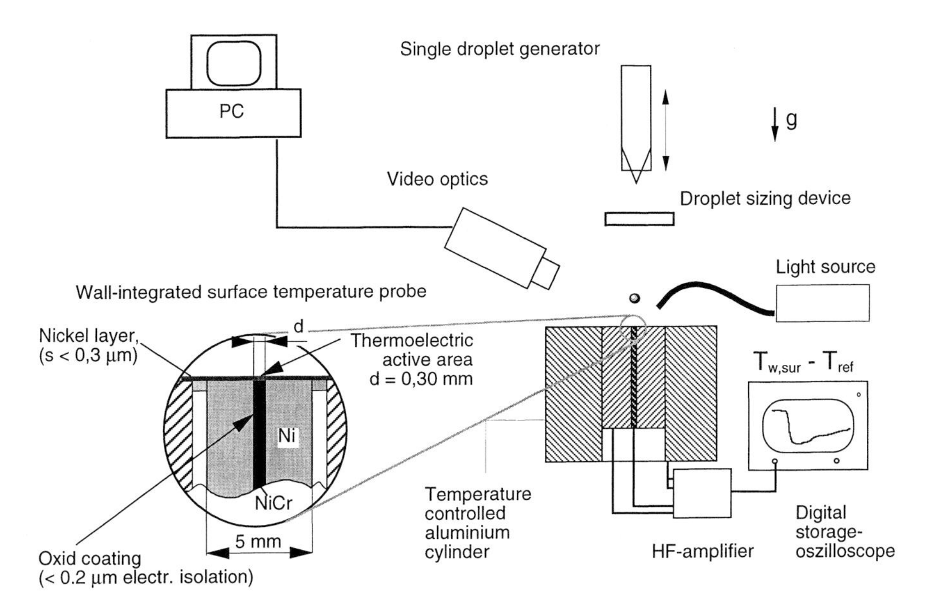

Figure 14.3: Experimental set-up.

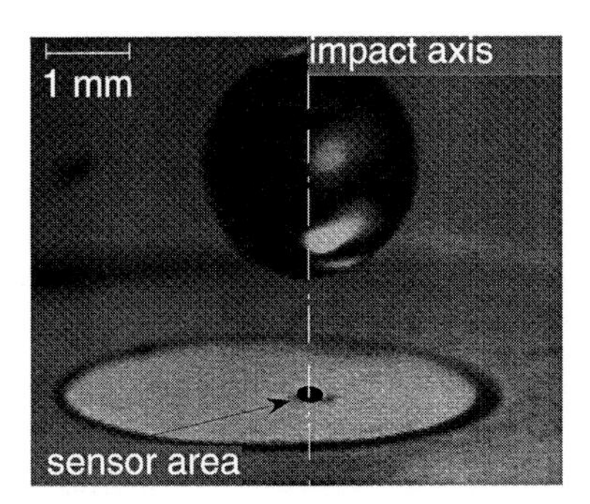

Figure 14.4: Surface temperature probe with approximating droplet.

Calibration experiments yielded excellent thermal properties with $(k\rho c)$ close to those of technical materials. Thus, the measured temperature describes the conditions occurring at the surface of practical applications, which is not the case if $(k\rho c)$ value of probe is low. The value of stainless steel (1.4301) at 400 K is $\sqrt{k\rho c} = 8000\ \mathrm{Ws^{0.5}\ m^{-2}\ K^{-1}}$, whereas that of the probe is $9000\ \mathrm{Ws^{0.5}\ m^{-2}\ K^{-1}}$. It can be seen in Figure 14.4 that the drilling of the small hole (diameter 0.3 mm) for the thermocouple produced a sensor area, which lies slightly eccentric in the mounting probe material. In the experiments, the impact axis is exactly adjusted to the sensor spot (Figure 14.4).

14.3 Results and Discussion

14.3.1 Boiling regimes of impacting droplets

The predominant parameter is the wall temperature T_w. It dominates both the heat transfer and the hydrodynamics related to the collision. Figure 14.5 shows measurements for isopropanol droplets with $d_0 = 2.6$ mm impinging on a NiCr surface for three different boiling regimes. The different experimental conditions (I–IV) refer to Figure 14.1. At $T_w = 300\,°C$, which means a wall far above Leidenfrost temperature ($T_{Leid} \cong 160–200\,°C$), the decay of surface temperature is weak and the interaction time is 20 ms. Since the Weber number is 27, the droplet is reflected elastically with only two secondary droplets occurring at the end of the collision.

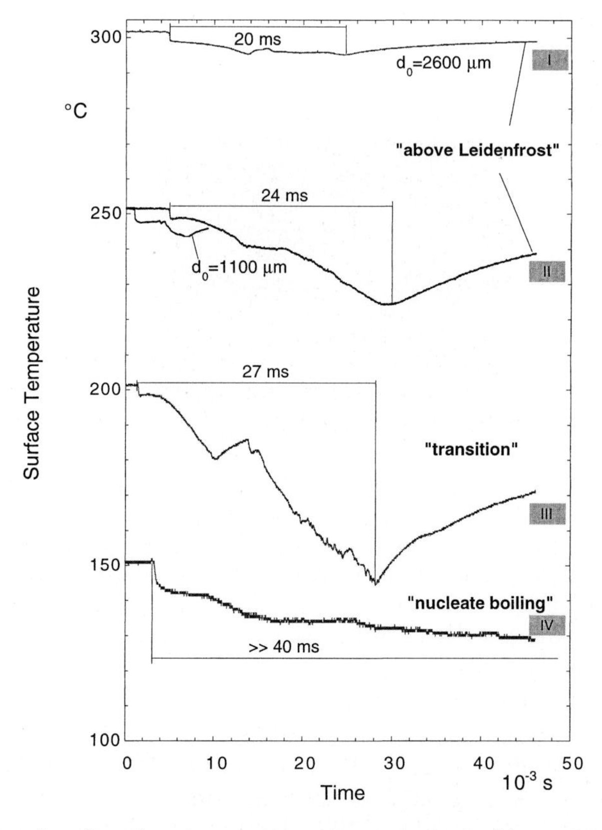

Figure 14.5: Local wall surface temperature due to impact of isopropanol droplets with a diameter of 2.6 mm ($We = 27$) on a NiCr surface: determination of different boiling regimes.

If T_w gets closer to the Leidenfrost region, as shown in case II ($T_w = 250\,^\circ C$), the droplet is also reflected elastically. But since the vapour layer between the wall and droplet fluid is thinner, the friction in the boundary layer increases, and so does the interaction time ($t_{int} = 24$ ms). So, as the isolating vapour layer is thinner, the heat transfer rate becomes higher. The temperature drop reaches 25 K. This is, in the "film boiling regime", a relatively high value due to the size of the drop and its long interaction time. Smaller droplets produce a much weaker decay in surface temperature because of their interaction time being much shorter. The second curve shown for $T_w = 250\,^\circ C$ was produced by a 1.1 mm diameter droplet. The curve is shifted slightly to the left so that its characteristics can be seen. The temperature decay at the beginning of the collision is slightly greater.

This effect seems not to be large but is reproducible. Measurements have shown that heat transfer rate averaged over the interaction time increases for smaller droplets [8], while the effect of Weber number was not pronounced.

At $T_w = 200\,^\circ C$, the transient surface temperature exhibits a "transitional" behaviour. The temperature decay reaches 60 K. The Leidenfrost temperature is no longer exceeded. Subsequently, the onset of nucleate boiling is observed.

The "nucleate boiling" regime is represented by the curve for $T_w = 150\,^\circ C$ in Figure 14.5. After the impact, the droplet sticks to the wall. Big vapour bubbles occur and prevent elastic rebound of the droplet. No vapour layer can be observed and the droplet remains at the wall until it is evaporated completely.

14.3.2 Dependencies of Leidenfrost temperature

Mostly, Leidenfrost temperature is said to be a function of the participated materials only. With impacting droplets that is not the case. An analysis of the surface temperature courses in Figure 14.5 shows that the Leidenfrost temperature is also a function of the droplet diameter. Although the temperature decay due to smaller droplets is generally steeper, the overall heat removal is lower since interaction time and covered wall area are reduced.

Subsequently, the minimum of transient surface temperature course is higher than for larger droplets. Because the local wall temperature under the drop determines the thickness of the vapour layer, it follows that in practice the Leidenfrost temperature is also a function of the droplet size. Figure 14.5 at $T_w = 250\,^\circ C$ may serve as an example.

The smaller droplet ($d_0 = 1.1$ mm) only has 8% of the mass of a 2.6 mm droplet und its interaction time is one-fourth (6 to 24 ms). Subsequently, the decay of surface temperature is only 8 K compared with 25 K. If these facts are considered in respect to a specific initial wall temperature that is close to the Leidenfrost temperature of a sessile drop, it is obvious that for big droplets, the local surface temperature sinks below a value that is necessary to maintain the vapour layer. For smaller droplets, that is not the case. The fact that the value of Leidenfrost temperature depends on the droplet dynamics leads to the so called "Dynamic Leidenfrost Phenomenon". In the present study with $d_0 = 1.1$–2.6 mm, a wall temperature above 250 °C is considered as far above Leidenfrost temperature.

14.3.3 Phase transition during collision

Transient liquid-vapour phase-change takes place in the collision zone at the wall. Simultaneous recording of both transient surface temperature and 8-frame video series provides exact interpretation of both the heat transfer and the hydrodynamics [9]. The surface temperature plots show a characteristic behaviour and allow statements about wetting, phase transition and relative thickness of the vapour layer. The video images with equal time spacing provide an overview of the hydrodynamics and clarify the relationship between changes in transient surface temperature and spreading or contraction period. In case of big diameters ($d_0 \approx 2$ mm), it must be remembered that there could be an oscillation of the fluid due to droplet generation. However, this is a question of damping and droplet mass and is no longer observed with smaller droplets. The results for smaller droplets are similar to those of the Figures 14.6–14.8, but are not presented here because of graphical quality. Even with the larger droplets, high contrast video imaging is nearly impossible when using a Cranz-Schardin camera [10].

14.3.4 Wall temperature far above Leidenfrost temperature ($T_w = 300$ °C)

In Figure 14.6 a, the transient surface temperature is depicted for a wall temperature of $T_w = 300$ °C. In Figure 14.6 b, simultaneously recorded video frames are shown. The dashed lines in the diagram indicate each moment that one of the eight video images was taken. The camera system was triggered

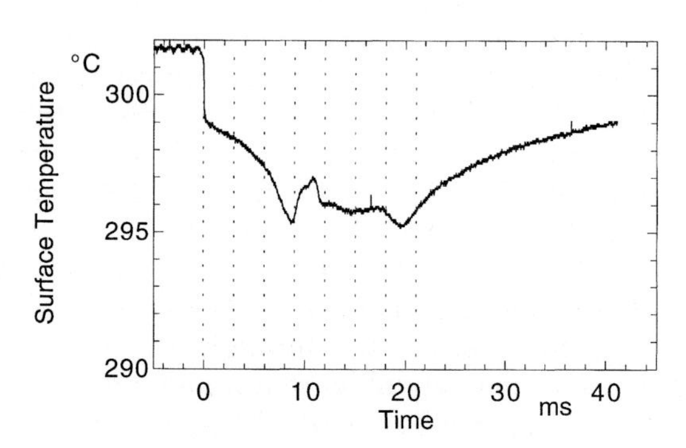

Figure 14.6 a: Local wall surface temperature during droplet impact at $T_w = 300$ °C (isopropanol, $d_0 = 2.6$ mm, $We = 27$, NiCr surface).

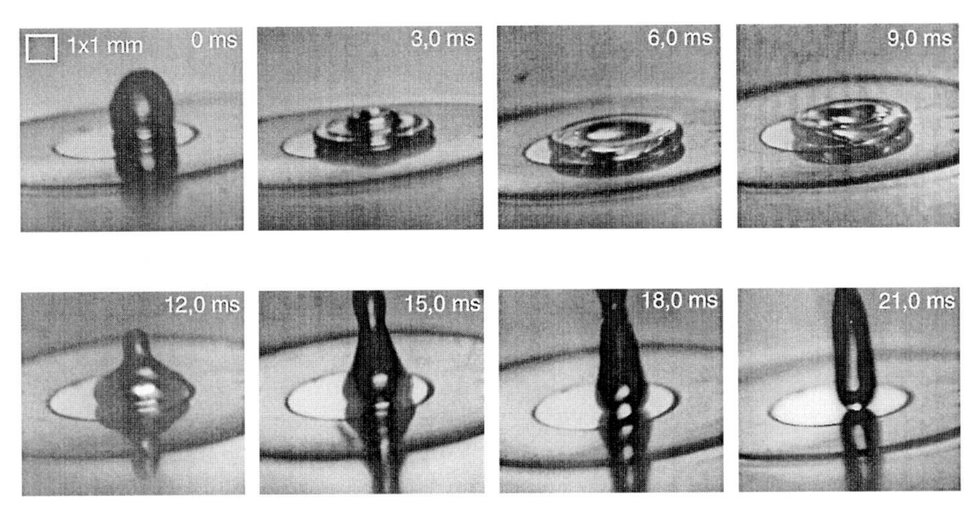

Figure 14.6 b: High speed video images of impacting droplet at $T_w = 300\,°C$ (isopropanol, $d_0 = 2.6$ mm, $We = 27$, NiCr surface).

by the signal of the surface temperature probe. The trigger delay of the system is 0.2 ms, which is approx. 1% of the duration of collision in the investigated case.

The basic deformation process of impacting droplets at wall temperatures far above Leidenfrost temperature will be considered only briefly. The process has been investigated previously for moderate Weber numbers. Excellent high speed images have been already provided (see [11]). The focus in the present study is put on the combination of "internal" information from the nearly inaccessible area between fluid and wall, and "external" overview from video imaging.

The duration of collision in Figure 14.6 is 20 ms and can be broken down into four parts. After an extremely steep decay of about 3 K at $t = 0$ ms, the curve shows a linear section with a temperature gradient of 250 K/s until $t = 6$ ms. This linear section during the spreading stage indicates a vapour layer of nearly constant thickness, covering the active sensor area. At $t = 9$ ms, the flattened droplet already forms a contracting torus. Surprisingly, the surface temperature increases between $t = 9$ and 12 ms. This is very probably due to a lower pressure underneath the center of the droplet. It can be seen as well in the images of Figure 14.6 b as in those of other researchers (e.g. [11]) that the fluid lamella reaches its thinnest stadium in this interval. As a consequence, the lower pressure causes that the thickness of the vapour layer grows rapidly over the active sensor zone. Heat transfer through that thicker vapour space is obviously lower than heat conduction between wall surface and metal body: the surface temperature rises for a duration of 3 ms.

At the next image ($t = 12$ ms), the center of fluid mass is already moving rapidly away from the wall. The pressure increases and the vapour is pushed away in a radial direction. Between $t = 12$ and 20 ms, the temperature in general takes a linear course again and it follows that the thickness of the vapour layer remains nearly constant. At $t = 20$ ms, the droplet leaves the surface.

14.3.5 Wall temperature slightly above Leidenfrost temperature ($T_w = 250\,^\circ$C)

The transient surface temperature in the case of $T_w = 250\,^\circ$C is plotted in Figure 14.7 a. The corresponding video frames are depicted in Figure 14.7 b. These images show only slight differences compared to Figure 14.6 b. However, if the temperature plot is considered, the interpretation is that Leidenfrost temperature is no longer clearly exceeded. This is derived from the collision duration, which is 25% longer. Moreover, the much stronger decrease of surface temperature indicates changes in heat transfer regime. The drop of surface temperature reaches 25 K, which has to be compared to 6 K in Figure 14.6 a. Because of the lower temperature difference between wall and fluid, the vapour layer stays thinner. Subsequently, friction losses during spreading and contraction stage increase.

The overall heat transfer is several times higher compared to the $T_w = 300\,^\circ$C case. At $t = 9$ ms, we observe again an increase of surface temperature. It shows similar behaviour as in Figure 14.6 a (duration 3 ms, temperature rise 1 K). A comparison of Figure 14.6 and Figure 14.7 demonstrates the specific features of the high resolution surface temperature probe: its signal exhibits remarkable changes in heat transfer, although the differences of hydrodynamics are only weak.

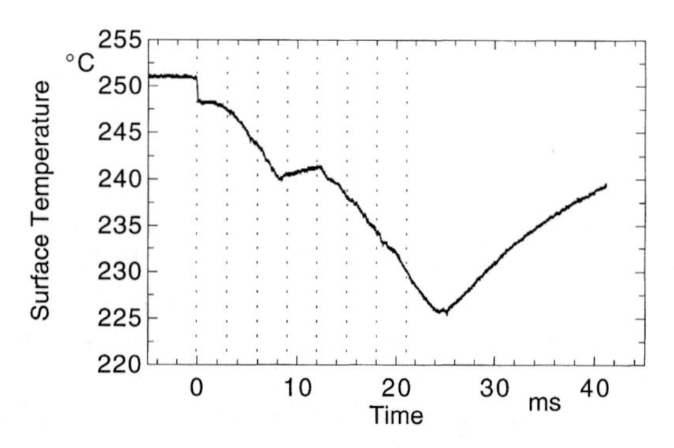

Figure 14.7 a: Local wall surface temperature during droplet impact at $T_w = 250\,^\circ$C (isopropanol, $d_0 = 2.6$ mm, $We = 27$, NiCr surface).

Figure 14.7 b: High speed video images of impacting droplet at $T_w = 250\,^\circ C$ (isopropanol, $d_0 = 2.6$ mm, $We = 27$, NiCr surface).

14.3.6 Wall temperature near to Leidenfrost temperature ($T_w = 200\,^\circ C$)

The trend towards longer interaction time continues in the case of $T_w = 200\,^\circ C$. In Figure 14.8 a, we observe the onset of transitional boiling and the droplet leaves the wall after 28 ms. The surface temperature drop equals 64 K. At the end of collision, the temperature plot shows a behaviour typical for transitional boiling. The enlarged section shows a rough course. An increasing number of vapour bubbles is mixing the fluid next to the wall. A stable vapour layer no longer exists. Finally, the droplet does not lift off the wall but tears off, as can be seen in the last video image of Figure 14.8 b. The rest of the fluid mass remains at the wall and evaporates in the nucleate boiling regime at a surface temperature below 150 °C.

14.3.7 Detection of boiling delay and fluid superheat

Under precisely adjusted geometrical conditions, i.e. the probe axis and droplet motion axis are aligned perfectly (see also Figure 14.4), the measured surface temperature course shows a characteristic behaviour. Figure 14.9 may serve as example. Begin, progression and end of collision are easily detected. Please note that Figures 14.6 a, 14.7 a and 14.8 a show these stages as well, but the vertical scale there is unfavourable for a more detailed consideration.

In the initial stage of the collision, the surface temperature shows a very steep decay with a temperature gradient up to 100 000 K/s. This period only lasts some microseconds and ends at a temperature some few degree below the

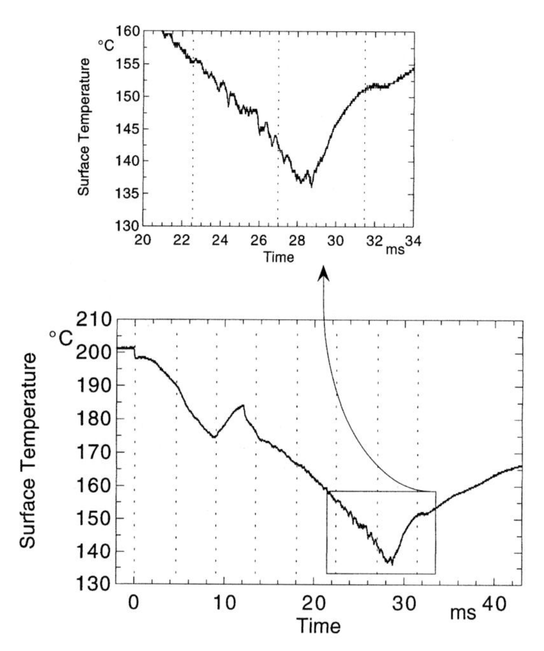

Figure 14.8 a: Local wall surface temperature during droplet impact at $T_w = 200\,^\circ$C (isopropanol, $d_0 = 2.6$ mm, $We = 27$, NiCr surface).

Figure 14.8 b: High speed video images of impacting droplet at $T_w = 200\,^\circ$C (isopropanol, $d_0 = 2.6$ mm, $We = 27$, NiCr surface).

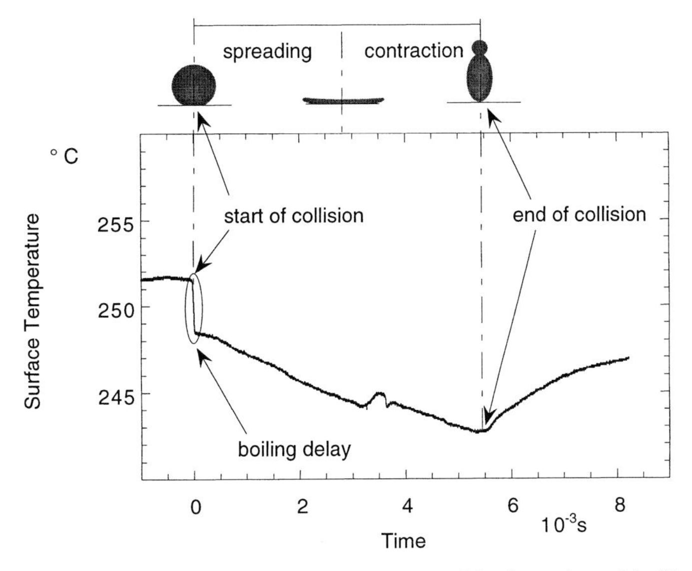

Figure 14.9: Local wall surface temperature at $T_w = 250\,^\circ$C: detection of boiling delay (isopropanol, $d_0 = 1.1$ mm, $We = 55$, NiCr surface).

initial temperature. This is the period of boiling delay. The subsequent decrease of surface temperature shows a much more moderate behaviour. Three different stages can be distinguished:

- initial stage with boiling delay,
- spreading stage with phase transition and
- contraction stage with phase transition.

14.3.8 "Contact Temperature Model"

From a simple theoretical model (see Figure 14.10), the superheat temperature of the fluid layer in the contact zone before the onset of evaporation can be estimated. A comprehensive description of that calculation and related considerations can be found in Wruck [12].

Assuming both the wall and the impinging fluid to be semi-infinite solid bodies, the contact temperature can be derived from the following equation [13, 14]:

$$\frac{T_{\mathrm{contact}} - T_{\mathrm{fluid,0}}}{T_{\mathrm{wall,0}} - T_{\mathrm{fluid,0}}} = \frac{b_{\mathrm{wall}}}{b_{\mathrm{wall}} + b_{\mathrm{fluid}}} \quad \text{with } b = \sqrt{(k\rho c)}\,. \tag{2}$$

In the present situation, the contact temperature is measured and the unknown temperature of the wall-facing fluid layer of the impinging droplet can be

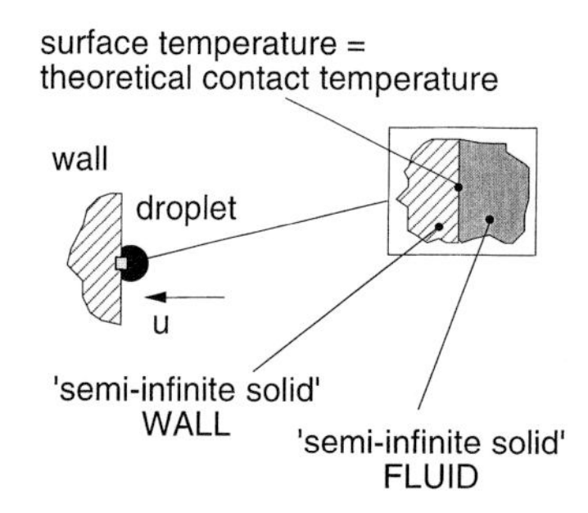

Figure 14.10: Model of "Contact Temperature".

calculated by rearranging Equation (2). In a first step, it is assumed that the thermal properties of the liquid are the same as with quasi-stationary heating. In the case depicted in Figure 14.9, the relevant values are as follows (Table 14.1):

Table 14.1: Values for calculating Equation (2).

$T_{contact}$	248.3 °C
$T_{wall,0}$	251.5 °C
$b_{sensor\ (=wall)}$	9000 $Ws^{0.5}\ m^{-2}\ K^{-1}$
$b_{fluid,\ 185\,°C}$	450 $Ws^{0.5}\ m^{-2}\ K^{-1}$

From Equation (2), an impingement temperature of $T_{fluid,0} = 185\,°C$ in the outer droplet shell can be calculated iteratively. Since this value, which is far above the saturation temperature, and a numerical estimation of the heat transfer to the wall-facing side of the droplet yields $T_{fluid,0} = 30\,°C$, the preconditions have to be re-examined.

While the contact temperature is measured and out of question in the considered calculation, the thermal properties of the liquid during very rapid heating are subject to a more detailed analysis. It appears that thermal properties undergo a remarkable change during very rapid heating of a liquid.

To our knowledge, this has not been measured before, but the result is consistent to theoretical considerations of Stephan [15]. Measuring $T_{contact}$ and calculating b_{fluid} by rearranging Equation (2), one yields a highly variable thermal effusivity (Figure 14.11).

With a value of $b_{fluid} = 120$, which is measured for rapid heating from 30 to 250 °C (see Figure 14.11), an analytical calculation of the fluid temperature in the contact zone is performed. Figure 14.12 shows three temperature traces at 0.1, 1 and 10 μs after the beginning of the collision. It displays which part of the fluid is superheated and therefore is subject to sudden phase-change after the boiling delay stage.

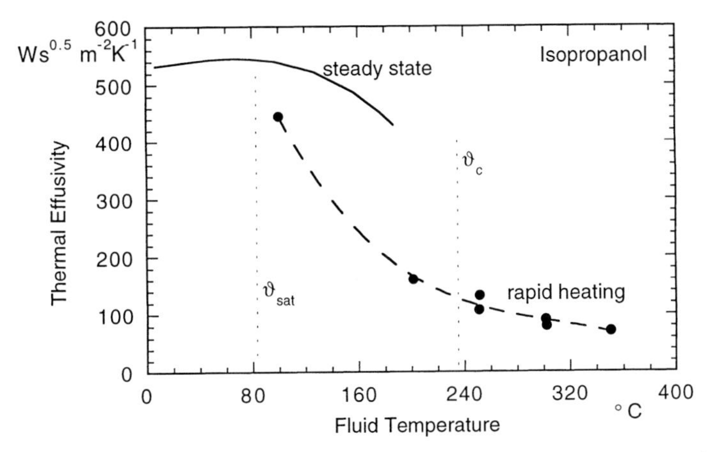

Figure 14.11: Thermal effusivity: effective value in case of rapid heating and steady state values.

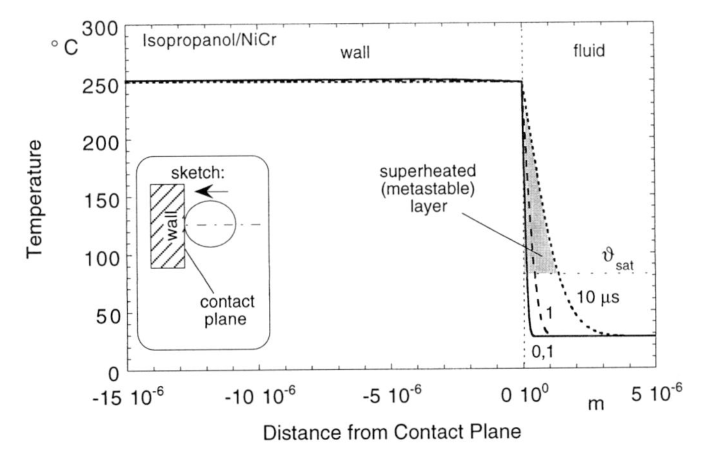

Figure 14.12: Calculated temperature in the contact zone (thermal properties of fluid: effective values after Figure 14.11).

The "Contact Temperature Model" predicts that very high temperatures occur in the liquid droplet during the short period of boiling delay. The highest value, $T_{contact}$, is far above the boiling temperature ($T_b = 82\,°C$) and also above a value of 200 °C published by Avedisian as the limit of superheat of isopropanol in case of quasi-stationary heating [16]. Furthermore, $T_{contact}$ is even above the critical temperature of isopropanol ($T_c = 235°C$). During the boiling delay period, the liquid exists in a highly superheated state.

Thus, the model provides a physical interpretation, in which the existence of metastable liquid is taken into account. It is free of inconsistency and explains the characteristic course of surface temperature during droplet-wall contact. It clarifies that a highly superheated liquid layer, which only exists for some 10 microseconds, flashes spontaneously and forms the well-known vapour "cushion" between the droplet and the wall.

14.4 Summary and Conclusions

The impact process of isopropanol droplets on hot metal walls under ambient conditions is analysed. Four different boiling regimes are identified due to varying wall temperature. Detailed information about hydrodynamics and phase transition are provided by a newly developed high resolution surface temperature measurement technique. Characteristic courses of transient wall surface temperature are discussed together with simultaneously recorded high speed video images. For droplets with moderate impact Weber numbers, three different stages during the collision can be distinguished:

- initial stage with boiling delay,
- spreading stage with phase transition and
- contraction stage with phase transition.

The detection of boiling delay is discussed by the model of "Theoretical Contact Temperature". This model, which is applied to the initial moment of droplet-wall contact, leads to new results in the field of superheated liquids. In the short period of boiling delay, a very high liquid superheat is detected in the wall-facing fluid layer. Rapid heating (more than 100 000 K/s) goes along with a remarkable change in thermal properties of the fluid.

Acknowledgements

This study was supported by the Deutsche Forschungsgemeinschaft by a grant from the SPP 725 (Transiente Vorgänge in mehrphasigen Systemen mit einer oder mehreren Komponenten).

We gratefully acknowledge the co-operation of Prof. F. Obermeier (now TU Freiberg, formerly Göttingen) when his co-workers Dr. A. Kubitzek and T. Jonas employed a Cranz-Schardin video camera system additionally to our apparatus (Figures 14.6 b, 14.7 b, 14.8 b). The experiments were conducted at Lehrstuhl für Wärmeübertragung und Klimatechnik, RWTH Aachen.

List of Symbols

c	specific heat [J kg^{-1} K^{-1}]
d	droplet diameter [m]
k	thermal conductivity [W m^{-1} K^{-1}]
p	pressure [N m^{-2}]
$\dot{Q}$	heat flow rate [W]
r	radial coordinate [m]
t	time [s]
T	temperature [K]
u, v	velocity [m s^{-1}]
We	droplet Weber number (see Equation (1))
$\sqrt{k\rho c}$	thermal effusivity [W s$^{0.5}$ m^{-2} K^{-1}]

Greek Symbols

a	thermal diffusivity, $k/\rho c_p$ [m^2 s^{-1}]
μ	dynamic viscosity [kg m^{-1} s^{-1}]
ρ	density [kg m^{-3}]

Subscripts

b	boiling
con	contact
d	droplet
int	interaction
ref	reference
sat	saturation
sur	surface
w	wall
0	initial value

References

[1] F. Papetti, S. Golini: *The Spray-Wall Interaction Model.* Remark in European KIVA Users Group Letter. Ed.: T. Baritaud (Inst. Francais du Petrole): December, 1993.

[2] M. Andreani, G. Yadigaroglu: *A 3-D Eulerian-Lagrangian Model of Dispersed Flow Film Boiling Including a Mechanistic Description of the Droplet Spectrum Evolution. I. The Thermal-Hydraulic Model.* Int. J. Heat Mass Transfer **40**(8) (1997) 1753–1772.

[3] T. Jonas, A. Kubitzek, F. Obermeier: *Transient Heat Transfer and Break-Up Mechanisms of Drops Impinging on Heated Walls.* Experimental Heat Transfer, Fluid Mechanics and Thermodynamics, Conference, Brussels, 1997.

[4] A. Karl, K. Anders, A. Frohn: *Experimental Investigation of Surface Roughness During Droplet-Wall Interactions.* Experimental Heat Transfer, Fluid Mechanics and Thermodynamics, Conference, Brussels, 1997.

[5] J.D. Bernardin, C.J. Stebbins, I. Mudawar: *Effects of Surface Roughness on Water Droplet Impact History and Heat Transfer Regimes.* Int. J. Heat Mass Transfer **40** (1997) 73–88.

[6] J.D. Bernardin, C.J. Stebbins, I. Mudawar: *Mapping of Impact and Heat Transfer Regimes of Water Impinging on a Polished Surface.* Int. J. Heat Mass Transfer **40** (1997) 247–267.

[7] H. Hüppelshäuser, U. Renz: *Messung der Oberflächentemperatur und Wärmestrommessung mit hoher zeitlicher Auflösung.* Sensor-Kongreß, Nürnberg (Germany) 13.–16. Mai 1991.

[8] S. Unverzagt: Diplomarbeit am Lehrstuhl für Wärmeübertragung und Klimatechnik, RWTH Aachen, 1995.

[9] J. Kubitzek: Co-operative investigation at Lehrstuhl für Wärmeübertragung und Klimatechnik, RWTH Aachen, 1996 (Please see acknowledgements).

[10] B. Stasicki, G.E.A. Meier: *A Computer Controlled Ultra High-Speed Video Camera System.* Proc. of 21th Int. Congress on High-Speed Photography and Photonics, Taejon, Korea, 1995.

[11] S. Chandra, C.T. Avedisian: *On the Collision of a Droplet with a Solid Surface.* Proc. Roy. Soc. Lond **432** (1991) 13–41.

[12] N. Wruck: *Transientes Sieden von Tropfen beim Wandaufprall.* Shaker-Verlag, Aachen, Germany. Dissertation, RWTH Aachen, 1998.

[13] H.D. Baehr, K. Stephan: *Wärme- und Stoffübertragung.* Springer-Verlag, Berlin etc., Germany, 1994, p. 163.

[14] H.S. Carslaw, J.C. Jaeger: *Conduction of Heat in Solids.* Clarendon Press, Oxford, 1959.

[15] K. Stephan: Private conversation with Prof. Dr.-Ing. Karl Stephan, Institut für Technische Thermodynamik und Thermische Verfahrenstechnik, Universität Stuttgart, Germany, 1998.

[16] C.T. Avedisian: *The Homogeneous Nucleation Limits of Liquids.* J. Phys. Chem. Ref. Data **14**(3) (1985) 695–729.

Boiling and Condensation

15 Model-Based Design, Control, and Evaluation of Transient Boiling Experiments

Joachim Blum and Wolfgang Marquardt *

15.1 Introduction

For the design of many technically relevant boiling processes, it is important to know the heat transfer for time-varying boiling surface temperature, here referred to as transient boiling. Applications of transient pool boiling include steel production, emergency cooling of nuclear reactor rods, or cooling of microelectronic components.

Boiling heat transfer is described by the so-called boiling curve (Figure 15.1, top left).

It characterizes the relationship between the transferred heat flux q and the wall superheat $\Delta T = T_w - T_{sat}$, i.e. the difference between the boiling surface temperature and the fluid's saturation temperature. Starting at low superheats, the curve passes through the nucleate (positive slope), transition (negative slope), and the film boiling (positive slope) region. We refer the reader to [1] for an extensive review of boiling fundamentals.

It is the aim of this joint project with a research group at the Technical University of Berlin [2] to quantify and understand the differences in pool boiling heat transfer between steady-state and transient conditions. For systematic investigations, a simple experiment representative of technically relevant boiling configurations is essential, allowing to vary the important process parameters in a controlled fashion. Controlled parameters here comprise the wall superheat, the speed of heating or cooling, pressure and saturation temperature. While the macroscopic heat transfer is described by the boiling curve – implying a uniform mean wall superheat across the boiling surface –, the physical process of boiling locally involves significant spatial and temporal heat flux fluctuations. The fluctuations are brought about by alternating vapor and liquid contacts with the boiling surface. To interpret and understand the boiling curve, it is therefore necessary to resolve the heat flux and temperature fields with high spatial and temporal resolution (Figure 15.1, top right).

* RWTH Aachen, Lehrstuhl für Prozeßtechnik, Turmstr. 46, D-52056 Aachen, Germany

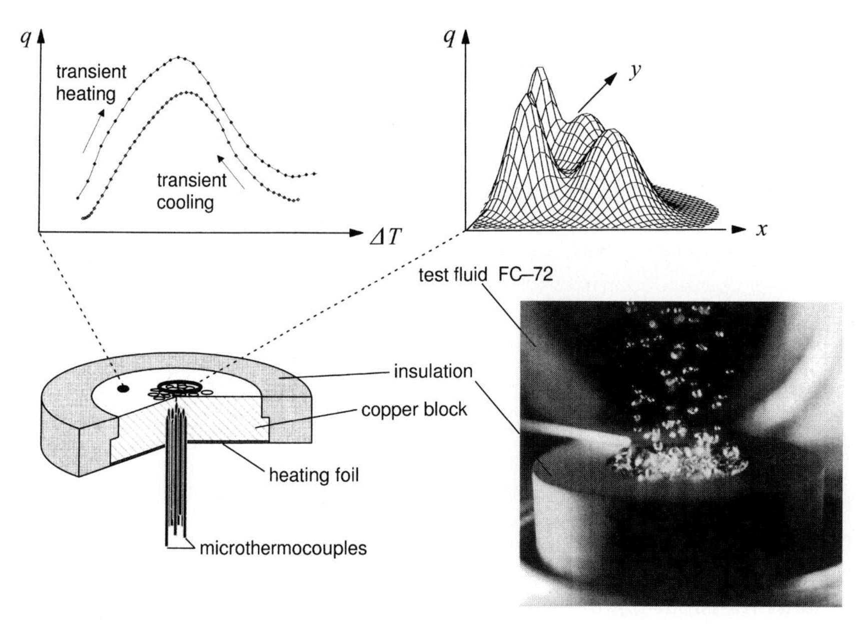

Figure 15.1: Test heater for measuring the macroscopic and local pool boiling heat transfer of FC-72.

Figure 15.1 also shows the core of the apparatus: the fluorinert FC-72 boils on a horizontally positioned copper block heated electrically from below. For temperature measurements, miniature thermocouples are galvanized into the copper block with their tips located just below, but not directly on the boiling surface in order to not disturb the local boiling process. As will become apparent, the outlined goals are only feasible by sensibly integrating the experimental work and a model-based analysis. While [2] treats the experimental and boiling-specific aspects in greater detail, we here focus on the model-based techniques, which help solve the following main tasks of an integrated experimental design:

- *Homogeneity of the boiling state:* There are two conflicting demands on the design of the test heater. On one hand, the heater should possess enough thermal inertia to avoid heterogeneous boiling states, i. e. the simultaneous existence of nucleate, transition, and film boiling on the same boiling surface [3]. On the other hand, the heater should be as thin as possible to achieve fast transients. The requirements on the thickness of a test heater, determined by means of a numerical stability analysis, will be given in Section 15.2. In addition, the influence of the heater insulation, the electric heating, and the thermocouples on the homogeneity of the temperature field near the boiling surface is assessed by detailed finite element simulation.
- *Temperature control:* To enable measurements along the entire boiling curve, operation of the apparatus in transition boiling must be stabilized.

Beyond mere stabilization, the control concept must allow for the surface temperature to follow a specified reference trajectory in transient experiments. Section 15.3 sketches the two-step development of a control concept involving a stability analysis and dynamic simulation of the controlled apparatus.

- *Measurement evaluation:* To infer heat flux and temperature at the boiling surface from heater-interior measurements, a suitable estimation algorithm must be provided. The solution of this so-called inverse heat conduction problem, whose main characteristic is its high sensitivity to measurement noise, is addressed in Section 15.4.

For a more detailed description of the project than here and in [2], we refer the reader to [4–9].

15.2 Test Heater Design

On thin test heaters such as wires or metal foils, the simultaneous existence of more than one boiling state has been observed [3]. In general, local and sufficiently strong disturbances of a homogeneous temperature field are needed to initiate heterogeneous boiling states, i. e. small dry spots on the boiling surface in nucleate boiling or liquid contacts of the surface in film boiling. This seed then spreads in a wave-like fashion across the boiling surface. For reproducibility of the results as well as to ensure the technical relevance of the experiment, heterogeneous boiling states must be ruled out.

The existence of such temperature waves is made possible by the interaction of heat transport near the boiling surface and the highly nonlinear characteristic of heat removal by boiling. While the conditions for initiating and the stability of heterogeneous boiling states on thin heaters can be determined analytically, for thick heaters, they so far must be checked by dynamic simulation. In a nucleate boiling scenario, for example, dry spots of varying extent and temperature are imposed as initial conditions of the simulation problem. The evolution of the temperature field then reveals whether this dry spot spreads or collapses. The results show that spontaneous dry spots of approx. 8 mm diameter are necessary to initiate heterogeneous boiling on the heater of Figure 15.1. Dry spots of this size have never been spotted on thicker heaters. Heterogeneous boiling in our experiments – or even thinner heaters of down to a few millimeters, as the simulations show – is therefore highly unlikely.

While the simultaneous existence of different boiling states is to be avoided during experiments, heterogeneous wetting and heat flux patterns still exist on a microscopic scale in homogeneous boiling. The stability of these patterns determines the boiling phenomenology and thus the macroscopic heat transfer characteristics. It can be studied using the same stability analysis techniques. The results show that the interaction of many tightly packed small dry spots, which, e.g., form during high heat flux nucleate boiling at the foot of growing bubbles, can lead to the propagation of vapor blanketing, i.e. film boiling, across the whole boiling surface – even on thick heaters. Macroscopically, this process appears as a spatially uniform transition from nucleate to film boiling. According to this novel interpretation, the maximum of the boiling curve corresponds to the stability limit, at which, microscopically, the initial dry spot distribution grows together. As mentioned, a critical factor influencing this stability limit is the thickness of the heated wall, with thinner walls tending to develop vapor blanketing at lower heat fluxes. Indeed, experimental studies of other research groups confirm our theoretical findings regarding the influence of wall thickness on the maximum heat flux.

Macroscopic temperature non-uniformity can also be introduced through non-ideal insulation of the heater, placement of miniature thermocouples near the boiling surface, or steps of the cross-sectional area of the test heater (cf. Figure 15.1). Detailed finite-element models of the copper heater are simulated using the commercial code ANSYS. The simulations reveal that, due to the high thermal conductivity of copper, cross-sectional temperature distortions stay well below the uncertainty of the temperature measurements of about 0.3 K.

15.3 Control Concept

The negative slope of the boiling curve prevents steady-state measurements in the transition boiling region for fixed heat flux. Stable operating conditions along the entire boiling curve can be accomplished in one of two ways: passive stabilization by means of fluid heating or active stabilization by feedback control of the surface temperature. We here choose feedback control, which, as a systematic analysis reveals, is both the most flexible approach and requires the least experimental effort to realize. The control loop is sketched in Figure 15.2. By measuring the wall temperature just below the boiling surface, local temperature fluctuations are smoothed, thus improving the performance of the controller which aims at controlling the mean surface temperature. The control law itself is realized on a PC. The actual temperature and the setpoint temperature can be matched by adjusting the electric heating of the copper block. Steady-state experiments are characterized by piecewise constant setpoint temperatures, also referred to as regulatory control; once the temperature settles on the setpoint, both the temperature and the heating voltage are recorded. In transient experiments, the surface temperature must follow a reference trajectory. This type of control problem is referred to as servo control. Here, the reference

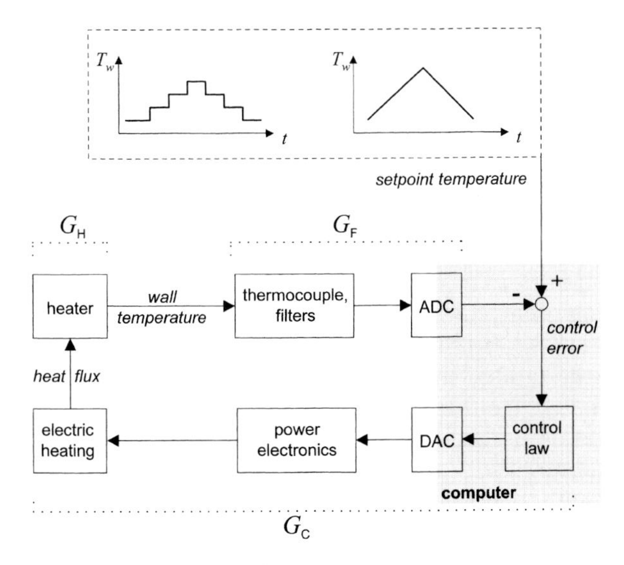

Figure 15.2: Temperature control loop elements.

trajectories are chosen as ramp functions. However, any setpoint function can be realized.

The development of the control concept involves two steps. First, the structure of the controller, i.e. the control law, the measurement location as well as admissible parameterizations of the controller, are determined. This is equivalent to the question of stability of the scheme. In a second step, control performance is optimized by dynamic simulation of the controlled experiment.

15.3.1 Stability analysis

As opposed to previous studies, the approach of this project [7, 8] allows to study the influence of all relevant parameters on the stability of an experiment: the dimensions and the material of the heater, the location of the temperature sensors, the control law and parameters, or the heat transfer to the boiling fluid. Only systems with a uniform mean temperature across the boiling surface are to be considered here.

The analysis is based on frequency domain transfer functions of all control loop components. The stability of the control loop is determined by the location of the roots of the so-called characteristic equation:

$$1 + G_H(s, \boldsymbol{p})G_C(s, \boldsymbol{p})G_F(s, \boldsymbol{p}) = 0\,, \tag{1}$$

in which G_H, G_C, and G_F are the transfer functions of the heater, the controller, and the additional filters in the loop (cf. Figure 15.2). They depend not only on the complex frequency s, but on the parameter vector $\boldsymbol{p}$. A system is asymptotically stable if all roots of the characteristic equation possess negative real parts. The stability bounds then are characterized by purely imaginary roots. An example of stability bounds for our test heater is given by the solid line in Figure 15.3. The variation parameters are the slope of the boiling curve and the gain of the PI-controller. While in nucleate and film boiling only an upper stability bound for the controller gain exists, there is an additional lower stability bound in the transition boiling region. If the controller gain drops below the lower bound in Figure 15.3, unstable oscillations of the wall temperature arise; beyond the upper limit stable, small-amplitude limit cycle oscillations of the temperature occur. In addition, a maximally stabilizable slope exists. Experimentally observed stability limits are shown for comparison. In view of significant experimental difficulties in determining these bounds [8], the agreement must be regarded as excellent. We here have exemplarily analyzed one of the realized test heaters and its control scheme. For a systematic derivation of this optimal configuration, we refer the reader to [7].

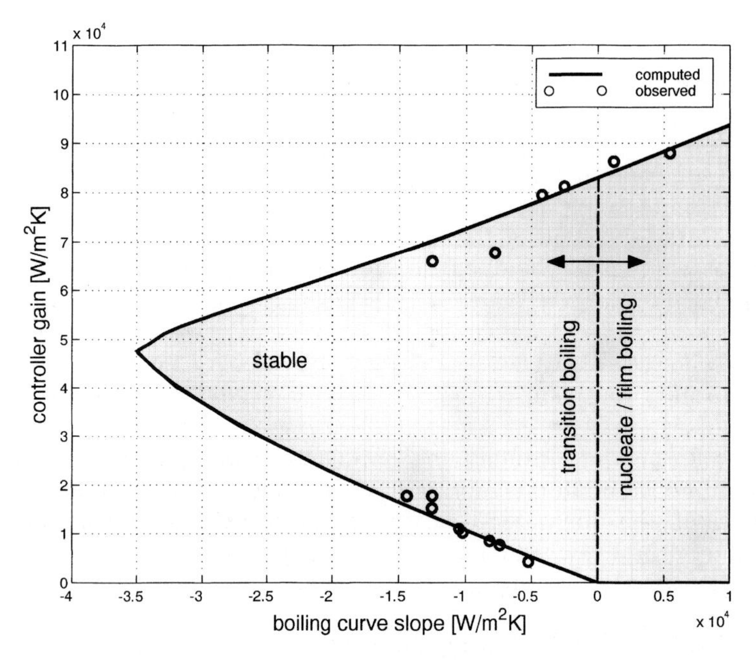

Figure 15.3: Comparison of computed and experimentally observed stability limits [8].

15.3.2 Dynamic simulation

In real experiments, two factors can possibly degrade control performance. One is the presence of measurement noise, the other the available heating power. Acceptable values of the controller parameters have to be determined by dynamic simulation of the controlled apparatus. In general, high controller gains lead to short response times, but also result in high sensitivity to measurement noise; low gains effect smoother temperature trajectories, but sluggish response to setpoint changes. To avoid deterioration of control performance by integrator windup when hitting the limits of the available heating power, an anti-reset-windup scheme was added to the PI-controller. In the simulations, the boiling curve is approximated by the boiling curve of FC-72 obtained in earlier experiments. Normally distributed noise is added to the simulated measurements. For regulatory control, setpoint changes are accommodated within 2–3 s. The servo controller allows close following of the reference trajectory for arbitrarily fast transients within the capacity of the electrical heating. Since no active cooling system is available, faster heating than cooling transients can be accomplished. The simulated control responses show close agreement with the experimentally recorded ones; very little fine-tuning had to be carried out on site.

15.4 Measurement Evaluation

The evaluation of the measurements focuses on the estimation of both the boiling curve and the spatially resolved heat flux distribution from temperature recordings below the boiling surface. In the past, simple extrapolation techniques have been used for the evaluation of transient boiling experiments. While these methods work for slow transients and one-dimensional problems, faster transients and multi-dimensional problems (cf. Figure 15.6) require more efficient algorithms. In the following, the nature of this estimation problem, a novel interpretation as a uniform basis for algorithm benchmarking, the derivation of an optimal solution algorithm, and finally exemplary results are briefly introduced.

15.4.1 Inverse heat conduction problems

Mathematically, solving the inverse heat conduction problem (IHCP) aims at minimizing the mean squared error of the estimates. This error includes two contributions. The first, referred to as an algorithm's sensitivity, stems from fluctuations in the estimates introduced through measurement noise. The second error is generated by the estimation procedure itself, i.e. the algorithm would not deliver correct estimates even if there were noise-free measurements. This error is inherent to the algorithm and is called an algorithm's bias. In IHCP, the bias and the sensitivity are unfavorably related to each other: If one is decreased (e. g. by adjusting the tuning parameters), the other increases and vice versa. IHCP therefore belong to the class of so-called ill-posed problems. Figure 15.4 illustrates this connection by the example of a transient boiling curve estimation [4].

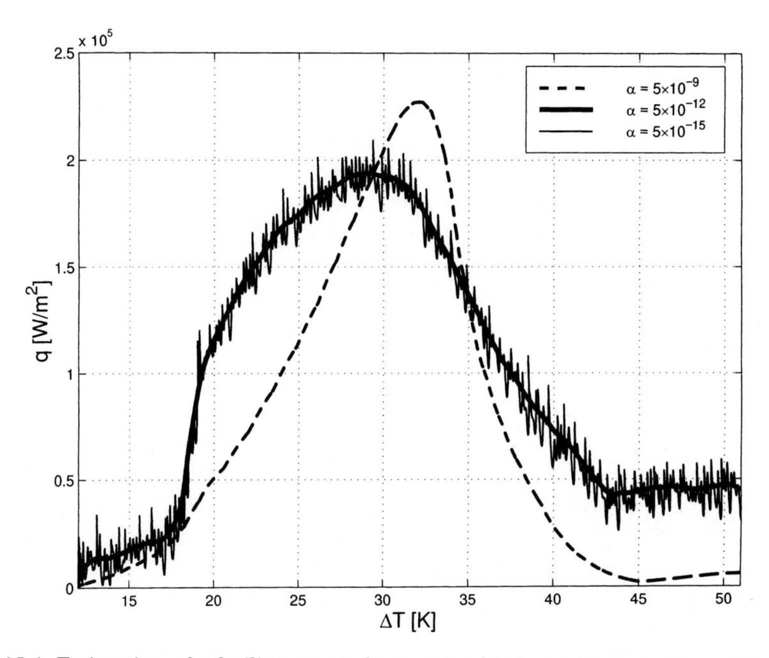

Figure 15.4: Estimation of a boiling curve for various bias-sensitivity settings [6].

Depending on the setting of the tuning parameter, the estimates can exhibit strong fluctuations (thin line: high sensitivity, low bias), be significantly damped, i. e. wrong (dashed line: low sensitivity, high bias), or closely approximate the true heat transfer characteristics (thick line). It is obvious that the tuning of an IHCP solution algorithm is critical for the success of the estimation. Although techniques for automated tuning are available and have been implemented here, in practice suitable parameters can be found by repeated data evaluations as well. Starting from oscillatory estimates, an optimal boiling curve reconstruction is obtained with the moving average of the oscillations. Some of the boiling curves recorded in the course of this joint project are shown and discussed in [2].

15.4.2 Frequency response interpretation

If the heat equation and its boundary conditions are spatially discretized, we obtain one or more independent heat fluxes as the unknown inputs. After Laplace transformation, we interpret the IHCP as a transfer system relating the vector of estimated heat fluxes $\hat{\boldsymbol{Q}}$ to the vector of the true heat fluxes $\boldsymbol{Q}$ and the measurement noise vector $\boldsymbol{N}$:

$$\hat{\boldsymbol{Q}}(s) = \boldsymbol{G}_{\mathrm{Q}}(s)\boldsymbol{Q}(s) + \boldsymbol{G}_{\mathrm{N}}(s)\boldsymbol{N}(s)\,. \tag{2}$$

$\boldsymbol{G}_{\mathrm{Q}}$ is referred to as the $(n_{\mathrm{M}} \times n_{\mathrm{M}})$ signal transfer function matrix, $\boldsymbol{G}_{\mathrm{N}}$ as the $(n_{\mathrm{M}} \times n_{\mathrm{M}})$ noise transfer function matrix. The number of measurements n_{M} must equal the number of unknown heat fluxes for the existence of a unique solution. In constructing any IHCP solution algorithm, we are ideally aiming for the following properties of the frequency response:

$$\boldsymbol{G}_{\mathrm{Q}} = \boldsymbol{I} \qquad \forall s = j\omega\,, \tag{3}$$

$$||\boldsymbol{G}_{\mathrm{N}}||_2 = 0 \quad \forall s = j\omega\,. \tag{4}$$

Here, $\boldsymbol{I}$ is the identity matrix and $||\cdot||_2$ the Euclidean norm of the transfer function matrix. Figure 15.5 exemplarily shows the frequency response magnitudes of G_{Q} and G_{N} for low and high bias in a one-dimensional problem. According to Figure 15.5, IHCP solution algorithms are filters passing low-frequency components of the true boundary heat flux signal while rejecting high-frequency components. The wider the passband ω_{E} of the filter (or the lower its bias), the higher is the amplification of measurement noise (the sensitivity) indicated by the shaded area.

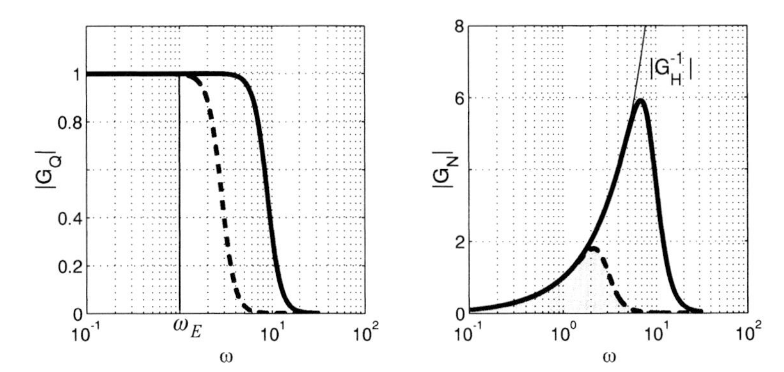

Figure 15.5: Frequency response magnitude of the transfer functions G_{Q} (left), G_{N}, and G_{H}^{-1} (right) for (– –) high bias/low sensitivity and (–) low bias/high sensitivity [6].

15.4.3 Observer algorithm

Observers are a well-known concept from systems and control theory allowing for the estimation of unknown or unmeasurable system quantities. In the context of IHCP, an observer can be derived to achieve any desired transfer characteristics $\boldsymbol{G}_{\mathrm{Q}}$ of the estimator. The resulting algorithm, whose transfer characteristics are given by:

$$\hat{\boldsymbol{Q}}(s) = \boldsymbol{G}_{\mathrm{N}}(s)\boldsymbol{T}^{*}_{\mathrm{M}}(s)\,, \tag{5}$$

will directly compute heat flux estimates from the noise-corrupted measurements $\boldsymbol{T}^{*}_{\mathrm{M}}$. It can be shown that the signal, noise and heater transfer function matrices are connected through

$$\boldsymbol{G}_{\mathrm{N}}\boldsymbol{G}_{\mathrm{H}} = \boldsymbol{G}_{\mathrm{Q}}\,. \tag{6}$$

The numerically stable inversion of $\boldsymbol{G}_{\mathrm{H}}$, necessary for the computation of $\boldsymbol{G}_{\mathrm{N}}$, poses the main difficulty in the derivation of the solution algorithm. Stable procedures for one- and multi-dimensional problems are described in [6, 9]. To obtain an algorithm, which can be implemented as a computer program, $\boldsymbol{G}_{\mathrm{N}}(s)$ is converted to discrete-time filter equations. If the IHCP is solved off-line, phase lag of the estimates can be removed by suitable forward-backward-filtering in time. One approach to compare different IHCP solution algorithms is to determine which algorithm is the least sensitive for a given bias. Quantitative measures of the sensitivity and the bias, for one-dimensional problems, are the passband of G_{Q} and the area under $|G_{\mathrm{N}}|$ marked in Figure 15.5. Comparisons with other solution methods show that the observer exhibits the best bias-sensitivity tradeoff. Due to its recursive discrete filter structure, the observer tops other algorithms with regard to computational efficiency as well. Data records on the order of a thousand time steps can be evaluated in fractions of a second. Provided the low measurement noise level achieved in our experiments, the algorithm is able to resolve heat flux fluctuations of up to several hundred Hertz. Although here only linear IHCP are considered, the observer can also be derived for nonlinear problems such as for temperature-dependent material properties.

15.4.4 Results

At last, we exemplarily and briefly present first results regarding the estimation of the local heat flux in Figure 15.6. The tips of a bundle of miniature thermocouples are located at a distance of approx. 20 μm below the boiling surface, spaced from 0.1 to 0.3 mm apart. Figure 15.6 shows the heat flux deviations from the mean surface heat flux associated with three of the sensors, derived by solution of a multiple heat flux IHCP in three space dimensions. It is

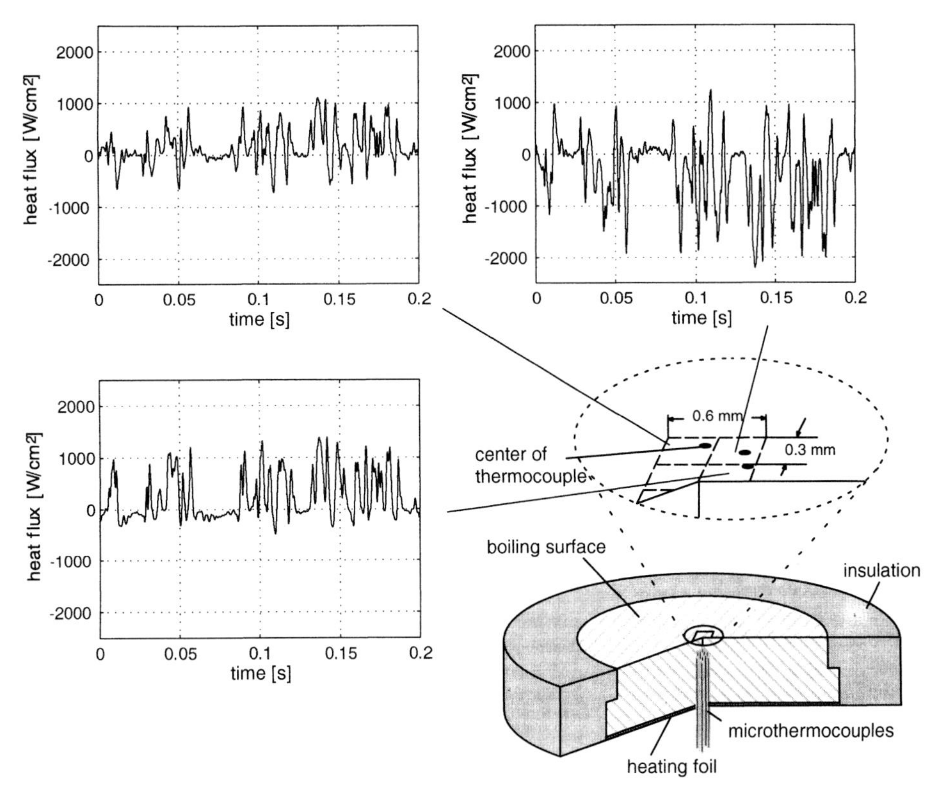

Figure 15.6: Local heat flux oscillations as deviations from the mean surface heat flux.

assumed that the surface heat flux is constant across the titles outlined in Figure 15.6. The shown oscillations around the nominal average heat flux indicated arbitrarily by zero are quite dramatic and undermine that boiling is a strongly heterogeneous process on a microscopic scale: While the heat flux maximum of the macroscopic boiling curve lies well below $100\,\mathrm{W/cm^2}$ [2], bursts on the order of $2000\,\mathrm{W/cm^2}$ frequently occur locally. Apparently, one tile contacts vapor more frequently, indicated by negative heat flux due to local insulation, another tile transfers peaks of high heat flux, which very well could be brought about by liquid contact and vapor generation, while the third tile displays intermediate wetting characteristics. For more detailed results as well as boiling curve estimates see [2, 4, 5].

15.5 Conclusions

Here and in [2], an integral approach for the measurement of transient pool boiling heat transfer, intelligently combining experiments and model-based techniques, has been outlined. The scheme allows the conduction of boiling experiments along the entire boiling curve and, for the first time, with arbitrarily specified temperature transients within the physical limits of the test heater and its electric heating. A new solution algorithm for the inverse heat conduction problem enables the reliable and efficient estimation of the boiling curve for fast transients and the reconstruction of high-frequency fluctuations of the local surface temperature and heat flux field.

References

[1] V. Carey: *Liquid-Vapor Phase-Change Phenomena.* Hemisphere, 1992.

[2] R. Hohl, H. Auracher: *Transient pool boiling experiments.* This book (Chapter 16).

[3] S. Zhukov, V. Barelko, A. Merchanov: *Wave processes on heat generating surfaces in pool boiling.* Int. J. Heat Mass Transfer **24** (1980) 47–55.

[4] R. Hohl, H. Auracher, J. Blum, W. Marquardt: *Pool boiling heat transfer experiments with controlled wall temperature transients.* In: G. Celata, P. Di Marco, A. Mariani (Eds.): Proceedings of the 2nd European Thermal Science and 14th UIT National Heat Transfer Conference, Rome, 1996, pp. 1647–1652.

[5] R. Hohl: *Mechanismen des Wärmeübergangs beim stationären und transienten Behältersieden im gesamten Bereich der Siedekennlinie.* Ph.D. thesis, Technical University Berlin, 1998.

[6] J. Blum: *Modellgestützte experimentelle Analyse des Wärmeübergangs beim transienten Sieden.* Ph.D. thesis, RWTH Aachen University of Technology, 1998.

[7] J. Blum, W. Marquardt, H. Auracher: *Stability of boiling systems.* Int. J. Heat Mass Transfer **39** (1996) 3021–3033.

[8] J. Blum, W. Marquardt, R. Hohl, H. Auracher: *Theory and experimental validation of boiling systems stability.* In: Proceedings of the Engineering Foundation Conference Convective Flow and Pool Boiling, Irsee, 1997.

[9] J. Blum, W. Marquardt: *An optimal solution to inverse heat conduction problems based on frequency domain interpretation and observers.* Numerical Heat Transfer, Part B **32** (1997) 453–478.

16 Transient Pool Boiling Experiments

Reiner Hohl and Hein Auracher *

16.1 Introduction

Although there are many technically relevant transient boiling processes, little is known about how boiling heat transfer characteristics under transient conditions differ from steady-state behavior. Several studies have suggested that different boiling curves describe the heat transfer for steady-state or transient heating and cooling experiments. Bui and Dhir [1] carried out pool boiling experiments with saturated water at atmospheric pressure and found that the critical heat flux is lower for transient cooling than for transient heating or for steady-state experiments. Similar results were obtained by Maracy and Winterton [2] and Peyayopanakul and Westwater [3]. The latter varied the cooling rate in quenching experiments with liquid nitrogen by varying the thickness, i.e. thermal inertia, of the heater. However, other researchers have established the opposite effect: Rajab and Winterton [4] found that steady-state boiling consistently caused higher transferred heat fluxes than heating or cooling experiments for a given wall superheat.

Systematic experiments on the influence of temperature transients on the boiling heat flux have not been carried out yet, mainly because of the lack of both a proper control concept and a reliable algorithm for the evaluation of transient measurements. The demands on a suitable experimental infrastructure for reproducible and systematic transient boiling experiments are (see [5] for more details):

- A control concept allowing for the surface temperature to follow a given reference trajectory along the entire boiling curve in the heating and cooling mode.
- Thermal inertia: The experimental heater should on one hand have enough thermal inertia to avoid heterogeneous boiling conditions on the

* Technische Universität Berlin, Institut für Energietechnik, Marchstr. 18, D-10587 Berlin, Germany

heated surface occurring on thin heaters [6]. On the other hand, its thermal inertia should be as low as possible as this favors faster transients.

- Data evaluation: The temperature sensors cannot be fixed on the heating surface but must be immersed in the heating block. Then, under transient conditions, heat flux and/or temperature at the boiling surface have to be determined by solving an Inverse Heat Conduction Problem (IHCP).

To account for the above points, an experimental apparatus, a control concept, and algorithms for the evaluation of transient measurements were developed and implemented as part of an integrated experimental design in interplay between simulation and experimental verification. Here, the "Institut für Energietechnik", Berlin, was responsible for the experimental set-up and test runs and the "Lehrstuhl für Prozeßtechnik", Aachen, accounted for the control concept, simulation and data evaluation algorithms.

16.2 Experimental

16.2.1 Experimental facilities

Figure 16.1 shows the apparatus with the test heater. It is located at the bottom of a stainless steel vessel (inner diameter: 209.1 mm, height: 332 mm). Four windows enable the observation of the boiling process. As test fluid FC-72 (C6F14, 3-M company) is used. It is kept at constant saturation temperature during the experiments by temperature-controlled condenser tubes. It can be cleaned by pumping it through a filter with a mesh size of 40 μm. Before starting experiments, the fluid is degassed by boiling of the fluid in the vessel allowing the noncondensible gases to vent through the condenser. The pressure is measured with a pressure transducer.

16.2.2 Test heater

The heating surface of the test heater 1 (Figure 16.1) is horizontally positioned. It consists of a 5 mm thick copper block, 20×20 mm squared in the bottom part and cylindrical with 18.2 mm diameter in the top part. It is insulated ($\lambda=0.22$ W/mK) and sealed by polyimide, a highly temperature-resistant synthetic. Heat losses through the polyimide block are small, hence one-dimensional heat conduction can be assumed (cf. [5]). Heating is accomplished by DC (20 V, max. 200 A) flowing through a foil made of a titanium/aluminium/vanadium alloy (thickness: 0.0125 mm, specific electric resistance: 126 μΩcm). The foil is pressed to the bottom of the heater with a 0.025 mm thick plate of muscowite

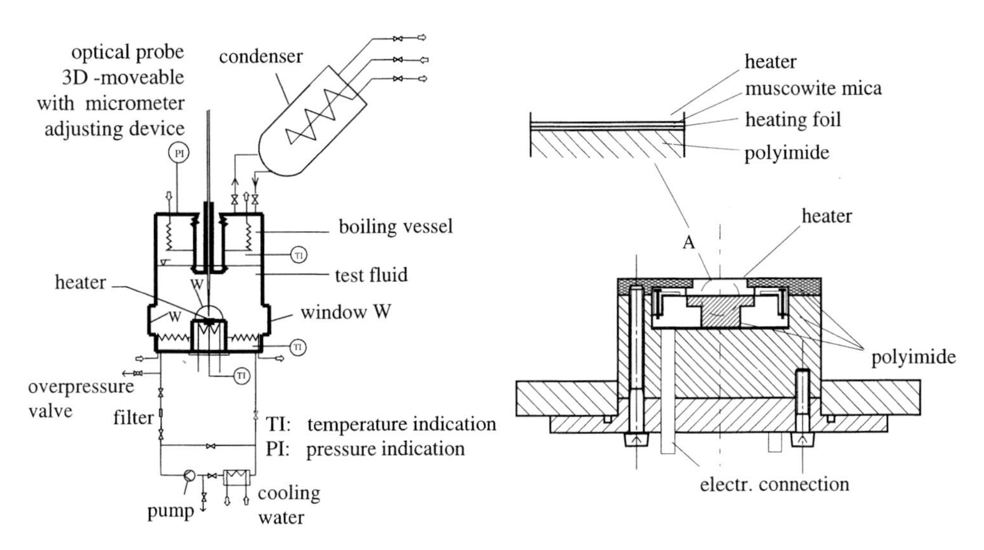

Figure 16.1: Experimental apparatus with test heater 1.

mica in-between for electrical insulation. The surface of the heater is electroplated with a 20 μm nickel-layer.

The thermocouples of 0.25 mm total thickness are implanted 1.3 mm beneath the heating surface by electroplating. One is used as sensor for temperature control, one as safety sensor against overheating and one to determine the surface temperature via a solution of the IHCP. A number of microthermocouples to measure the surface temperature fluctuations are also installed in heater 1. The details of this technique are presented in Section 16.4.3.

For the steady-state experiments described in Section 16.3, a similar heater (2) but with a somewhat higher thermal inertia has been applied (copper, 34 mm diameter, 10 mm thickness; see also [7–9]).

16.2.3 Optical probe

To investigate the liquid-vapor fluctuations above the heater, a fiber-optic probe has been developed. The set-up of the optical probe is based on an earlier development of Auracher and Marroquin [10]. It consists of a gradient index glass fiber with the end of the fiber formed into a conical shape (Figure 16.2). This is done by melting the tip under a microscope with a small burner. The fiber is then glued into small stainless steel tubes with increasing diameter to make it sufficiently stiff. The probe can be moved at different heights and different locations above the heater by means of a 3D-micrometer adjusting device (cf. Figure 16.1).

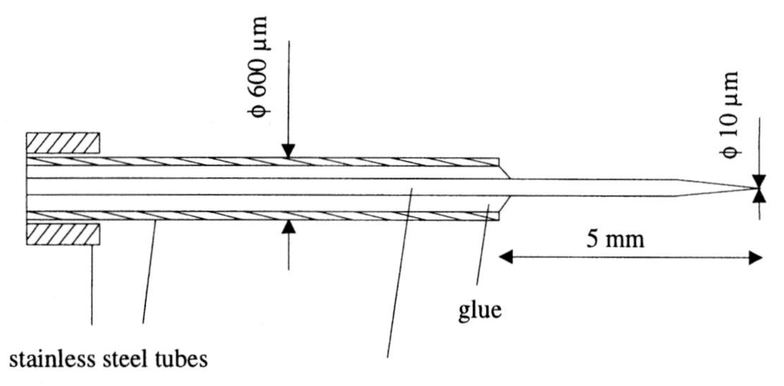

Figure 16.2: Scheme of optical probe.

16.2.4 Control concept

The wall temperature is controlled by measuring the temperature close to the boiling surface, comparing it to a setpoint value, which is either constant in steady-state experiments or time-varying in transient experiments, feeding the difference signal into the controller and adjusting the power of the electrical heating according to the controller output. The setpoint signal as well as the control law are implemented on a computer. More details are presented in [5].

16.2.5 Data acquisition and uncertainties

Data for heater temperature, system pressure and heating voltage are sampled with a data acquisition board for PC at a sampling rate of 200 Hz in steady-state experiments and 500 Hz in transient runs. To obtain a steady-state point, data are sampled for 2 seconds and then averaged. Data of probe measurements are sampled at a rate of 10 kHz for 60 seconds, and the sampling rate of the microthermocouple measurements (see Section 16.4) is 5050 Hz.

The uncertainty in temperature measurement is about ± 0.3 K. The uncertainty of calculated heat flux, taking into account the entire evaluation system, yields ± 1.5 W/cm^2. However, as for all experiments, apparatus, heater and data acquisition remained unchanged, measurement uncertainties concerning the effect of wall-temperature transients are better represented by the reproducibility, which is smaller than ± 0.5 W/cm^2 (cf. [11]).

16.3 Results of Steady-State Experiments

In this section, some important results of the optical probe measurements under steady-state boiling conditions are presented. They give important information on the mechanism of transient boiling as discussed in Section 16.5.

16.3.1 The boiling curve

In Figure 16.3, the steady-state boiling curve of FC-72, measured with heater 2, is plotted. All experiments were performed with saturated test fluid at a temperature of 333 K. The saturation temperature is calculated from the system pressure, and the heat flux is set equal to the electrical heat input. Knowing the heat flux, the heater surface temperature is obtained by linear extrapolation of the measured temperature inside the heater to the heater surface. The measured boiling curves for experiments with stepwise increasing and decreasing heater temperature were found to be identical.

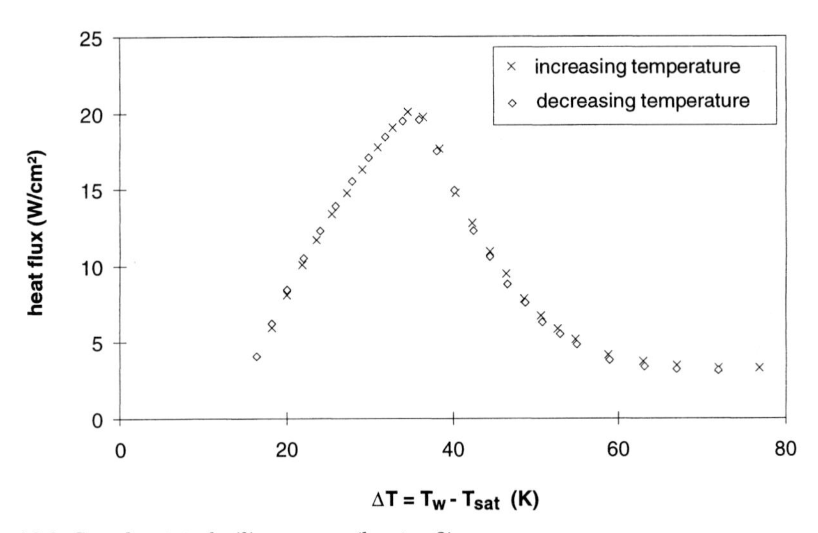

Figure 16.3: Steady-state boiling curve (heater 2).

16.3.2 Optical probe signals

Three series of measurements were carried out, one with the probe tip at the heater's center, a second one 8 mm from the center and a third one 15 mm from the center. It was found that no significant difference exists between the probe signals from different locations, indicating homogeneous boiling conditions on the heater surface. Measurements were carried out at wall superheats between 20 and 50 K at steps of 5 K and at 80 K superheat. The distances of the probe tip from the heater surface were 0.01, 0.03, 0.05, 0.1, 0.5 and 1 mm. The distance between probe tip and heater surface was determined with an accuracy of ±0.01 mm by measuring the signal rise as a result of light reflexion at the heater surface when the probe comes very close to it. In measurements at 0.01 mm nominal distance from the heater, it can be assured that the probe tip was fixed within a real distance between 0.005 mm and 0.02 mm from the heater surface.

In Figure 16.4 a the probe signals in nucleate boiling ($\Delta T = 25$ K), near CHF ($\Delta T = 35$ K), in transition boiling ($\Delta T = 45$ K) and in film boiling ($\Delta T = 80$ K) at 0.1 mm distance from the heater are plotted (cf. Figure 16.3). High signal level represents vapor, low signal level represents liquid phase. The number of vapor contacts increases with increasing wall superheat up to CHF. In transition boiling, fewer and longer vapor contacts dominate until in film boiling only few contacts are detected. At this height, the probe tip is situated between the minimum and maximum thickness of the wavy vapor film. When the probe tip approaches to 0.01 mm distance from the heater (Figure 16.4 b), it is mainly wetted by liquid in the nucleate boiling region up to CHF. In transition boiling again, longer vapor contacts are detected, which are interrupted by periods of very rapid liquid-vapor fluctuations. In film boiling, the probe tip is fully immersed in the vapor film and no liquid phase is detected. Apparently, the probe signal for liquid phase appears to be rather "noisy". This is a result of vibration of the apparatus, which leads to changes in light reflexion from the heater surface back into the glass fiber.

In measurements at a distance of 0.01 mm, the probe might of course influence the boiling mechanism. Still, in all boiling regimes except film boiling, distinct vapor or liquid contacts of various durations are clearly detected. So, apparently, the probe tip is still significantly smaller than the "vapor-bubbles" we want to detect, and the distortion effect is probably weak. The electrical response time of the system for 90% signal rise is about 0.05 ms. Obviously, this is not a limiting factor for the measurement system since the fastest signal changes due to the probe wetting mechanism took place within a time interval of not less than appr. 0.1 ms.

The data processing of the signals and results on the void fraction at different heights above the surface as well as on the distribution of vapor and liquid contact times are presented in [12]. Here, only some typical results on the mean vapor contact frequency are shown because they are important for the interpretation of transient behavior in boiling.

16.3.2.1 Mean vapor contact frequency

Figure 16.5 shows the mean vapor contact frequency at the probe tip. The number of vapor contacts increases when the probe is moved towards the heater surface and reaches frequencies as high as 500 to 600 Hz at about 0.05 mm from the heater. It seems that, at about these distances, the probe detects bubbles from several nucleation sites before they coalesce to bigger vapor masses resulting in a decrease of the frequency at larger distances. As the high num-

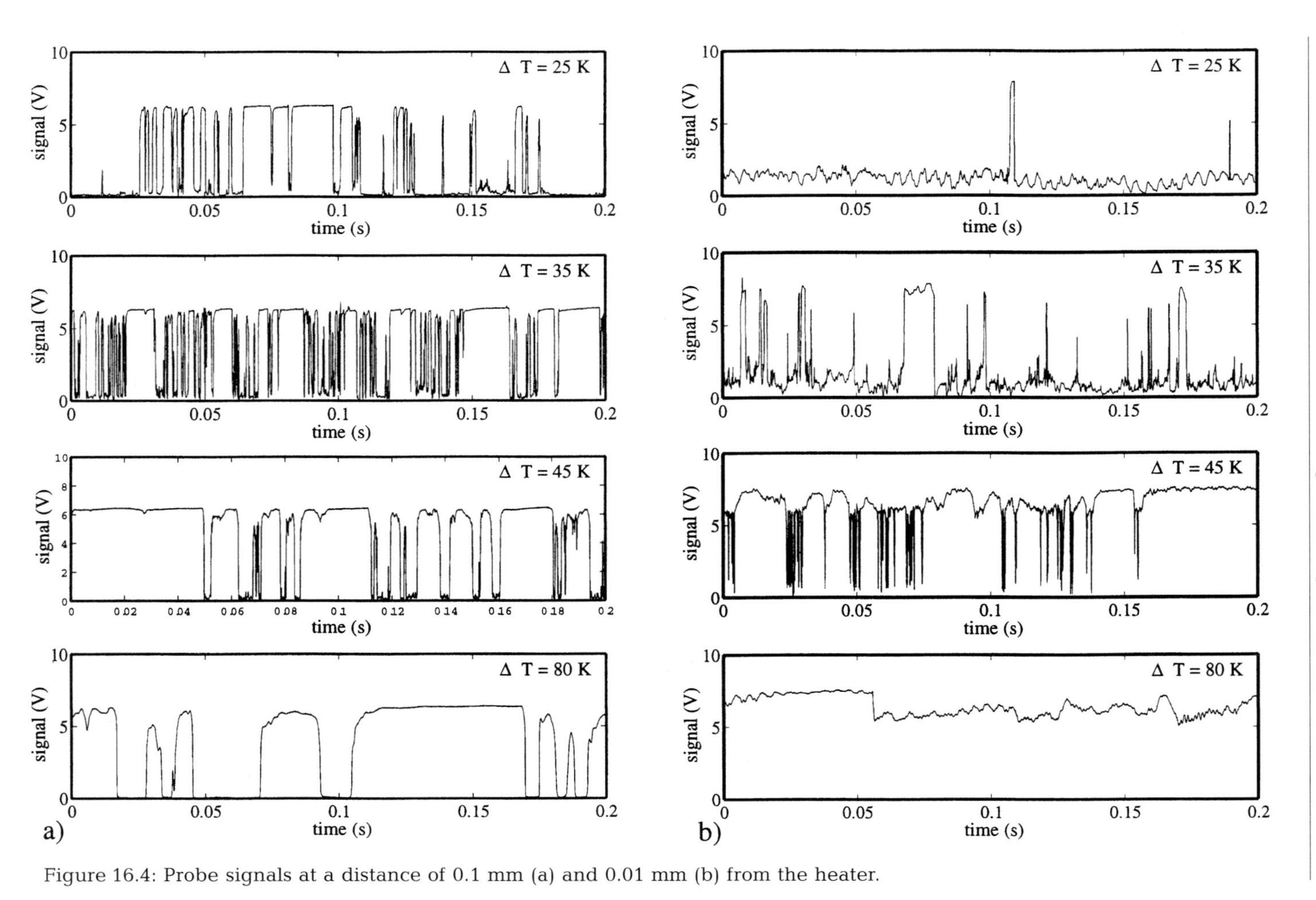

Figure 16.4: Probe signals at a distance of 0.1 mm (a) and 0.01 mm (b) from the heater.

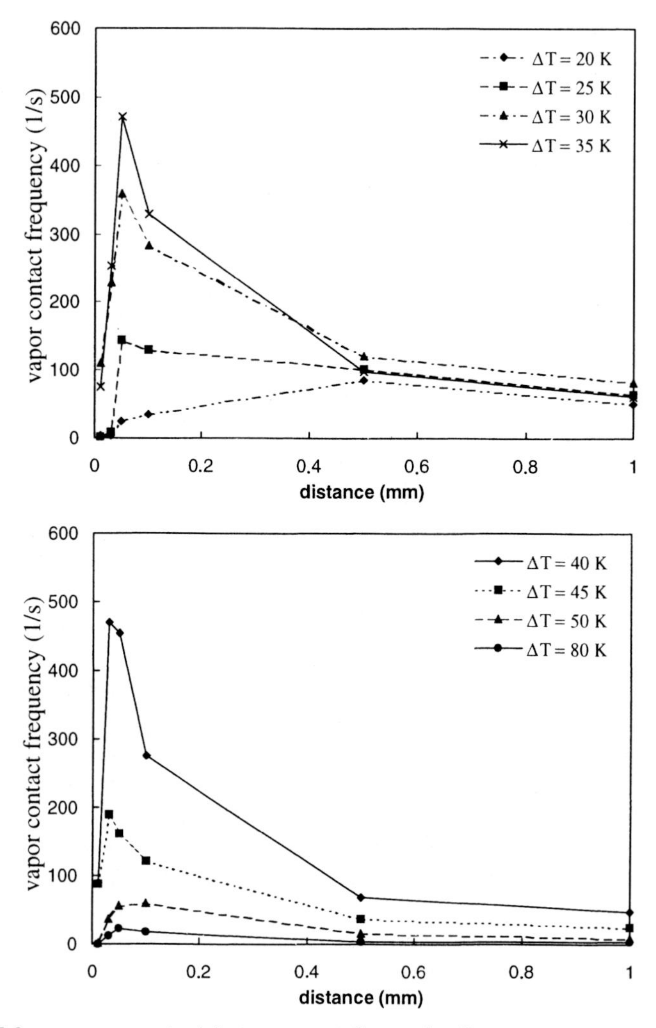

Figure 16.5: Mean vapor contact frequency at the probe tip.

ber of vapor contacts indicate a highly agitated two-phase flow, the possibility that the probe detects the same vapor masses not only once can, however, not be excluded. A further approach of the probe tip to the surface results in a decrease of the frequency. An explanation could be that on the one hand the effect of multiple detection becomes more and more insignificant, and on the other the probe detects bubbles from less nucleation sites than at larger distances.

In transition boiling, liquid contacts during a "wetting period" become so short that it is not possible to clearly detect all of them. Consequently, the liquid contact frequency and therefore as well the vapor contact frequency is underestimated. Here, it is more reasonable to look at the distribution of the vapor detection frequency and the distribution of vapor and liquid contact times as presented in [9] and [12].

In transient boiling, the probe measurements were not successful due to the unavoidable thermal expansion and contraction of both the probe and the heater. It turned out that, during a transient run, the distance between probe tip and heater surface varied within a distance, where the strongest changes in contact frequencies and other quantities occur.

16.4 Results of Transient Experiments

16.4.1 Boiling curves for transient heating

Boiling curves for transient heating with up to 50 K/s nominal temperature change per second of the heating surface measured with heater 1 (Figure 16.1) are depicted in Figure 16.6. These curves represent heat fluxes and temperatures at the heater surface determined by solving the Inverse Heat Conduction Problem (cf. [5]). For comparison, the steady-state boiling curve is also plotted. Like the boiling curve in Figure 16.3, it has been measured with stepwise increasing and decreasing temperature and again no hysteresis was observed. The experiments were carried out in a way that nucleation sites were already active at the start of heating up (ΔT appr. 15 K) to avoid the disturbance effect of boiling incipience.

In the ideal case, the temperature change with time of the heating surface should be kept constant by the control system during the entire transient period. However, the constant heating rates given in Figure 16.6 are nominal values, which deviate slightly from the real ones depending on the transient velocity. In the first part of the heating period, the surface temperature does not follow precisely the set-point velocity of the control system due to the thermal inertia of the heater. In the subsequent period the effect is vice versa. More details are presented in [11].

The heat flux increases strongly with increasing heating rate. At 50 K/s, the critical heat flux (CHF) is by a factor 4 higher than the one in the steady-state case. The superheat at CHF does not change remarkably with the heating rate. On the other hand, the minimum heat flux of film boiling (MHF) does not exhibit a clear behavior. With growing heating rates, it moves first to smaller and than to higher superheats. In some transients, no distinct minimum is observed. The heat flux in film boiling during the heating up mode is significantly higher than in steady-state heating. After switching off the heating power to start the cooling mode somewhere between a superheat of 60 and 70 K, the heat flux drops down to at least the value of the steady-state case (see Figure 16.7).

It can easily be proven that the behavior shown in Figure 16.6 is not due to an evaluation uncertainty or error. A simple energy balance reveals that the

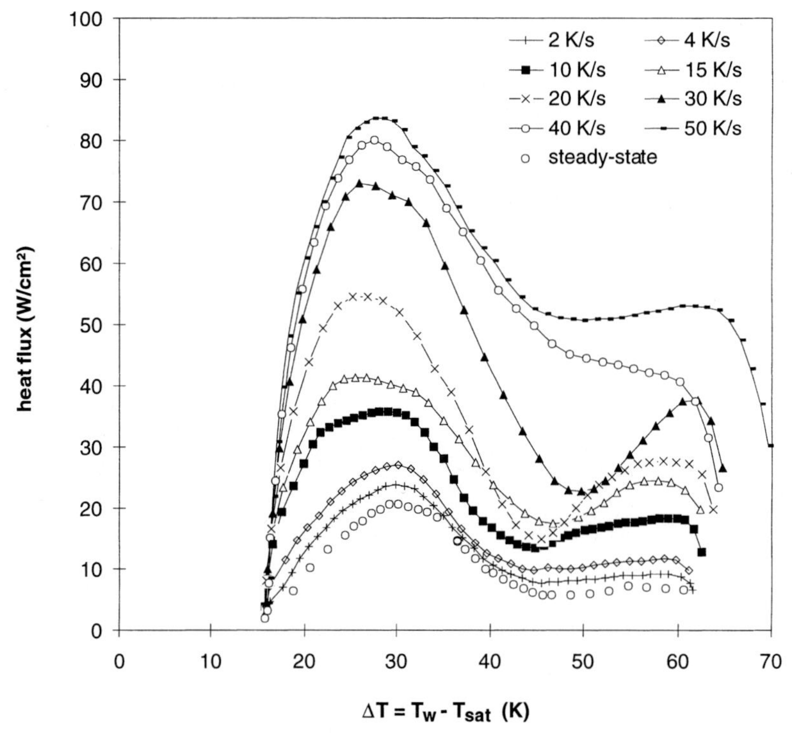

Figure 16.6: Boiling curves for transient heating.

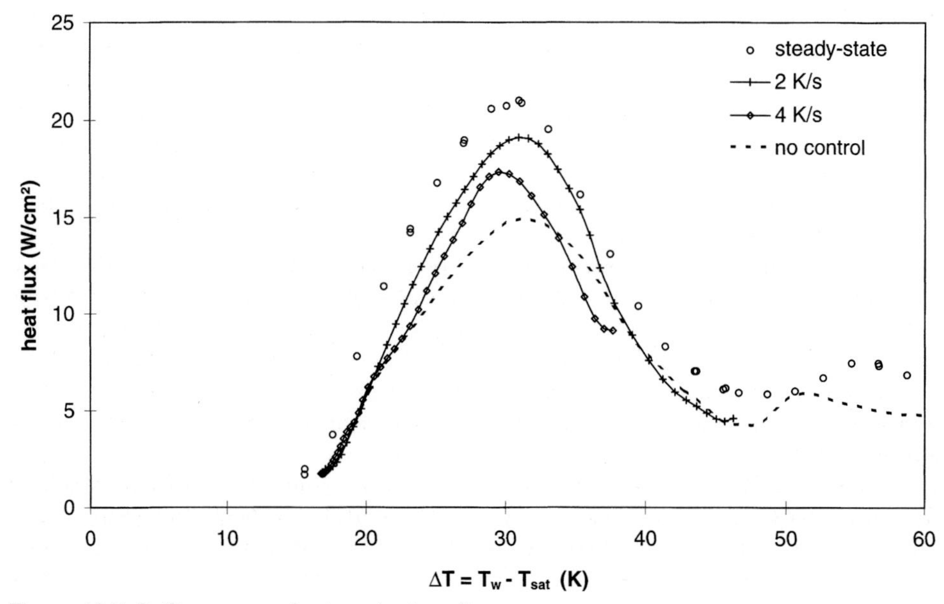

Figure 16.7: Boiling curves for transient cooling.

total heat input during the heating period between nucleate and film boiling is equal to the one accumulated in the heater plus the heat supplied to the fluid, which follows from a timely integration of the transient boiling curve. An estimate shows that an error due to a storage of heat in the insulation around the heater is negligibly small. Hence, the following conclusions can be drawn: 1. the inverse heat conduction problem is solved properly, and 2. the boiling curve characteristic in transient heating shown in Figure 16.6 is due to a physical effect of the boiling mechanism.

16.4.2 Boiling curves for transient cooling

Due to the thermal inertia of the heater, the cooling rate is limited. Constant cooling rates were only possible for 2 K/s along the entire boiling curve and for 4 K/s between transition and nucleate boiling (see Figure 16.7). With zero heat input, the cooling rate was about 2 K/s in film boiling and 6.8 K/s in the CHF-region. Obviously, the transient cooling effect is contrary to the transient heating behavior. The faster the cooling rate the smaller the heat flux. Keys for a physical explanation of the transient boiling mechanism are the results obtained by the optical probe (Section 16.3) and data for the temperature fluctuations on the heating surface presented in the following.

16.4.3 Temperature fluctuations at the heating surface

To detect temperature fluctuation due to the wetting mechanism, several micro-thermocouples are imbedded in heater 1. Both wires of a Ni-CrNi-thermocouple (0.025 mm diameter each) with a polyester coating for electrical insulation are fixed in the copper block by electroplating and connected at the surface by a 20 μm Ni-layer, again by an electroplating process. 16 of these pairs were implanted within an area of 1 mm^2 near the center of the heater. During the sophisticated procedure, only 6 elements survived. Their relative location is shown in Figure 16.8. Another problem occurred during the first test measurements: the thermal inertia of the 20 μm Ni-layer prevented a pronounced temperature signal well above the noise band. Therefore, the Ni-layer was grinded down to a smaller thickness but obviously not uniformly because the average amplitudes of some thermocouples were found to be different. Despite of these experimental problems, some conclusions can be drawn from the measured temperature fluctuations on the boiling mechanism under steady-state and transient conditions.

In Figure 16.9, the simultaneously measured temperature fluctuations of thermocouples 5 and 6 during a period of 0.05 s are presented for a steady-state and a transient heating run (50 K/s) at different locations on the boiling curves, which are depicted in Figure 16.6. In steady-state nucleate boiling (A, a), temperature drops of appr. 0.5 K during about 1 ms are observed, and during a sub-

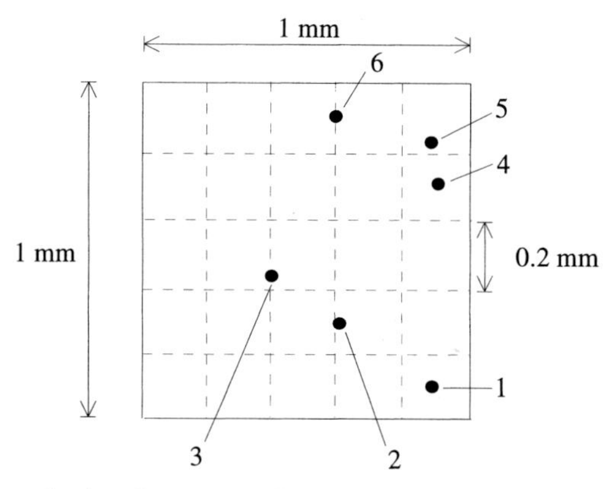

Figure 16.8: Location of microthermocouples.

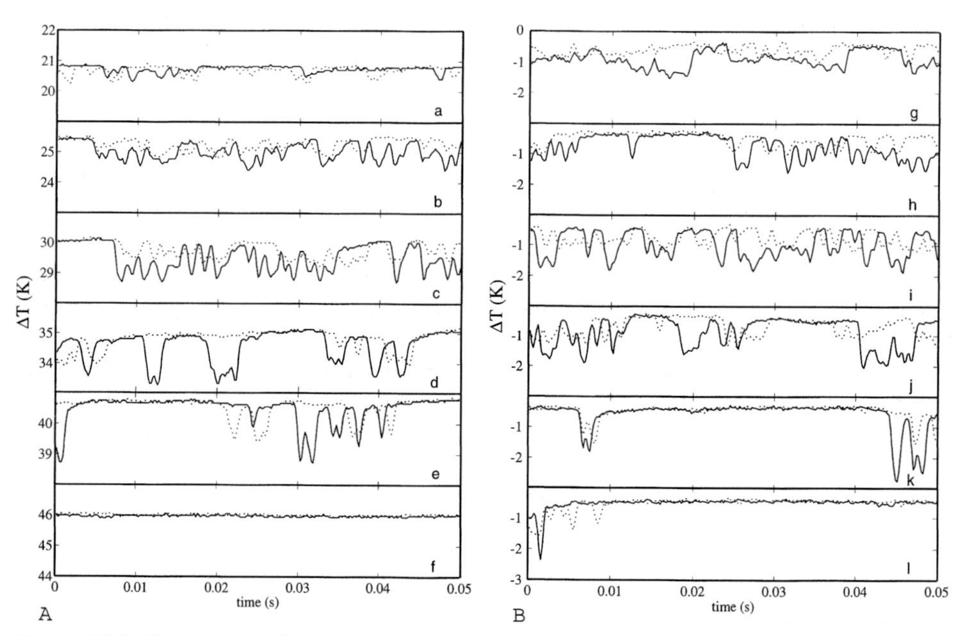

Figure 16.9: Temperature fluctuations in steady-state (A) and transient heating with 50 K/s (B); corresponding boiling curves see Figure 16.6; __TH5. ...TH6 (see Figure 16.8); heat fluxes q [W/cm^2]: a: 10.7, b: 17.8, c: 21.7, d: 17.6, e: 8.6, f: 6.1. Temperature difference ΔT_m [K]: g: 20.7, h: 24.9, i: 29.9, j: 35.2, k: 40.6, l: 45.9.

sequent period of 1 ms, the initial temperature level is reached again. In intensive nucleation (A, b), the temperature drops are more pronounced and so fast that "individual events" are not detected. In the CHF-region (A, c), a further increase in the amplitudes is observed and longer vapor contact times occur (upper level). This trend continues towards higher superheats (transition boiling; A, d and e), and in film boiling (A, f), no liquid contacts are observed.

To simplify the comparison of steady-state and transient results, the temperature fluctuations of the transient run (Figure 16.9B) are presented as follows: During the period of 0.05 s plotted in Figure 16.9, the temperature rise of a heating transient with nominal 50 K/s is about 2.5 K. This rise was obtained more precisely by determining an instantaneous arithmetic mean temperature difference ΔT_{m} from all 6 thermocouples and a linearization of ΔT_{m} with time during the total interval of 0.05 s. Then, the corresponding instantaneous mean temperature difference ΔT_{m} is subtracted from the measured signal to obtain only the temperature changes due to the wetting mechanism. The temperature differences ΔT_{m} listed in the caption of Figure 16.9 are those reached after 0.025 s of heating. These values enable to determine the corresponding average location on the 50 K/s-boiling curve in Figure 16.6.

Except for the signal-characteristics in nucleate boiling (Figure 16.9 A, a and B, g), no qualitative difference between the steady-state and the transient mode can be seen. Also standard statistical analyses yield no indication of a distinct difference in the temperature fluctuations. It can therefore be concluded – under the reservation that the presently available number of data are not high enough for a well established statistical statement – that no significant difference in the wetting mechanism exists under steady-state and transient conditions at the same wall superheat. A similar conclusion holds for the comparison with cooling transient temperature fluctuations, which are not presented in this report.

At least, if we look at the nucleation mechanism, the above conclusion is not surprising. Assume the bubble departure frequency be 100 Hz. Then, during the period between two bubbles, the temperature rise of the heater surface is only 0.5 K at the fastest transient of 50 K/s. Hence, in a first approach, we may argue that since vapor generation and wetting mechanism are much faster processes than the change of the average temperature with time during a transient, the vapor/liquid structure close to the heater surface region is not strongly influenced by the transient. It is therefore acceptable to rely also on the steady-state optical probe results if we search for a physical explanation of the heat transfer behavior in transient boiling.

16.5 Conclusions on the Physical Mechanism of Transient Boiling

A comparison of probe signals (cf. Figure 16.5) and microthermocouple signals enables an estimate of the nucleation site density. The cross-correlation function of signals detected by thermocouples 4 and 5 during nucleate boiling reveals no distinct maximum. Consequently, taking into account the size of a sensor and the distance between both, one may conclude that a microthermocouple can only detect nucleation sites within an area of appr. 0.01 mm^2 around it. Thermocouple 5, e.g., detects near CHF a long term average bubble frequency of 300 Hz (several and longer runs than depicted in Figure 16.9). Frequencies of this order of magnitude are measured by the optical probe at distances of appr. 0.03 mm above the surface (cf. Figure 16.5). If the probe distance is reduced to 0.01 mm, the measured frequencies decrease by a factor of about 3.

Even if, at that distance, the probe detects only one nucleation site (a further evaluation of all our measurements let us conclude that it is a higher number), it would detect bubbles from 3 nucleation sites at a distance of 0.03 mm. Consequently, the probe records at that distance the same number of bubbles as the microcouple on a surface area of 0.01 mm^2. The resulting minimum nucleation site density near CHF is 3×10^8 per m^2. Regarding the uncertainty of this rough estimate, we may conclude that the nucleation site density is between 10^8 and 10^9 per m^2. This is in the order of magnitude of Pinto et al's. [13] results, who observed 4×10^7 nucleation sites per m^2 in propane boiling at a 5 W/cm^2 heat flux, much below CHF.

The high density of active nucleation sites and the resulting highly turbulent two-phase boundary layer let us conclude that the strong increase of heat flux in heating transients and vice versa in a cooling process is mainly due to the intensive two-phase convection heat transfer from the wall to the bulk. Note that this is most likely not primarily because of a change in the two-phase structure (i.e. nucleation site density, bubble frequency, void fraction distribution etc.) near the wall, which is not significantly affected by temperature transients as discussed in the previous section. It is mainly because the instantaneous temperature gradient in the fluid at the heater surface is increased or decreased during the transient modes. Hence, the highly turbulent microconvective transfer of energy is intensified if it takes place in a temperature field with steeper gradient caused by transient heating and vice versa with weaker gradient in the cooling mode. Clearly at this stage, we have no direct experimental indication for steeper or weaker temperature gradients during the transients. Furthermore, our measurements of the two-phase structure near the wall and its behavior during transients are not precise enough to exclude an error in the above statement. Therefore, more precise measurements along the line presented in this report should be carried out to support this conclusion. Besides, an estimate reveals that unsteady-state heat conduction to the liquid cannot explain the strong increase of heat flux during a heating transient.

16.6 Conclusions

- Boiling curves measured under steady-state conditions with stepwise increasing and decreasing temperature do not exhibit a hysteresis.
- In transient heating, the heat flux is higher than in the steady-state case along the entire boiling curve. The higher the heating rate (K/s) the more pronounced is the heat flux increase. At the highest measured heating rate of 50 K/s, the CHF is by a factor 4 larger than the one under steady-state conditions. In transient cooling, the result is vice versa.
- Data for the mean vapor contact frequency at different heights above the heater surface, measured by a miniaturized optical probe, as well as data for temperature fluctuations on the surface, measured by microthermocouples, let us assume that modified heat transfer in transient cases is due to a convective effect in the boundary layer. The enhancement or reduction of heat transfer is most likely not primarily due to a change in the two-phase structure and turbulence near the wall, but mainly due to an increase or decrease of the instantaneous temperature gradient during the transients. However, these conclusions must be verified by more precise measurements.

Acknowledgements

The financial support of the "Deutsche Forschungsgemeinschaft" (DFG) is highly acknowledged. We also thank 3M Deutschland GmbH for providing us with test fluid FC-72.

References

[1] T.D. Bui, V.K. Dhir: *Transition boiling heat transfer on a vertical surface*. J. Heat Transfer **107** (1985) 756–763.
[2] M. Maracy, R.H.S Winterton: *Hysteresis and contact angle effects in transition pool boiling of water*. Int. J. Heat Mass Transfer **31** (1988) 1443–1449.
[3] W. Peyayopanakul, J.W. Westwater: *Evaluation of the unsteady-state quenching method for determining boiling curves*. Int. J. Heat Mass Transfer **1** (1978) 1437–1445.
[4] I. Rajab, R.H.S. Winterton: *The two transition boiling curves and solid-liquid contact on a horizontal surface*. Int. J. Heat Fluid Flow **11** (1990) 149–153.

[5] J. Blum, W. Marquardt: *Model-based design, control, and evaluation of transient boiling experiments.* This book (Chapter 15).
[6] S. Zhukov, V. Barelko, A. Merchanov: *Wave processes on heat generating surfaces in pool boiling.* Int. J. Heat Mass Transfer **24** (1980) 47–55.
[7] R. Hohl, H. Auracher, J. Blum, W. Marquardt: *Pool boiling heat transfer experiments with controlled wall temperature transients.* Proc. 2nd Europ. Therm. Sci. Conf., Rome, ETS Pisa **3** (1996) 1647–1652.
[8] J. Blum, R. Hohl, W. Marquardt, H. Auracher: *Controlled transient pool boiling experiments – methodology and results.* Proc. Eurotherm Sem. No. 48, Pool Boiling, Paderborn, Germany, ETS Pisa, 1996, pp. 301–310.
[9] R. Hohl, H. Auracher, J. Blum, W. Marquardt: *Identification of liquid-vapor fluctuations between nucleate and film boiling in natural convection.* In: Proc. Engng. Found. Conf. Convective Flow and Pool Boiling, Irsee, Germany, 1997.
[10] H. Auracher, A. Marroquin: *A miniaturized optical sensor for local measurements in two-phase flow.* Proc. 10th Brazilian Congr. of Mech. Engng., Rio de Janeiro, 1989, pp. 13–16.
[11] R. Hohl: *Mechanismen des Wärmeübergangs beim stationären und transienten Behältersieden im gesamten Bereich der Siedekennlinie.* Fortschritt-Berichte VDI, Reihe 3, Nr. 957, VDI Verlag GmbH, Düsseldorf, 1999.
[12] R. Hohl, H. Auracher, J. Blum, W. Marquardt: *Characteristics of liquid-vapor fluctuations in pool boiling at small distances from the heater.* Proc. 11th Int. Heat Transfer Conf., Kyongjy, Korea, Vol. **2** (1998) 383–388.
[13] A. Pinto, D. Gorenflo, W. Künstler: *Heat transfer and bubble formation with pool boiling of propane at a horizontal copper tube.* Proc. 2nd Europ. Therm. Sci. Conf., Rome, ETS Pisa **3** (1996) 1653–1660.

17 Boiling Delay and Boiling Fronts in Pure Liquids and in Liquid Mixtures

Jovan Mitrovic and Jürgen Fauser *

17.1 Introduction

It is well known that formation of a new phase within the homogeneous mother phase does not occur under saturation conditions. Pure liquids, for example, can reach considerable superheats prior to vapour formation (see e.g. [1]). Usually, boiling is initiated not in the bulk but at the interface, across which the superheated liquid interacts with its surroundings, say a heated wall. Once started at a certain point, the vapour generation spreads along the wall surface, forming a boiling front. This changes the heat transfer mode from single-phase convection of superheated liquid into a stable boiling mode. Hereby, a transition into nucleate as well as into film boiling is possible.

In comparison to common boiling, e.g. nucleate boiling, processes taking place during these transients have scarcely been studied [2, 3]. According to this situation, the corresponding knowledge is very limited. The experiments, the results of which are summarised in this article, shall contribute to understand these processes and the boiling events associated with them. Reported are the maximum attainable liquid superheats under heterogeneous conditions and the velocities of boiling fronts, propagating along heated wall surfaces. Further, heat transfer during transition from free convection of superheated liquid into stable boiling modes is determined, and the conditions required for a direct jump from liquid free convection into film boiling are specified. Also, representative results on the time evolution of wall temperature in the region of propagating fronts are analysed.

* Universität Paderborn, Institut für Energie- und Verfahrenstechnik, Lehrstuhl für Thermische Verfahrenstechnik und Anlagentechnik, Warburger Straße 100, D-33098 Paderborn, Germany

17.2 Test Conditions and Measuring Procedure

The experiments have been performed by using an apparatus schematically shown in Figure 17.1. The boiler and the condenser adjust the desired liquid state that is controlled via pressure and temperature. The outer surface of a tube, arranged horizontally in the test vessel, serves as heating surface. The storage tank allows variation of the level of the free liquid surface above the test tube. A small heater, arranged immediately below the test tube, at one end of its heated length, is used as boiling starter. All components of the apparatus interacting with the test fluid are made of stainless steel except the gaskets and the test tube.

The test tube (basic material copper, outside diameter 18 mm, wall thickness 2.5 mm, heated length 180 mm) is provided with a cartridge heater, and is constructed with cold ends that largely suppress axial heat losses. By means of a copper plating technique, three thermocouples are embedded in the wall of the copper tube, approximately 0.4 mm below its outer surface. The distances of the thermocouple junctions (TC 1, TC 2, and TC 3) from the end of the tube, where the boiling front is started, are 50, 90 and 130 mm, respectively. The tube is coated (galvano plating) with an about 10 μm thick layer of copper or nickel. Its surface is carefully polished; the roughness parameters of the surface, R_a (arithmetic mean deviation) and R_q (root-mean-square deviation), are

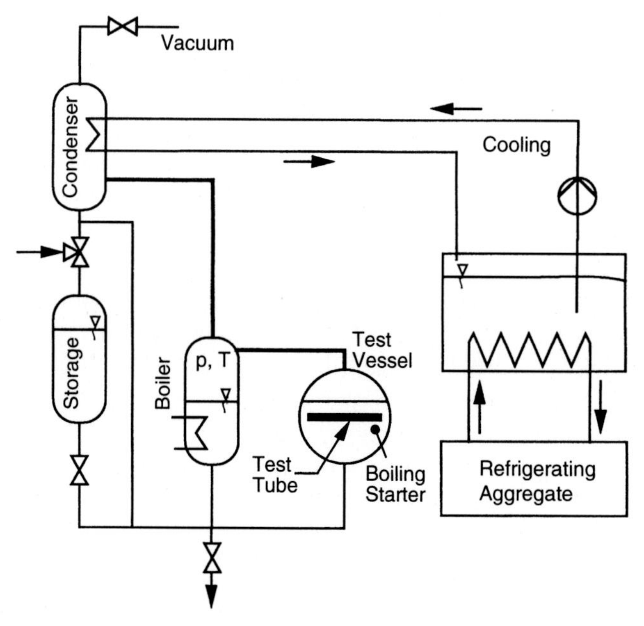

Figure 17.1: Flow sheet of the test facility. The test tube is horizontally arranged in the test vessel, about 40 mm below the free liquid surface.

0.050 μm and 0.070 μm for the copper, and of 0.017 μm and 0.023 μm for the nickel coating, respectively. The test tube is such arranged in the test vessel that the thermocouples are positioned near the upper tube vertex.

The experiments are performed with pure liquids (alcohols, heptane, Frigen R11) as well as with mixtures of methanol and isopropanol (50 mol-% methanol), and of methanol and heptane (20 mol-% methanol). They are undertaken along isobars at pressures of 0.01 to 0.35 MPa. The free liquid surface was usually 40 mm above the tube. During an experimental run, the tube is heated at a constant electric power. In different runs, different heat inputs are applied. Because of careful preparation of the tube surface, bubble formation is first suppressed and the liquid layer interacting with the test tube becomes more and more superheated as the time passes. The heat is transferred from the tube to the liquid by free convection only and after a certain time period – usually about one minute – a steady-state is reached. Then, a few vapour bubbles are generated on the starting heater. These bubbles rise, "touch" the test tube, and initiate boiling on it. The boiling manifests itself as a two-phase front propagating along the tube length.

The front velocity is determined from pictures, taken with a video camera. The tube length, observed with the camera, is about 120 mm long. The wall temperatures are normally recorded every 0.02 s, but some data are also taken every 0.007 s. This temperature, corrected for heat conduction within the tube wall, represents the actual temperature of the liquid layer immediate on the tube surface.

17.3 Maximum Attainable Liquid Superheat

The liquid superheat ΔT, reported in the following, is the difference between the actual temperature T of the liquid layer and the saturation temperature T_{sat},

$$\Delta T = T - T_{sat} \,. \tag{1}$$

The highest liquid superheats ΔT_{max} reached experimentally under free convection conditions are summarised in Table 17.1. These superheats are considerably lower than those at the spinodal, the limit of liquid superheat, which correspond to homogeneous nucleation and which are not reported here. This is in agreement with theoretical notions of various authors, who assume minute gas or vapour rests entrapped in holes on the heated surface to act as bubble nuclei, thus reducing the maximum superheat under heterogeneous conditions. For example, after Thormählen [4], the maximum liquid temperature is achieved, when the radius of the so-called critical bubble r_e equals half the diameter D of the mouth of such a hole:

$$r_e = \frac{D}{2} \,. \tag{2}$$

Table 17.1: Measured maximum liquid superheat ΔT_{max} and the radius of the critical bubble r_e, obtained by Equation (3). The radius r_e is largely independent of pressure; the average values $\bar{r}_e$ are determined over the whole pressure range (ΔT_{max} in K, r_e in μm). Mixture A: Methanol-Isopropanol (equimolar, x_M=0.5), Mixture B: Methanol-Heptane (x_M=0.2).

Surface	Copper								Nickel				Copper			
Fluid	Isopropanol		Methanol		Heptane		Frigen R11		Isopropanol		Frigen R11		Mixture A		Mixture B	
p, MPa	ΔT_{max}	r_e	ΔT_{max}	r_e	ΔT_{max}	r_e	ΔT_{max}	r_e	ΔT_{max}	r_e	ΔT_{max}	r_e	ΔT_{max}	r_e	ΔT_{max}	r_e
0.01	93.5	0.058	100.2	0.054	–	–	–	–	113.5	0.029	–	–	110.5	0.035	–	–
0.02	79.3	0.060	78.0	0.059	87.7	0.069	–	–	98.5	0.031	–	–	111.3	0.022	102.1	0.034
0.05	61.9	0.059	78.4	0.038	68.1	0.049	–	–	90.4	0.022	104.2	0.016	81.8	0.032	76.4	0.041
0.10	44.7	0.067	52.8	0.063	58.0	0.044	68.5	0.036	78.2	0.021	90.9	0.017	79.5	0.020	66.3	0.034
0.20	34.8	0.062	47.6	0.042	–	–	55.7	0.029	56.5	0.024	70.3	0.017	56.4	0.027	48.5	0.037
0.30	–	–	–	–	–	–	42.8	0.036	–	–	61.4	0.016	–	–	–	–
$\bar{r}_e$, μm		0.061		0.051		0.054		0.034		0.026		0.017		0.027		0.037

According to Equation (2), r_e is a function of surface roughness only and should be the same for all liquids. On the other hand, the critical bubble is in equilibrium with the superheated liquid and its size can be obtained from the equilibrium conditions:

$$r_e = \frac{2\sigma}{p_V - p_L}, \quad p_V = p_{sat} \exp\left(\frac{p_L - p_{sat}}{\rho_L RT}\right), \tag{3}$$

where σ, ρ_L, and R denote the surface tension, the liquid density, and the gas constant, respectively; p_V is the pressure inside the bubble, p_L the adjusted system pressure, and p_{sat} the common saturation pressure, fixed by the temperature T of the liquid.

We used Equation (3) to prove the idea stated in Equation (2). The values thus obtained are listed in Table 17.1. As seen, the theoretical values of r_e for a given fluid-surface pair are largely independent of the pressure. The average value $\bar{r}_e$, determined over the whole pressure range, is for pure hydrocarbons of the same order as the roughness of the heated surface. Consequently, smaller values are obtained at the nickel-coated than on the copper-coated test tube. However, R11 shows smaller values of $\bar{r}_e$ than the hydrocarbons. This might be ascribed to a better wetting ability of the refrigerant, resulting in a stronger liquid superheat.

When mixtures are used as test liquids, higher superheats are reached than with their pure components, which results in lower values of $\bar{r}_e$. The reason for this is not clear yet. One is inclined to speculate about the change of liquid structure in mixtures in the region interacting with the solid wall. Depending on these interactions, the adsorption layer, forming on the heating surface, is expected to assume properties that strongly differ from those in the bulk. This could lastly affect the attainable liquid superheat.

17.4 Velocities of the Boiling Fronts

17.4.1 Basics

At present, only few studies deal with the propagation of boiling fronts along heated surfaces [2, 3, 5]. Depending on the initial liquid superheat, Zhukov and Barelko [2] describe different mechanisms of front propagation, leading to generation of slow or fast fronts. As they report, at low superheats, slow fronts form; their velocity increases at a relatively small rate as the superheat rises. At a certain superheat, the front velocity increases almost jumpwise, reaching a value about twice as much as just before. It increases then continuously again with liquid superheat, but now the slope of the velocity is much steeper. This is the region of the fast fronts.

Avksentyuk and Ovchinnikov [3] treat fast fronts like shock waves. During the propagation of a fast front, the heating surface is considered to play only a minor role. The energy, required for vapour formation on the front boundary, is mainly provided by the superheated liquid. The expression for the front velocity u thus arrived at is:

$$u = \dot{m}\left(\frac{2}{\rho_L \rho_V}\right)^{1/2}, \tag{4}$$

where ρ_V and ρ_L are the densities of the phases. The mass flux $\dot{m}$ of vapour leaving the interface depends in a complex manner on pressure and liquid superheat.

17.4.2 Front velocities in pure liquids

17.4.2.1 Copper-coated tube surface

Slow moving fronts. In the region of slow fronts, the front velocity depends on liquid superheat and pressure. It rises with increasing both these parameters. For an easier illustration and comparison of the velocity data, the measured liquid superheat ΔT is divided by the superheat at the spinodal ΔT_{sp}, thus defining a nondimensional quantity S:

$$S = \frac{\Delta T}{\Delta T_{sp}} = \frac{T - T_{sat}}{T_{sp} - T_{sat}}. \tag{5}$$

For present purpose, values of T_{sp} (spinodal temperature) are taken from Avedisian [6] or are calculated by homogeneous nucleation theory.

As an example, Figure 17.2 shows the front velocity u vs. S for isopropanol as test fluid. The velocity increases with S almost linearly; its pressure dependence is largely contained in the quantity S. At higher values of this quantity, the front velocity changes abruptly, becoming larger. This is indicated in the figure by the arrows. Above these values of S, fast fronts develop, which are described in the next subsection.

The front velocities obtained with isopropanol can be approximated by:

$$u = m(S - S_0), \tag{6}$$

where $m = 2.59$ and $S_0 = 0.15$.

The velocity of slow moving fronts is fluid specific, but with other fluids the same qualitative behaviour as with isopropanol is observed. Also, Equation (6) may be used to represent their velocities. The values of the parameters m are 1.711, 3.130, and 1.667, whereas S_0 is determined as 0.159, 0.186, and 0.240 for methanol, heptane, and Frigen R11, respectively.

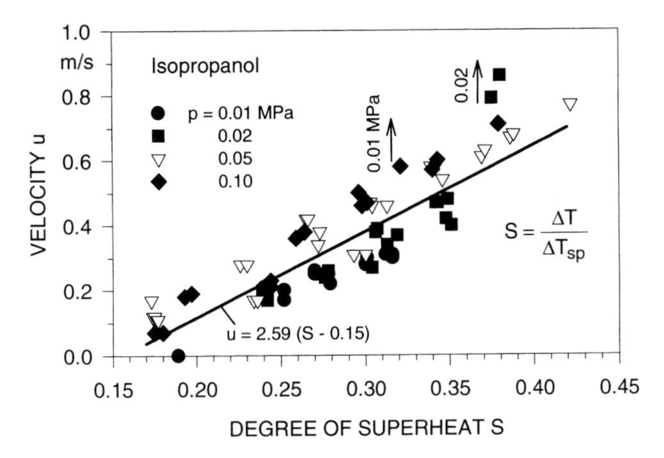

Figure 17.2: Velocity of slow propagating fronts, approximated by a simple function of S.

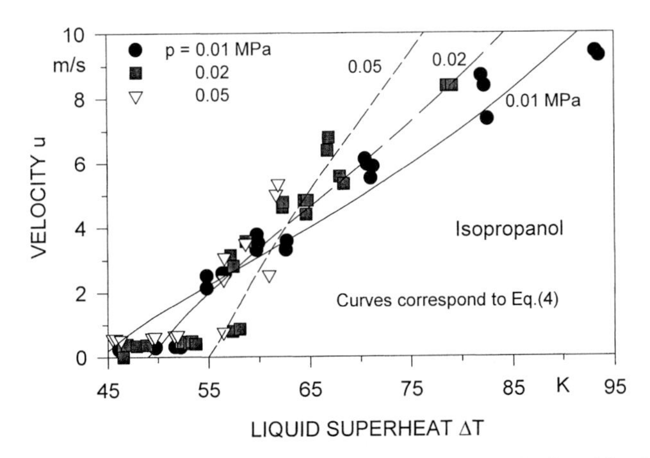

Figure 17.3: Velocity of fast propagating fronts. Curves are calculated by Equation (4).

Fast moving fronts. Figure 17.3 shows the velocities of fast propagating fronts obtained with isopropanol. The front velocity rises strongly with increasing superheat. At p=0.01 MPa, a value of about 10 m/s is reached at a superheat of 92 K. Although the heated surface is carefully prepared, no higher liquid superheats could be attained and, consequently, no faster boiling fronts could be generated. This is associated with spontaneous, uncontrolled nucleation on the heated surface. At a pressure higher than 0.05 MPa, the minimum liquid superheat for fast fronts to occur, could not be achieved.

The experimental velocities of fast propagating fronts are used to validate Equation (4). Values obtained from this expression are represented by lines in the figure. As may be seen, the calculated velocities u agree satisfactorily with the experiments. The same qualitative behaviour is also found with heptane.

However, larger deviations are observed with methanol, where the calculated values of u lie some 30% below the experimental data.

To compare the results of different fluids, we introduce a cavitation number Cav:

$$Cav = \frac{2(p_{\mathrm{sat}} - p_{\mathrm{L}})}{\rho_{\mathrm{L}} u^2}, \tag{7}$$

which measures the maximum pressure difference corresponding to liquid superheat in terms of kinetic energy of liquid at the velocity of the front.

In Figure 17.4, the number Cav is plotted over the molar rate of evaporation $\dot{n}$ in the front region. This flux is calculated for a flat interface according to Walton [7]. The symbols represent experimental data at $p = 0.02$ MPa. Different fluids give almost the same dependence of the cavitation number Cav on the evaporation rate $\dot{n}$. Fast boiling fronts establish with heptane only at high values of $\dot{n}$, whereas, with isopropanol, this is already the case at low evaporation rates. Under isobaric conditions, as shown in the figure, an increase of $\dot{n}$ leads to a decrease of Cav. This qualitative behaviour is also observed at other pressures. Higher pressures, however, result in larger cavitation numbers. This is illustrated by the lines representing averaged experimental data at different pressures.

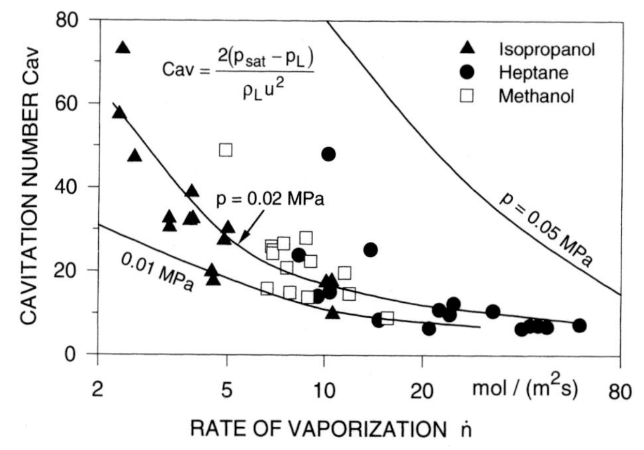

Figure 17.4: Nondimensionalised presentation of fast fronts for different test fluids. The symbols represent the data at $p = 0.02$ MPa; they scatter around a single curve. For reason of clearness, instead of direct data, curves are shown at other pressures.

17.4.2.2 Nickel-coated tube surface

Some results of measurements performed on the nickel-coated tube are shown in Figure 17.5. The front velocity of isopropanol is plotted over the degree of superheat at various pressures. As may be observed in this figure, the velocities of slow fronts on the nickel-coated test tube are at a low pressure almost equal to those on the copper surface. However, on average, the fronts are slower on the nickel than on the copper surface.

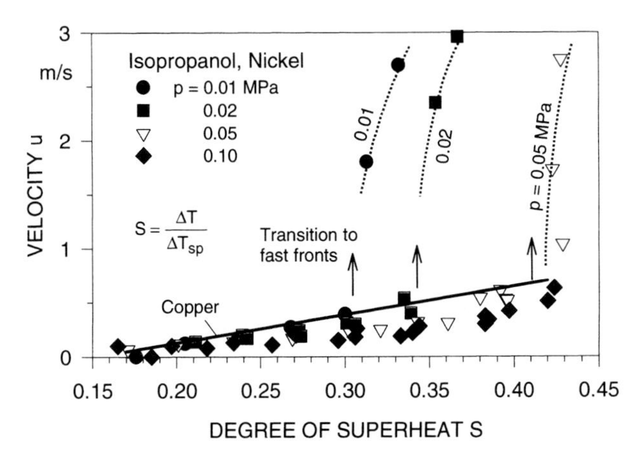

Figure 17.5: Front velocity on a nickel-coated surface. On average, slower fronts establish at the nickel than at the copper surface. Arrows indicate transition from slow to fast fronts.

The reason for this behaviour might lie in different adsorption properties of isopropanol on nickel and copper as well as in the different surface roughness of the two coatings. The nickel coating is smoother and a lower density of nucleation sites on the tube surface is generally expected, leading to a lower probability of bubble formation and thus to slower boiling fronts in comparison to the copper surface. Also the lower thermal conductivity of nickel should be mentioned in this connection.

As is further seen in Figure 17.5, the minimum superheat, necessary to generate boiling fronts on the nickel surface, coincides with that on the copper surface (see Figure 17.2). However, the superheat, above which fast fronts form, as indicated by arrows in Figure 17.5, is practically the same as for the copper-coated tube. When the refrigerant R11 is used as test liquid, the velocities of slow and fast fronts, observed within the range of our measurements (0.01 MPa $<p<$0.35 MPa, $S<0.6$), are almost the same on both surfaces.

To sum up, our results with pure liquids confirm the validity of Equation (4) and indicate that the formation of fast moving fronts is scarcely affected by the properties of the surface coating. Only in the region of slow fronts, there is an effect of heating surfaces on front velocity. This effect seems stronger at higher pressure, but it is generally weaker for liquids having higher wetting abilities.

17.4.3 Liquid mixtures

Besides single component liquids, also two binary liquid mixtures are examined in this study: an equimolar mixture of methanol and isopropanol that is in the thermodynamic view almost ideal, and a mixture of methanol and heptane with a molar fraction of methanol $x_M=0.2$. The phase behaviour of the latter mixture is far from ideal.

With these mixtures, the same qualitative behaviour of the front velocities is found as with pure liquids. There are slow fronts at low superheats, which become fast propagating above a particular superheat value. Representative results of the velocity measurements of slow moving boiling fronts in the methanol-isopropanol mixture are shown in Figure 17.6. As may be seen, the front velocities are almost independent of the pressure, as is the case with pure liquids. For reason of comparison, the data of the mixture components are reproduced by solid lines in this figure.

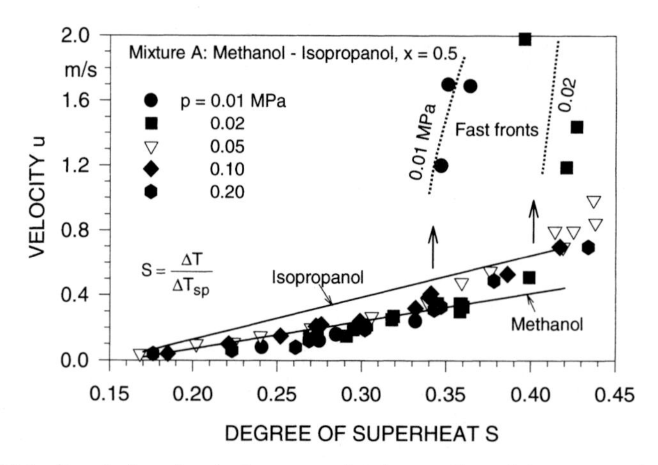

Figure 17.6: Velocity of slow fronts in an equimolar methanol-isopropanol mixture. Solid lines represent average velocity values of the pure components.

The observed front velocities of the mixture are partly less than those of pure methanol, the mixture component with the lower front velocity. This can be linked to mass transfer associated with phase transition in the mixture, resulting in a reduction of bubble kinetics and a slowing down of front propagation. With increasing superheat, the effect of mass transfer on bubble generation lowers and the front velocity reaches values that are between those of the pure mixture components. However, the processes of mass transfer are still present and are seemingly the reason why, in comparison to the pure components, the formation of fast fronts in mixtures is shifted towards higher superheats.

Basically, the same qualitative behaviour is observed with the methanol-heptane mixture. The difference worthy of mentioning is that in this mixture, the front velocities are always lower than those of the pure mixture components. This is probably due to stronger mass transport effects rooting in the larger difference of equilibrium compositions of the phases.

17.5 Wall Superheat and Heat Flux Histories in the Front Region

17.5.1 Transition into a stable boiling mode

As is known from numerous experiments, at low liquid superheats, boiling inception is followed by a transition into nucleate boiling. Figure 17.7 shows a typical temperature history taken directly (symbols) with the thermocouples in the tube wall during such a transition. The time scale is so adjusted that at $t=0$, the boiling front reaches the position of the thermocouple located in the middle of the test tube (TC 2).

For $t<0$, heat transport from the tube to the superheated liquid is by free convection, and the applied steady-state heat flux q_{st} of 11.9 kW/m^2 yields a superheat on the tube surface of about 37 K. The slightly lower temperatures of the thermocouples TC 1 and TC 3 are mainly due to variation of heat transfer conditions at the tube circumference. Note that the thermocouples are not arranged at the same circumferential position but at 30 deg apart, the mid-thermocouple (TC 2) being at the top of the tube. After the leading line of the boiling front has passed the positions of the thermocouples, a transition from free liquid convection into nucleate boiling occurs and the wall temperature decreases monotonously, approaching a steady-state value of 24 K. Together with the front velocity, the measured wall temperature represents a set of primary data for further evaluations.

The temperature distribution within the tube wall is calculated from the energy equation. Assuming the temperature profile in the wall to move along

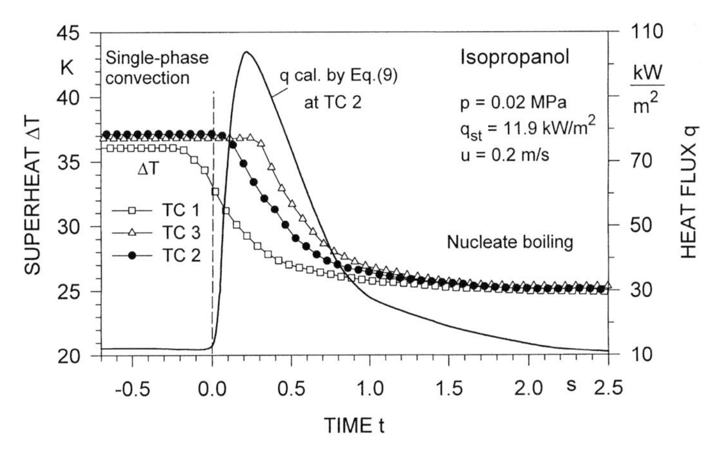

Figure 17.7: Temperature and heat flux history during transition into nucleate boiling. The thermocouples TC 1, TC 2, and TC 3 are 40 mm apart, TC 2 being in the middle of the heated tube length. The boiling front moves along the tube from TC 1 towards TC 3. The heat flux is represented by a solid line, averaging the calculated values.

the tube length at a constant (front) velocity u, and neglecting heat conduction in angular direction, this equation becomes:

$$\frac{1}{a}\frac{\partial T}{\partial t} - \frac{1}{u^2}\frac{\partial^2 T}{\partial t^2} = \frac{\partial^2 T}{\partial r^2} + \frac{1}{r}\frac{\partial T}{\partial r}, \tag{8}$$

where the axial co-ordinate is replaced by $c + ut$; c being a constant, r and a are the radial co-ordinate and the thermal diffusivity.

To solve the equation, initial and boundary conditions have to be specified. In our case, boundary conditions are the measured temperature $T = T(t)$ at the known radial position and the steady-state heat flux at the contact surface between the copper tube and the cartridge heater. Steady-state temperature distributions both prior to and for large time after boiling inception are taken as temporary conditions.

A numerical solution of Equation (8) yields the temperature on the tube surface ($r = r_a$) as well as the transient heat flux:

$$q(t) = -\lambda \left.\frac{\partial T}{\partial r}\right|_{r=r_a}, \tag{9}$$

where λ represents the thermal conductivity of the copper tube. The effect of the nickel coating is disregarded. To exclude possible side effects, caused by the finite tube length, only the temperatures taken in the middle of the test tube (TC 2) are used for further data processing.

The solid line in Figure 17.7 represents the calculated values of $q(t)$ based on the given temperature and velocity data. The local heat flux increases very rapidly after the leading front line has passed the position observed (TC 2) and reaches some 0.2 to 0.3 s later a maximum value of about 105 kW/m^2, which is several times higher than the stationary one. During this time, the front line has moved about 40 mm along the tube. After passing the boiling front, the heat flux decreases continuously, approaching again its steady-state value, at $t = 2.5$ s.

Further details about front behaviour and more general conclusions can be obtained when the instantaneous heat flux is plotted vs. the wall superheat. Some of the results gained with isopropanol during transition into stable boiling modes are visualised in Figure 17.8.

The curve A–C in the figure, obtained from our measurements, characterises the steady-state heat transfer by free convection with superheated liquid. The curve represents at the same time the initial states for transient measurements. Some such states are fixed by black dots. The lines originating at these dots describe the corresponding transient boiling behaviour for steady-state heat fluxes indicated on the curves. As may be seen from this figure, at low superheats, e.g. at point A, immediately after boiling inception, the heat flux increases very rapidly accompanied by a small change of wall superheat. Since more energy is carried away from the heating surface than is supplied to the tube, the surface temperature decreases steadily. The transient heat flux di-

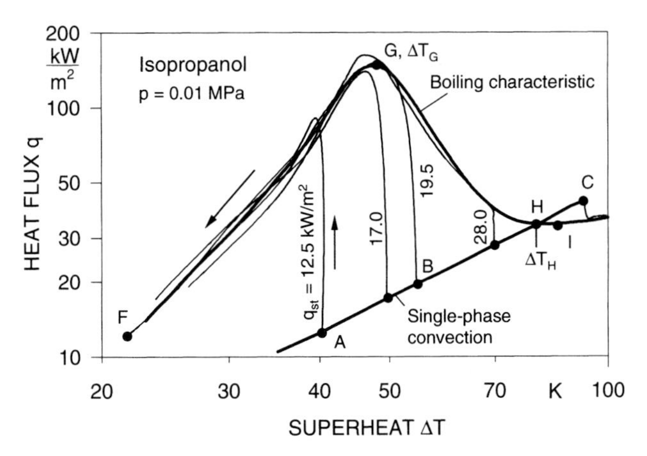

Figure 17.8: Transition from single-phase convection of superheated liquid into stable boiling modes. Initial states for transient experiments are fixed on the steady-state curve (A–C) for single-phase free convection.

minishes with decreasing superheat and a new steady-state is finally reached at point F, at which the superheat corresponds to the stable nucleate boiling mode.

The curve F–G–I, drawn through the transient boiling states at different heat fluxes, represents the "transient" Nukiyama curve. The value of the heat flux at point G represent the transient CHF. After boiling inception at initial liquid superheats higher than the corresponding critical superheat ΔT_G, the transient heat flux first increases, while the superheat decreases. Under these conditions, the mode of free convection heat transfer changes to the so-called transition boiling, before stable nucleate boiling is attained.

At point H, the heat transfer characteristic of single-phase convection and the transient Nukiyama curve cross each other. At initial liquid superheats higher than ΔT_H, e.g. at point C, the heat flux first decreases indicating that less energy is carried away from the tube than is supplied to it. Thus, a temperature increase of the test tube occurs, leading to transition into the state of stable film boiling.

The main features illustrated in Figure 17.7 and Figure 17.8 are also observed at other pressures and with other fluids, as reported earlier [8].

17.5.2 Relationship between boiling characteristic and front velocity

Further transient boiling characteristics (Nukiyama curves) are visualised in Figure 17.9 at different pressures of isopropanol. These characteristics are obtained from Equation (9) using the measured front velocities and temperature distributions. In addition, the figure contains the corresponding heat transfer characteristics of free convection (dashed lines) of the superheated liquid

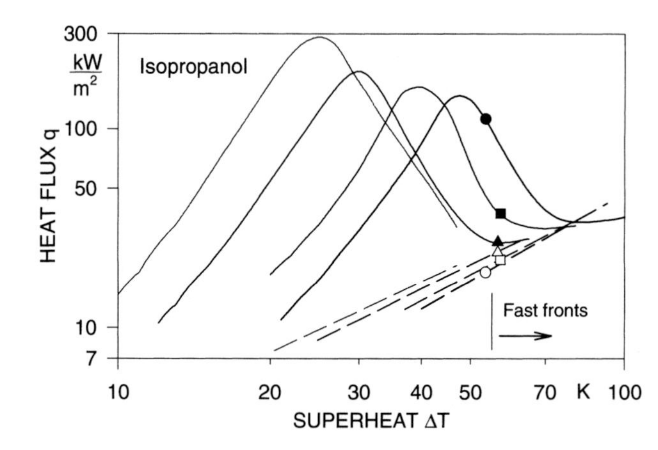

Figure 17.9: Region of fast propagating boiling fronts. Depending on pressure, the transition from slow to fast fronts also occurs at superheats below the Leidenfrost point.

phase. The empty symbols on the free convection curves mark the liquid superheats, above which fast boiling fronts establish. On the Nukiyama curves, these superheats are marked by black symbols. As mentioned above, no fast fronts could be generated in our experiments at the pressure $p=0.1$ MPa.

As is seen in Figure 17.9, the minimum superheat, needed for fast boiling fronts to form, are widely independent of pressure. This superheat almost coincidences with the Leidenfrost temperature at the pressure of 0.05 MPa. Thus, at this pressure, fast fronts form when boiling inception leads from single-phase convection directly into film boiling. Slow boiling fronts develop within the transition or nucleate boiling region. However, this is not the case at lower pressures. Namely, with decreasing pressure, the threshold value of liquid superheat to establish fast fronts is shifted towards the region of stable nucleate boiling.

The same qualitative behaviour is also observed with other liquids, as explained in more detail in an earlier paper [5]. Thus, one may conclude that the generation of fast fronts does not occur at the transition from single-phase convection into film boiling only, but also into the so-called transition or even into the nucleate boiling mode.

17.5.3 Temperature distribution near the front boundary

In the region of slow moving fronts, the temperature continuously decreases with time after boiling inception (see Figure 17.7). This is due to generation of individual bubbles in the front region. However, when fast fronts form, the temperature history shows a much complicated, an irregular behaviour. Sometimes, the temperature continuously increases after the front passes the thermocouple position. But, in most cases, first a temperature drop occurs. Some measure-

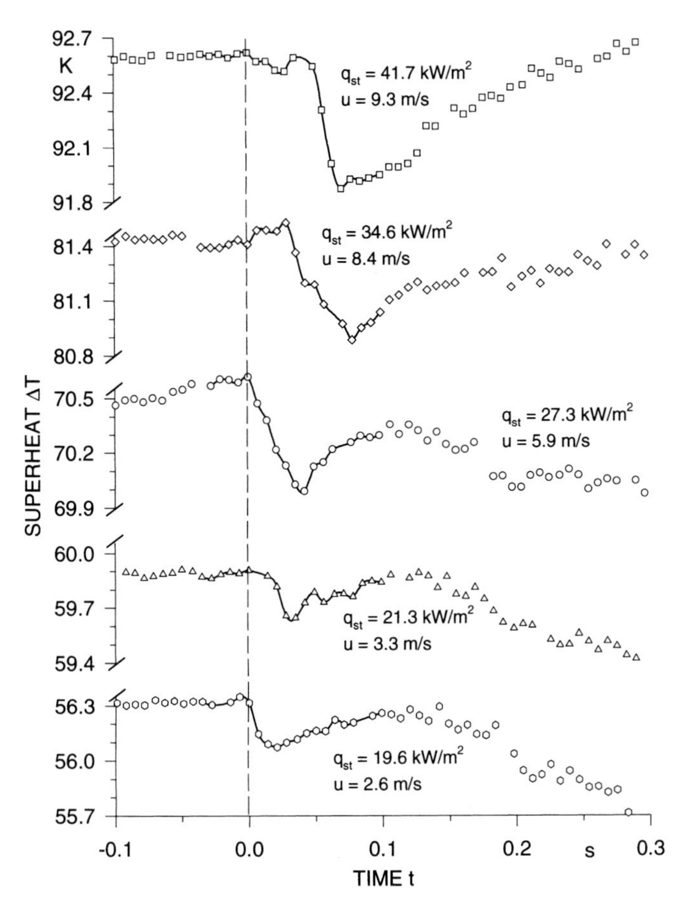

Figure 17.10: Superheat history in the case of fast propagating fronts. After the position of the thermocouple ($t=0$), the superheat minimum is followed either by a maximum or by a continuous rise of ΔT. Test fluid: isopropanol, $p=0.01$ MPa.

ments with isopropanol at a pressure of 0.01 MPa, illustrating the main features, are given in Figure 17.10. The experimental runs are ordered according to increasing steady-state heat fluxes and thus to increasing initial liquid superheats. Also, the corresponding front velocity u is recorded in each diagram.

The general tendency that may be followed from this figure is that the superheat ΔT shows a pronounced minimum followed by a soft maximum or a continuous rise. The latter establishes at a higher initial superheat. With increasing superheat, the temperature drop magnifies. It reaches, for example, at a superheat of 92.6 K (the most upper curve), a value of 0.7 K. However, in some experiments, at test conditions comparable to those in Figure 17.10, considerably less temperature drops are observed. This finding, supported by small irregularities along the superheat curves in the transition region, reflects the complex nature of boiling events occurring and their mutual interactions. In this connection, also the strong turbulence in the vapour phase and a possible partial rewetting of the heating surface should be mentioned.

The temperature increase in the transition region is due to the reduction of heat transfer by a thin vapour film left behind the sliding front. At steady-state heat fluxes up to about 30 kW/m^2, a change from single-phase convection to transition boiling occurs, and the vapour film, which forms on the heated surface, is not stable for a longer time. It collapses at about 0.15 s after the front leading line has passed the thermocouple. At heat fluxes above 30 kW/m^2, the vapour film remains stable and the temperature increases tending to steady-state film boiling.

At higher pressures, the temperature drop, occurring in the front, is not as strong and above 0.05 MPa even insignificant. The same behaviour and tendency show also methanol and heptane. However, the values of the temperature decrease with these fluids somewhat vary, e.g. with methanol, a drop of 1.5 K within 0.05 s was measured.

17.6 Summary

We studied experimentally boiling fronts propagating along the heated surface. The velocity of the fronts is shown to decisively depend on the initial liquid (wall) superheat. The energy necessary for the front propagation is withdrawn both from the superheated liquid and the heating wall. The wall heat flux in the front region reaches values that are much higher than the stationary ones.

The experiments have further revealed that, above a threshold superheat, a transition from single-phase free convection into the film boiling mode may occur. In this case, the transition always proceeds via fast propagating fronts.

Acknowledgement

The authors wish to acknowledge the financial support of the Deutsche Forschungsgemeinschaft (DFG).

References

[1] P. V. Skripov: *Metastable liquids.* John Wiley N.Y., 1974.

[2] S. A. Zhukov, V. V. Barelko: *Dynamic and structural aspects of the processes of single-phase convective heat transfer metastable regime decay and bubble boiling formation.* Int. J. Heat Mass Transfer **35** (1992) 759–764.

[3] B. P. Avksentyuk, V.V. Ovchinnikov: *Model of an evaporation front propagation in metastable liquid.* Proc. 2nd European Thermal-Sciences and 14th UIT National Heat Transfer Conference Rome, 1996, pp. 459–465.

[4] I. Thormählen: *Grenze der Überhitzbarkeit von Flüssigkeiten – Keimbildung und Keimaktivierung.* Fortschrittsberichte VDI, Reihe 3, Nr. 104, 1985.

[5] J. Fauser, J. Mitrovic: *Propagation of boiling fronts in superheated liquids.* Proc. Convective Flow and Pool Boiling Conference, Irsee, 1997.

[6] C. T. Avedisian: *The homogeneous nucleation limits in liquids.* J. Phys. Chem. Ref. Data **14** (1985) 695–728.

[7] A. J. Walton: *Three phases of matter.* Oxford University Press, New York, 1983.

[8] J. Fauser, J. Mitrovic: *Heat transfer during propagation of boiling fronts in superheated liquids.* Proc. EUROTHERM Seminar 48: Pool Boiling 2, Paderborn, 1996, pp. 283–290.

18 Vapor Condensation on a Thin Wire in a Transient Gas Expansion

Franz Peters *

Abstract

This paper deals with vapor condensation on a thin wire. Assuming the temperature of the condensed liquid, the wire allows the determination of the condensate temperature. In an analytical part, a remarkably simple solution is derived for this temperature. After that, the wire experiment in a piston-expansion tube is described. In a representative case of water vapor carried in nitrogen, it is shown that the predicted liquid temperature is in fact confirmed.

18.1 Introduction

A vapor diluted in a carrier gas may become saturated or even supersaturated by transient changes of state of the mixture. These may be temperature or pressure changes or combined ones occurring in an expanding gas. Homogeneous or heterogeneous nucleation and formation of a liquid phase follow. The rate of phase transition is determined by the rates, at which mass and heat (latent heat) are transferable to the condensation site and away from it, respectively. A vast body of literature exists on nucleation (e.g. Wilemski [1]), experimental as well as theoretical. Much less is available on the phase change at the droplet surface (e.g. Young [2]). Experimental work on small droplets is even scarce. As a forerunner to this work, we have investigated how small droplets grow and shrink in expansion and compression waves (Peters and Paikert [3], Rodeman and Peters [4]). We used a Mie-light scattering technique to resolve the growth and solved the according transfer equations for diffusional mass transfer and heat conduction. Very good agreement was found. A disadvantage of

* Universität Essen, Strömungslehre FB 12, Schützenbahn 70, D-45127 Essen, Germany

the droplet experiment is that the liquid temperature is not measurable, a crucial parameter in checking out transfer laws and coefficients. This was the incentive to study condensation onto a thin wire that assumes the temperature of the liquid film, which translates into the measurable electrical resistance.

In the next section, we show analytically how the liquid temperature simply depends on the ambient vapor conditions. Diffusion coefficient and thermal diffusivity enter the equation. With the experimental temperature at hand, it is then possible to determine these coefficients. It is not necessary to measure the mass flux or to know the temperature profile about the cylinder.

In the experimental section, we demonstrate how the wire experiment was integrated into the piston-expansion tube (pex-tube) previously developed for droplet experiments. An example of an experimental run is given with evaluation of the temperature for different diffusion coefficients.

18.2 Model of Condensation on a Thin Wire

Condensation of a vapor releases a considerable amount of latent heat, which has to be transferred away from the condensation site, otherwise the process quenches. In a condenser, there is normally a condensation surface like a wall, which absorbs the latent heat effectively. Whenever that is not the case, the heat has to be transferred back into the gas surrounding the condensation site. These situations may be reduced to the standard cases listed in Figure 18.1.

In each case, a vapor diffuses against the r-axis through a carrier towards the wall, cylinder or sphere surface where it condenses. The latent heat is conducted backwards along the r-axis into the gas space. The diffusion is driven by a vapor pressure difference between surroundings and liquid surface ($p^{\mathrm{v}}_{\infty}-p^{\mathrm{v}}_{\mathrm{l}}$), while the heat follows an opposed temperature gradient ($T_{\mathrm{l}}-T_{\infty}$).

Figure 18.2 depicts the according thermodynamic situation in the log (p)-T phase diagram. At the liquid surface, we have phase equilibrium, i.e. the respective state lies on the coexistence curve. The ambient state has a smaller

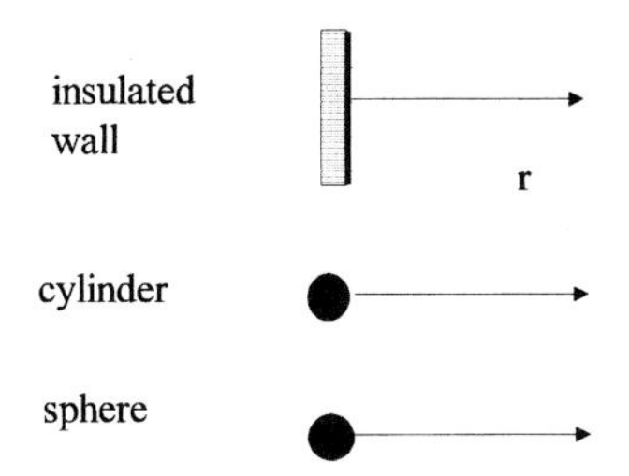

Figure 18.1: Standard situations of condensation, in which the latent heat needs to be conducted away from the condensation site into the ambient gas.

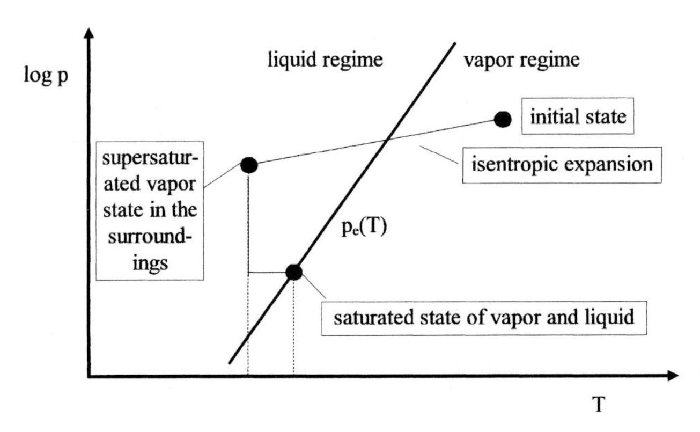

Figure 18.2: Thermodynamic situation for vapor condensation in a phase diagram.

temperature and a higher vapor pressure as compared to the equilibrium state. The ratio of actual vapor pressure over equilibrium vapor pressure is called the supersaturation:

$$S = \frac{p_\infty^v}{p_e(T_\infty)} . \tag{1}$$

When there is droplet condensation in the supersaturated state, case three of Figure 18.1 applies. In the present case, we study condensation onto a wire (cylinder) where we keep the supersaturation small enough such that droplet condensation by homogeneous nucleation is just prevented. Our way to reach the supersaturated state is by the indicated fast isentropic expansion realized by a moving piston (pex-tube) enlarging a given volume.

The existing work on droplet condensation presumes stationary transport processes. For fixed ambient conditions, this leads to the well known Maxwell's law saying that the radius squared grows proportionally with time. Interestingly, this restriction is not necessary, which is noticed in dealing with the cylinder case, in which no stationary solution exists to start with.

The 1-D transfer equations for temperature T and concentration c are:

$$\frac{\partial T}{\partial t} = a\left[\frac{\partial^2 T}{\partial r^2} + \frac{C}{r}\frac{\partial T}{\partial r}\right] \tag{2}$$

$$\frac{\partial c}{\partial t} = D\left[\frac{\partial^2 c}{\partial r^2} + \frac{C}{r}\frac{\partial c}{\partial r}\right] . \tag{3}$$

The parameter $C=0$ refers to the insulated wall, $C=1$ to the cylinder and $C=2$ to the sphere. The thermal diffusivity $a = k/\rho c_p$ and the diffusion coefficient D of the vapor in the carrier are taken to be constant in the range of operation (k is the thermal conductivity, ρ the total density and c_p the specific heat; c is the vapor density with respect to the total density). The liquid film wetting the solid cylinder has the radius r_l (a liquid cylinder is not conceivable). The ambient temperature is T_∞, the total pressure p_∞, the vapor pressure p_∞^v and the partial pressure of the carrier gas p_∞^g. Since there is no flow, the total pressure is constant and in the ideal gas:

$$c = \frac{R}{R^v} \frac{p^v}{p_\infty} . \tag{4}$$

Here R is the gas constant of the mixture which is put constant like a and D. Then Equation (3) may be written in p^v:

$$\frac{\partial p^v}{\partial t} = D \left[\frac{\partial^2 p^v}{\partial r^2} + \frac{C}{r} \frac{\partial p^v}{\partial r} \right] . \tag{5}$$

As mentioned above, the vapor pressure at the liquid/vapor interface equals the equilibrium pressure:

$$p^v(r_l) = p_e(T_l) \tag{6}$$

so that at the film ($r = r_l$)

$$\frac{\partial p^v}{\partial t} = \frac{\partial p_e(T_l)}{\partial t} = \left. \frac{\partial p_e}{\partial T} \right|_{T=T_l} \frac{\partial T_l}{\partial t} \tag{7}$$

with the differential equations

$$\frac{\partial T_l}{\partial t} = a \left[\frac{\partial^2 T}{\partial r^2} + \frac{C}{r} \frac{\partial T}{\partial r} \right] \tag{8}$$

$$\left. \frac{\partial p_e}{\partial T} \right|_{T=T_l} \frac{\partial T_l}{\partial t} = D \left[\frac{\partial^2 p^v}{\partial r^2} + \frac{C}{r} \frac{\partial p^v}{\partial r} \right] . \tag{9}$$

Observing that gradients of T and p have opposite signs, it then follows that

$$a T_l + \frac{D p_e(T_l)}{\left. \frac{\partial p_e}{\partial T} \right|_{T=T_l}} = \text{const.} \tag{10}$$

The only reference temperature is T_∞. Therefore, $C_1 = T_\infty$ and to satisfy the equation $C_2 = p_\infty^v$. Then:

$$\left. \frac{\partial p_e}{\partial T} \right|_{T=T_l} = \frac{D}{a} \frac{p_\infty^v - p_e(T_l)}{T_l - T_\infty} . \tag{11}$$

This result is remarkable. It shows that a system determined by T_∞, p_∞^{v}, $p_e(T)$, a and D fixes T_l, the liquid temperature. T_l is not only the same in all three cases, it is also constant in time. The solutions for the temperature and concentration profiles do not need to be known to find T_l. They may be stationary or not.

The solutions for the profiles are scattered in the literature (see: Tautz [5], Eckert and Drake [6], Özisik [7], Grigull und Sander [8]). In the sphere case, it turns out that the solution in fact converges rapidly against a stationary solution (with a stationary mass flux).

In the cylinder case, there is no stationary solution. The temperature distribution starts as a step function and converges for long times against a horizontal. The analytical solution is a combination of Bessel functions with x as the integration variable, τ as the Fourier number and R as the normalized radius:

$$R = \frac{r}{r_l}; \quad \tau = \frac{at}{r_l^2} \tag{12}$$

$$\frac{T - T_\infty}{T_l - T_\infty} = 1 - \frac{2}{\pi} \int_0^\infty \frac{J_0(x)Y_0(Rx) - J_0(Rx)Y_0(x)}{x(J_0^2(x) + Y_0^2(x))\exp(\tau x^2)} dx \,. \tag{13}$$

An analytical solution of this integral is not known. Its numerical evaluation is difficult because of its improper character for $x \to 0$. It is less an effort to solve the differential equations numerically with a finite difference method (Graßmann [9]), the result of which appears in Figure 18.3. The temperature curves rise with τ approaching the horizontal for $\tau \to \infty$.

The analytical approach via Equation (13) is still useful for determining the heat flux at the wall, which is the product of temperature gradient at the wall and heat conductivity k. First, the derivation of Equation (13) with respect to R at $R=1$ is taken:

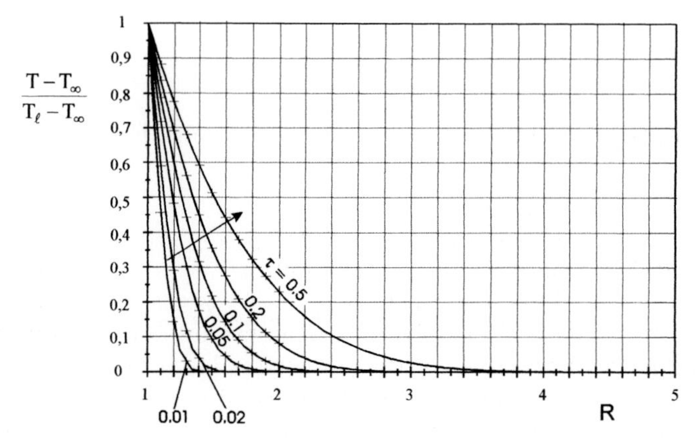

Figure 18.3: Temperature distribution in the vicinity of a cylinder.

$$\frac{\partial}{\partial R}\frac{T(r,t)-T_\infty}{T_l-T_\infty}\bigg|_{R=1} = -\frac{2}{\pi}\int\limits_0^\infty \frac{J_0(x)Y_1(x)-J_1(x)Y_0(x)}{(J_0^2(x)+Y_0^2(x))\exp(\tau x^2)}\mathrm{d}x\,. \tag{14}$$

Second, the Bessel functions are linearized for $x<x^*<<1$. This way, one dodges around the improper character of the integral, i.e.:

$$\frac{\partial}{\partial R}\frac{T(r,t)-T_\infty}{T_l-T_\infty}\bigg|_{R=1} = \frac{1}{\ln x^*}-\frac{2}{\pi}\int\limits_{x^*}^\infty \frac{J_1(x)Y_0(x)-J_0(x)Y_1(x)}{(J_0^2(x)+Y_0^2(x))\exp(\tau x^2)}\mathrm{d}x\,, \tag{15}$$

which is then easily evaluated by a mathematical program (like Mathcad). The instantaneous heat flux density at the film is then computed from:

$$\dot{q} = -k\frac{\partial}{\partial R}\frac{T-T_\infty}{T_l-T_\infty}\bigg|_{R=1}\frac{T_l-T_\infty}{r_l}\,. \tag{16}$$

For the temperature gradient, one gets for example –1.236 at $\tau=0.5$ and –0.166 at $\tau=10^5$. To become aware of the order of magnitude, we condense water at $\tau=10^5$, T_l–T_∞ =10 K; r_l=2.5 10^{-6} m; k=0.023 W/mK. We get for the heat flux density 1.5 10^4 W/m^2. The corresponding mass flux density follows by division through the latent heat of 2.5 10^6 Nm/kg. The result is 6 10^{-3} kg/m^2 s.

The presented analysis allows to evaluate the experiments and draw conclusions on the transfer coefficients. The given space of this paper is not sufficient to discuss the assumptions and implications in more detail. We just mention the Stefan flow problem, which means that the diffusive flux has to be corrected because the carrier gas is stagnant and not counterdiffusing. We went through the analysis of this problem starting with the fundamental equations as given in Landau und Lifschitz [10]. The outcome is that a correction is only required for small times which are experimentally irrelevant.

18.3 Experimental Set-Up

A pex-tube (piston-expansion tube) as outlined in Figure 18.4 is used to subject a vapor diluted in a carrier gas to a fast isentropic expansion from an initially undersaturated to a saturated state (Figure 18.2). The pex-tube was originally developed for the investigation of homogeneous nucleation with subsequent droplet growth. The droplets were detected by a laser Mie-scattering method included in Figure 18.4 but not used in the present experiment. Along with the droplet work the pex-tube was described several times. A comprehensive description is found in Peters and Rodemann [11].

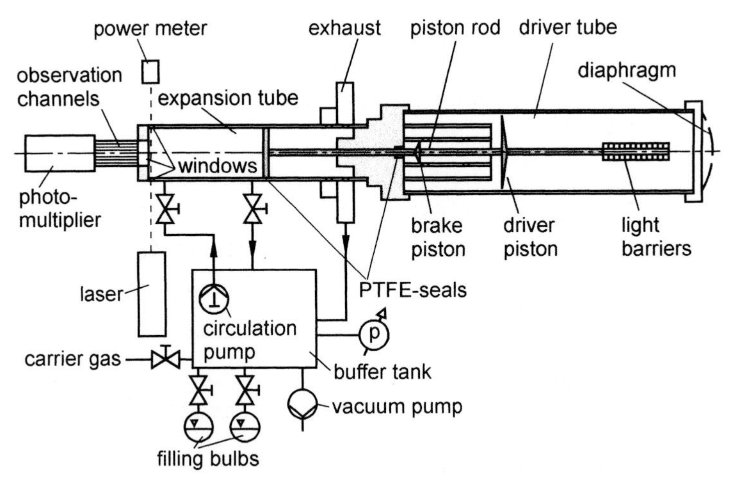

Figure 18.4: Pex-tube set-up. The thin wire is fixed where the laser passes the expansion tube. Light scattering is not used in the present experiments.

Here, we restrict to a brief outline and elucidate the adaption to the wire experiment. The desired vapor/gas mixture is prepared in the buffer tank and filled into the expansion tube. Initial partial pressures and temperatures are very well known before the expansion piston is displaced to the right enlarging the initial volume. Everything in the driver tube serves to facilitate the rapid piston motion, which takes less than 10 ms from start to stop. The state at the tail of the expansion stays constant for up to 50 ms. The state parameters are selectable by the initial conditions, the piston travel and the type of carrier gas. The supersaturation is chosen such that the critical value for substantial nucleation is not reached, while heterogeneous condensation onto the wire takes place.

The wire is simply from a spool of hot wire material (tungsten) of 5 μm diameter. It is mounted 6 mm away from the end wall of the tube between two prongs 48 mm apart as shown in Figure 18.5. The heat capacity of the wire is so small that the wire temperature follows the expansion temperature closely. At the expansion tail where the gas temperature levels off and the liquid film builds up, the wire temperature becomes certainly equal to the liquid temperature. Since the wire temperature is uniquely coupled to its resistance, the measurement of the latter determines the liquid temperature. We measure the resistance by a voltage drop at a constant current of 1 mA. Of course, a calibration of the wire is inevitable. A perfect linearity between temperature and resistance was found at coefficients around 2 K/Ω.

Two problems associated with the wire method were investigated. The first is the heating of the wire by the current during calibration. A Nußelt number calculation revealed that an overtemperature of 0.1 K is to be expected, which is negligible in view of the overall accuracy of a few tenth of a degree. The second is the heating of the wire by the prongs: while the wire follows the

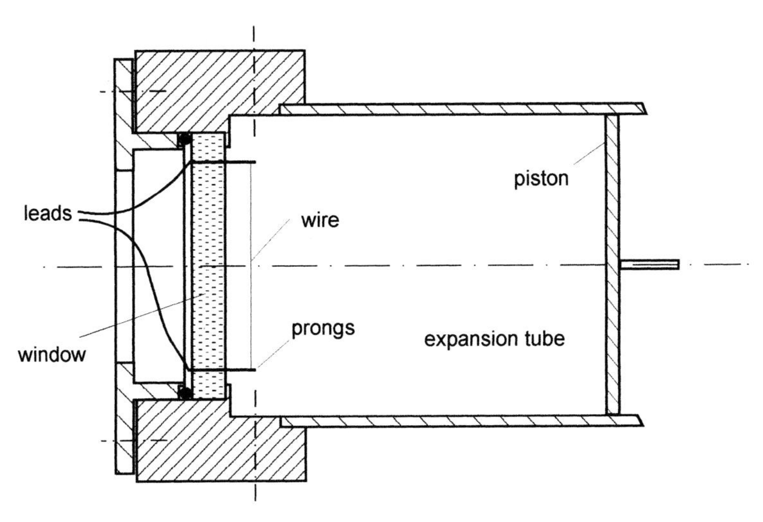

Figure 18.5: Mounting of the wire at the end wall of the expansion tube.

cooling of the gas, the prongs do not releasing their internal heat into the wire. Consequently, the wire assumes a temperature distribution instead of a constant temperature. The analysis of this problem (for the temperature distribution see Eckert and Drake [6]) made clear that, with the enormous aspect ratio of the wire of almost 10^4, the mean wire temperature may be off by a few tenth of a Kelvin at the most.

18.4 Experiments and Results

So far, experiments were conducted with water vapor in nitrogen as a carrier gas. The mixture was prepared in the buffer tank. The tank was evacuated and subsequently filled with water vapor and nitrogen up to desired partial pressures. The total pressure was established around atmospheric, while the initial temperature was elevated above atmospheric by electrical heating. The gas sample was circulated through tank and expansion tube to assure uniform mixing. After that, the expansion tube was separated from the rest by valves and the piston was moved by operating the driver unit. Two signals were monitored: the expansion pressure and the voltage drop at the wire. An analog/digital converter card in conjunction with a PC served to store and read out the signals. Figure 18.6 provides an example.

The upper curve represents the expansion pressure starting at 752.4 Torr and levelling off at 404.3. At closer inspection, one notices that three curves are plotted on top of each other with little differences. They associate with dry nitrogen with a water vapor partial pressure of 14.08 Torr and 20.21 Torr, respec-

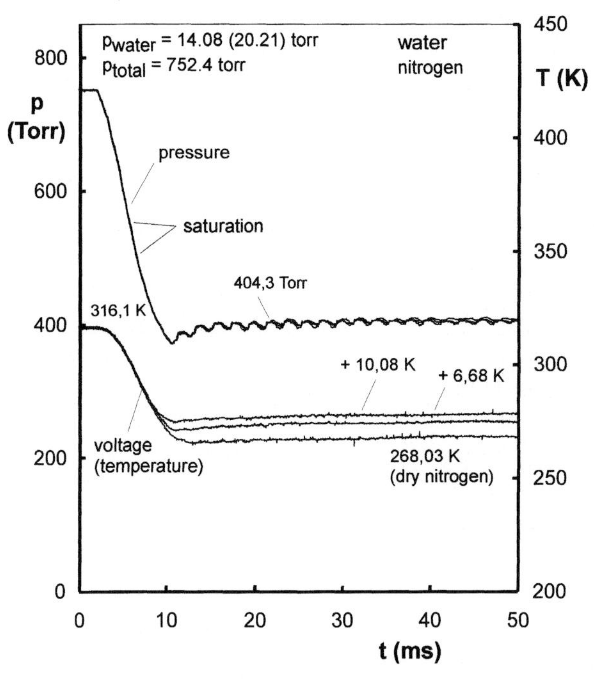

Figure 18.6: Experimental run showing expansion pressure (top) and voltage drop across the wire converted to temperature (bottom).

tively. The lower curves show the wire voltage converted to temperature, which can be read off at the right ordinate. It is seen that the temperature follows the expansion closely and, furthermore, the wire temperature takes a constant value as predicted by Equation (11). It is indicated where saturation is reached in the two cases. Shortly after saturation, the three curves deviate from each other. The bottom one belongs to dry nitrogen. When 14.08 Torr of water vapor are added, the temperature rises by 6.68 K, and when 20.21 Torr are added, the rise is 10.08 K.

18.5 Model Calculations

The main measuring result is the liquid temperature T_l at known ambient conditions. In order to calculate T_l from Equation (11), the equilibrium vapor pressure of water is taken from Sonntag und Heinze [12]. It should be noted that we are in principle in the solid regime of water where the vapor pressure of ice applies. However, since ice does not form on the time scale of the experiment the extended vapor pressure of the liquid matters. For $a = k/\rho c_p$, the density follows from the equation of state. The heat transfer coefficient k is determined according to Wilkes mixing rule (Reid et al. [13]). c_p, R and k are also calculated using mixing rules as described in Rodemann [14].

The diffusion coefficient follows from Fuller's equation (Reid et al. [13]):

$$D = 9.1546 \cdot 10^{-3} \frac{T^{1.75}}{p} , \tag{17}$$

in which p is the total pressure in Torr, while D comes out in cm^2/s. With these data, T_l may be computed from Equation (11). We use the T_l as the reference temperature in computing D and a. One could also use T_∞ or a mean value. It does not matter much because a and D have similar temperature dependencies and only the ratio of a and D counts. Alternatively Pruppacher's model is used (Pruppacher and Klett [15]), which is not quite correct as it holds for air rather than nitrogen. Another alternative is the Chapman/Enskog model (Hirschfelder et al. [16]), the application of which is given in Paikert [17].

In calculating the two cases, we get the following results:

	1. case	2. case
initial vapor pressure:	20.21 Torr	14.08 Torr
ambient temperature:	268.03 K	268.03 K
liquid temperature:	278.11 K	274.71 K
liquid temp. with Fuller model	278.62 K	275.21 K
liquid temp. with Pruppacher model	278.46 K	275.1 K
liquid temp. with Chapman/Enskog model	277.9 K	274.7 K

The result is quite satisfying. The analytical prediction of the liquid temperature is closely confirmed for all three diffusion models.

18.6 Conclusions

The liquid film temperature was analysed when condensation takes place on an insulated wall, a sphere or a thin wire. It turned out that this temperature is identical in all cases and a constant when the ambient conditions are constant. Notably, this result is independent on the temperature and concentration profile solutions whether they are stationary (sphere) or not (cylinder).

Transient condensation of water vapor on a thin wire was studied by means of a pex-tube. The liquid temperature was evaluated and compared to the analytical result. Good agreement was found. Different vapor diffusion models entering the calculations worked equally well.

The work will proceed with a detailed study of the temperature dependency of the diffusion coefficient since the models are different in this respect. Then we will deal with molecular transport, i.e. Knudsen numbers about and greater than one. In this range, we hope to get experimental verification of the socalled Schrage correction, which accounts for the fact that the transport is a nonequilibrium process.

Acknowledgement

This project is part of the "Schwerpunktprogramm Transiente Vorgänge in mehrphasigen Systemen mit einer oder mehreren Komponenten" funded by the Deutsche Forschungsgemeinschaft (DFG). The progress of our project was delayed for various reasons. Roughly 50% of the research has been completed.

References

[1] G. Wilemski: *The Kelvin equation and self-consistent nucleation theory.* J. Chem. Phys. **103** (1995) 1119–1126.
[2] J.B. Young: *The condensation and evaporation of liquid droplets at arbitrary Knudsen number in the presence of an inert gas.* Int. J. Heat Mass Transfer **36**(11) (1993) 2941–2956.
[3] F. Peters, B. Paikert: *Nucleation and growth rates of homogeneously condensing water vapor in argon from shock tube experiments.* Exp. Fluids **7** (1989) 521–530.
[4] T. Rodemann, F. Peters: *Measurement and interpretation of growth of binary droplets suspended in a water/n-propanol/nitrogen mixture by means of a pex-tube.* Int. J. Heat Mass Transfer **40**(14) (1997) 3407–3417.
[5] H. Tautz: *Wärmeleitung und Temperaturausgleich.* Verlag Chemie Weinheim, 1971.
[6] E.R.G. Eckert, R.M. Drake: *Analysis of Heat and Mass Transfer.* McGraw-Hill, New York, 1972.
[7] M.N. Özisik: *Heat Transfer.* McGraw-Hill, New York, 1985.
[8] U. Grigull, H. Sander: *Wärmeleitung.* Springer-Verlag, Heidelberg, 1990.
[9] A. Graßmann: *Zusammenstellung der Lösungswege für stationäre und instationäre Wärmeleitung in den Grundgeometrien Platte, Zylinder und Kugel.* Studienarbeit am Lehrstuhl für Strömungslehre der Universität Essen, 1997.
[10] L.D. Landau, E.M. Lifschitz: *Lehrbuch der Theoretischen Physik VI.* Akademie Verlag, Berlin, 1966.
[11] F. Peters, T. Rodemann: *Design and performance of a rapid piston expansion device for the investigation of droplet condensation.* Exp. Fluids **24** (1998) 300–307.
[12] D. Sonntag, D. Heinze: *Sättigungsdampfdruck- und Sättigungsdampfdichtetafeln für Wasser und Eis.* VEB Deutscher Verlag für Grundstoffindustrie Leipzig, 1982.
[13] R.C. Reid, J.M. Prausnitz, B.E. Poling: *The Properties of Gases and Liquids.* McGraw-Hill, New York, 1987.
[14] T. Rodemann: *Homogene Keimbildung und Tröpfchenwachstum in binären Dampfmischungen.* Diss. Universität Essen, Shaker Verlag, Aachen, 1997.
[15] H.R. Pruppacher, J.D. Klett: *Microphysics of Clouds and Precipitation.* D. Reidel, Dordrecht, 1980.
[16] J.O. Hirschfelder, C.F. Curtiss, R.B. Bird: *Molecular Theory of Gases and Liquids.* John Wiley and Sons, New York, 1964.
[17] B. Paikert: *Untersuchung der Kondensation und Verdampfung ruhender Tropfen in Gas-Dampf-Gemischen mit Hilfe eines Stoßwellenrohres.* Diss.Universität Essen, 1990.

Numerical Methods

19 Dynamics of Forced and Self-Excited Instabilities in Heterogeneously/ Homogeneously Condensing Flows through Nozzles and Steam Turbine Cascades

Günter H. Schnerr, Michael Heiler, Stephan Adam and Gunter Winkler *

Abstract

About 10% of the total electrical net power of steam turbine plants is produced in the last stages. Since condensation is observed here, it is obvious that the understanding and accurate modelling of steady two-phase flows and the control of unsteady two-phase phenomena is of great importance. Therefore, we concentrate on the following topics:

- self-excited instabilities in condensing nozzle and cascade flows,
- interaction of heterogeneous/homogeneous condensation,
- nucleating blade to blade flows,
- rotor/stator interaction in nucleating flows,
- interaction of vortex shedding with condensation.

Self-excited instabilities may cause frequency jumps up to a factor of two, whereas additional heterogeneous condensation may be able to suppress unsteadiness. The temperature wakes caused by rotor/stator interaction significantly influence the two-phase flow regime. The same is true for vortex shedding.

* Universität Karlsruhe (TH), Fachgebiet Strömungsmaschinen, Kaiserstr. 12, D-76128 Karlsruhe, Germany

19.1 Introduction

Steady and unsteady phenomena in steam turbine stages are subject of intensive research for several reasons. Such studies were conducted using nozzles and cascades of airfoil sections in atmospheric indraft wind tunnels or more recently in steam tunnels. Aside from homogeneously condensing steam or steam components, the sources of failure and losses in turbine bladings are manyfold, ranging from heterogeneous condensation caused by mechanical or chemical impurities of the steam or by the formation and movement of coarse water, by the movement of water films along bladings, by rotor/stator interactions with entrainment of water films to droplet breakup at stator trailing edges. Therefore, the purpose of the current research reported is to provide a better understanding of steady and unsteady two-phase flow phenomena in cascade flows.

19.2 Physical Model and Numerical Scheme

The homogeneous nucleation process is modelled by the classical nucleation theory of Volmer, Frenkel, and Zel'dovich [1]. Dealing with steam/carrier gas mixtures at very low partial vapor pressures, the Hertz-Knudsen law is well suitable for modelling the droplet growth. Simulating pure steam flows in turbines requires droplet growth expressions, which cover the whole Knudsen number range and take the temperature as a driving potential into account, e.g. see Gyarmathy [2], Young [3, 4], or Peters [5]. We herein use the model given by Gyarmathy because calculations using this formula show good agreement with nozzle flow experiments, which are particularly relevant to LP steam turbine conditions. Furthermore, this expression provides the lowest CPU-time requirements. The heterogeneous condensation caused by particles or chemical impurities of the mixture is modelled using the heterogeneous nucleation theory of Kotake and Glass [6–8]. After the particles are fully covered by liquid, the same droplet growth model as used for the homogeneous condensation is applied.

The numerical scheme applied solves the time-dependent 2-D Navier-Stokes equations in conservation form coupled to four additional equations for the homogeneous and heterogeneous condensation and two additional equations for turbulence closure [9]. Air is considered to be a perfect gas, whereas steam is treated as an imperfect gas. The system of equations for the two-phase flow is then solved with a MUSCL-type finite volume method on structured body-fitted grids applying explicit second order accurate time integration. More details of the physical model and numerical scheme can be found in [10–12].

19.3 Instabilities and Bifurcation

19.3.1 Steam/carrier gas mixture flows

19.3.1.1 Symmetric nozzle A1

The investigation of a slender nozzle with parallel outflow (nozzle A1 – isentropic outflow Mach number $M_{e,is}=1.2$) yielded a yet unknown oscillation mode. In wind tunnel experiments, we observed this new bifurcation where the flow oscillation is unsymmetric in perfect symmetric nozzles. Discontinuous bifurcation of the frequency dependence on the water vapor content in the supply, i.e. on the heat addition, leads to a tremendous increase of the frequency by a factor of two or more (Figure 19.1). Experimental investigations (Figure 19.2) and numerical solutions of the time-dependent Euler equations (Figure 19.3) show a very good agreement in the structure of the 2-D flow field and the dynamics of the process, e.g. the frequency and the stability limit. The numerical schlieren pictures in Figure 19.3 of one flow oscillation of each branch reveal the background for this frequency bifurcation. The sequence at left shows an oscillation of the lower branch, indicated by the left arrow in Figure 19.1. It represents, what is to be expected, a symmetric oscillation. The right sequence of an oscillation in the upper frequency regime shows the new unsymmetric mode. A complex system of oblique shocks forms periodically and moves through the nozzle opposite to the main flow direction. More details of the bifurcation in symmetric nozzles can be found in [13–16].

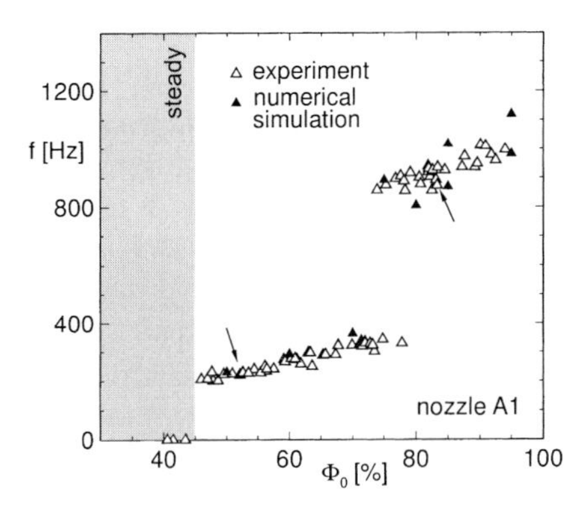

Figure 19.1: Frequency dependence on the reservoir relative humidity Φ_0 in nozzle A1, steam/carrier gas mixture; experimental results compared to numerical simulation; lower branch: experiment 287.5 K$\leq T_{01} \leq$305 K, numerical simulation 300 K; upper branch: experiment 285.5 K$\leq T_{01} \leq$295 K, numerical simulation 285 K, 295 K.

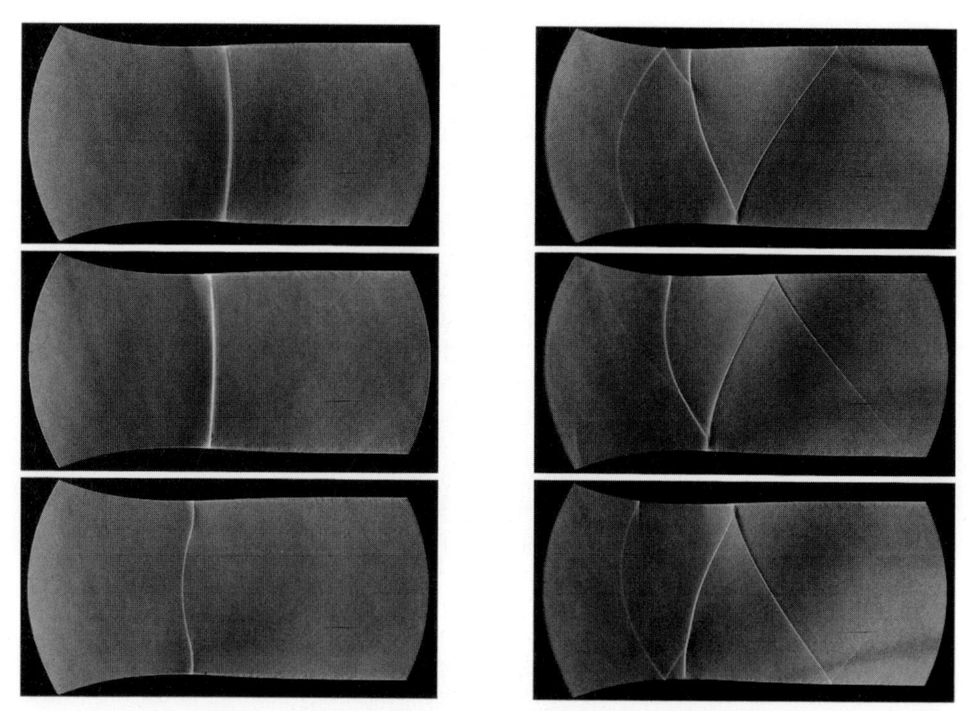

Figure 19.2: Experiment-bifurcation in nozzle A1, schlieren pictures of symmetric and unsymmetric flow oscillation, steam/carrier gas mixture; left: symmetric mode, f=225 Hz, T_{01}=292.8 K, p_{01}=0.1 bar, Φ_0=51.6%; right: unsymmetric mode, f=905 Hz, T_{01}=288.2 K, p_{01}=1.0 bar, Φ_0=82.0%.

19.3.1.2 Shifted nozzle A1A

In order to prove that the bifurcation is not restricted to a limited class of nozzle shapes, especially not to nozzles with very slender, and therefore very sensitive supersonic regions, we investigated a plane nozzle where the walls are formed by circular arcs. Furthermore, to prove that these phenomena are not dependent on a perfect symmetric geometry, the upper wall of the nozzle is shifted by 40°. The numerical schlieren simulations of Figure 19.4 show clearly the formation of the two typical wave patterns for identical conditions of the oncoming flow. Independent of whether the flow is quasi-symmetric ('symmetric') or not, the flow accelerates to supersonic exit Mach numbers $M_e>1$. Due to the supersonic boundary condition at the exit of nozzle A1A, the bifurcation is clearly a local instability in the most sensitive transonic throat region, which again supports the validity and accuracy of our experimental findings.

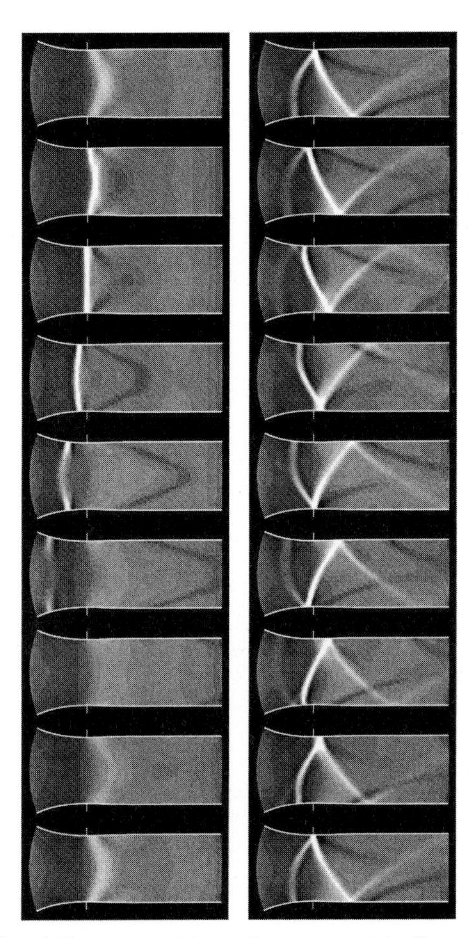

Figure 19.3: One cycle of the symmetric and unsymmetric flow oscillation in nozzle A1 for identical reservoir conditions $T_{01}=295$ K, $p_{01}=1.0$ bar, $\Phi_0=90.0\%$, steam/carrier gas mixture; numerically simulated schlieren pictures; left: symmetric $f=447$ Hz, right: unsymmetric $f=1068$ Hz; flow from left to right, time increases from top.

19.3.1.3 Linear cascade A1A

In contrast to boundaries formed by nozzle walls, the periodicity of cascades ahead and behind the bladings accounts for the interaction of the pressure and suction sides of the blades, which is a quite different and more sensitive configuration in bifurcation dynamics. As a first step, the blades are simply formed by circular arcs of 12.5% thickness ($d/c=0.125$). Without stagger, the cooling rate at the equivalent throat position is the same as that of the nozzle A1 where the bifurcation was detected first. Figure 19.5 shows one period of the 'symmetric' oscillation mode with moving normal shocks at left and one cycle of the unsymmetric oscillation with an additionally upward and downward moving oblique

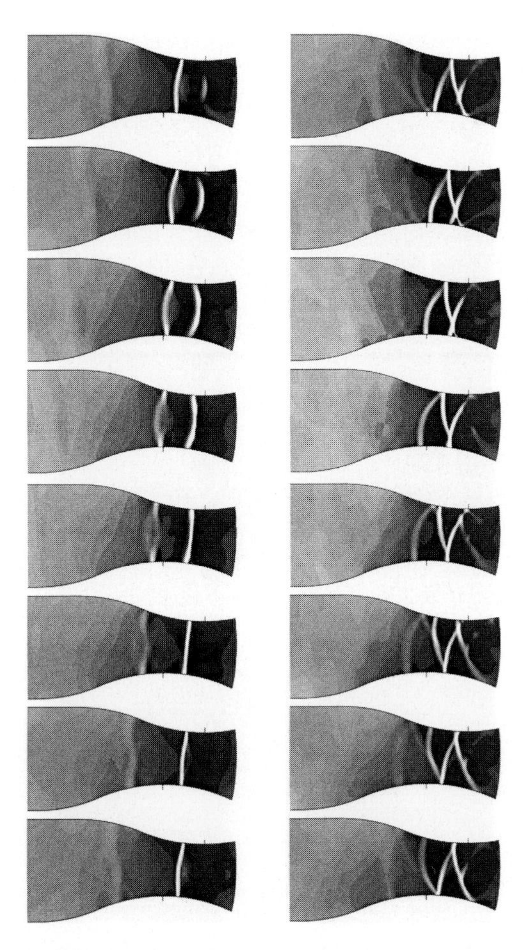

Figure 19.4: One cycle of the 'symmetric' and unsymmetric flow oscillations in the shifted nozzle A1A, shift angle $\beta=40^\circ$, for identical reservoir conditions $T_{01}=295$ K, $p_{01}=1$ bar, $\Phi_0=90\%$, steam/carrier gas mixture; numerically simulated schlieren pictures; left: 'symmetric' $f=854$ Hz, right: unsymmetric $f=1082$ Hz; flow from left to right, time increases from top.

shock system (the time starts from top) for a stagger angle $\beta_s=100^\circ$. As the main result, we can conclude that the existence of bifurcations is not prevented by periodic boundary conditions, i.e. in axial cascades. For more details, especially of the dispersed structure of the condensate of the different oscillation modes, see [10, 17, 18].

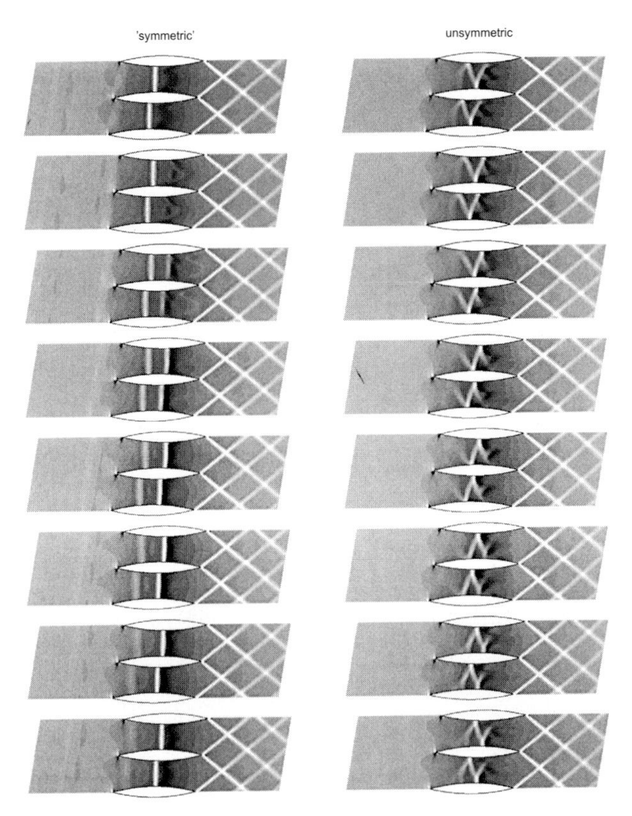

Figure 19.5: One cycle of the 'symmetric' and unsymmetric flow oscillations in the staggered cascade A1A, shift angle $\beta_s = 100^\circ$, for identical reservoir conditions $T_{01} = 295$ K, $p_{01} = 1$ bar, $\Phi_0 = 100\%$, steam/carrier gas mixture; numerically simulated schlieren pictures; left: 'symmetric' $f = 835$ Hz, right: unsymmetric $f = 1202$ Hz; flow from left to right, time increases from top.

19.3.2 Pure steam flows

19.3.2.1 Symmetric nozzle Ba1

So far, the investigation of the appearance of the unsymmetric mode has been restricted to steam/carrier gas mixture (moist air) flows. To prove that this instability is not dependent on this special fluid, we now change it. Because of its importance in turbine flows, we here consider pure steam. The drastical increase of the vapor pressure in conjunction with the use of pure steam has three consequences. At first, due to the lower Knudsen number, the Hertz-Knudsen droplet growth model is not accurate anymore. Therefore, the Gyarmathy model is applied. Secondly, using the classical nucleation theory, an additional isothermal correction has to be added and, furthermore, especially for

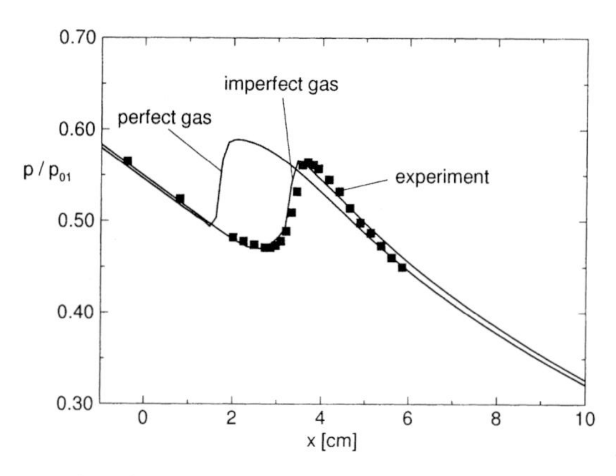

Figure 19.6: Pressure ratio p/p_{01} along the axis in nozzle BA1, comparison between numerical simulation of perfect, imperfect gas behavior and experiment; pure steam, reservoir conditions T_{01}=373.15 K, p_{01}=0.7839 bar.

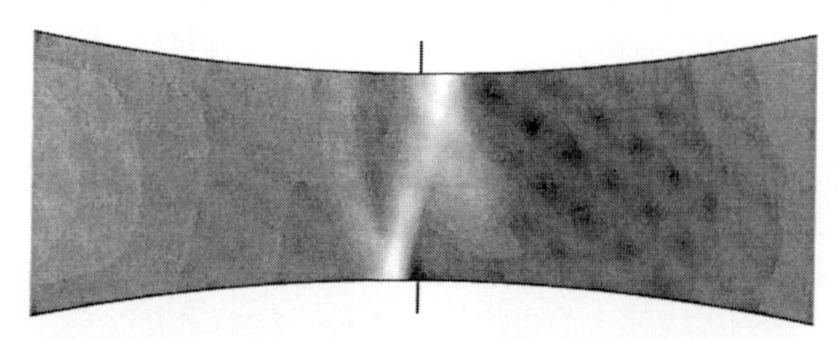

Figure 19.7: Unsymmetric flow oscillation in nozzle BA1; numerically simulated schlieren picture, pure steam, reservoir conditions T_{01}=363.0 K, p_{01}=0.7839 bar, frequency f=1820 Hz; flow from left to right.

higher pressures, imperfect gas effects have to be taken into consideration. Figure 19.6 depicts that adapting all these considerations allows an accurate modelling of the condensing pure steam flow.

Decreasing the reservoir temperature T_{01} from T_{01} = 373.15 K (see Figure 19.6) to T_{01}=363.0 K causes the flow to become unsteady showing the well known symmetric oscillation mode with a frequency f=1100 Hz [19]. However, when a small unsymmetric disturbance is introduced, the flow field rearranges and the unsymmetric oscillation mode appears and remains there. Figure 19.7 shows one instantaneous picture of the unsymmetric oscillation cycle. The frequency increases from f=1100 Hz to f=1820 Hz. Therefore, we conclude that the unsymmetric oscillation mode also appears in pure steam flows, which further underlines the fact that this instability is caused by the interaction of the homogeneous condensation with the transonic flow.

19.4 Heterogeneous/Homogeneous Condensation in Nozzle A1

For the investigation of the influence of heterogeneous condensation, we use the nozzle A1. Again a steam/carrier gas mixture is considered. This facilitates the understanding of the effects in conjunction with heterogeneous condensation because the effects due to purely homogeneous condensation in the nozzle A1 have already been investigated in detail.

19.4.1 Steady flow

Figure 19.8 depicts the steady heterogeneously/homogeneously condensing flows for four different particle concentrations n. On the left side the schlieren pictures, on the right side the corresponding pressure distributions p/p_{01}, homogeneous nucleation rates J_{hom} and condensate mass fractions g/g_{max} along the nozzle axis are shown. For a particle concentration $n_1=3\cdot10^{10}$ 1/m^3, no heterogeneous condensation effect can be detected. In increasing the particle concentration to $n_2=1\cdot10^{13}$ 1/m^3, a noticeable amount of heterogeneous condensate is formed. Nevertheless, the overall flow field is not influenced significantly. The shock position moves only a little bit downstream. However, when the particle concentration is further increased to $n_3=1.25\cdot10^{13}$ 1/m^3, heterogeneous condensation starts to dominate. The shock is weakened and delayed far downstream compared to the case of purely homogeneous condensation. For $n_4=5\cdot10^{13}$ 1/m^3, the flow is completely dominated by heterogeneous condensation. No noticeable homogeneous condensate could be formed. Due to the strong subsonic heat addition, a continuous pressure distribution, with only some weak waves, develops. This variation shows that the effect of the heterogeneous condensation is to weaken or to suppress completely the phenomena arising in connection with spontaneous heat addition (e.g. formation of shocks, observation of unsteadiness) due to homogeneous condensation.

19.4.2 Unsteady flow

Now, we show the aforementioned effect that heterogeneous condensation can suppress unsteadiness. For this reason, we consider a purely homogeneously condensing flow with reservoir conditions (see caption of Figure 19.9) causing a symmetric oscillation with a frequency $f=410$ Hz. Figure 19.9 depicts the effect of adding particles. Until the particle concentration attains a value of $n=1\cdot10^{11}$ 1/m^3, the frequency remains nearly constant. Then, further increase in the particle concentration results in a decrease in the frequency to approximately a value of=200 Hz for $n=1\cdot10^{12}$ 1/m^3. For $n=2\cdot10^{12}$ 1/m^3, finally the flow becomes steady due to the strong subsonic heat addition.

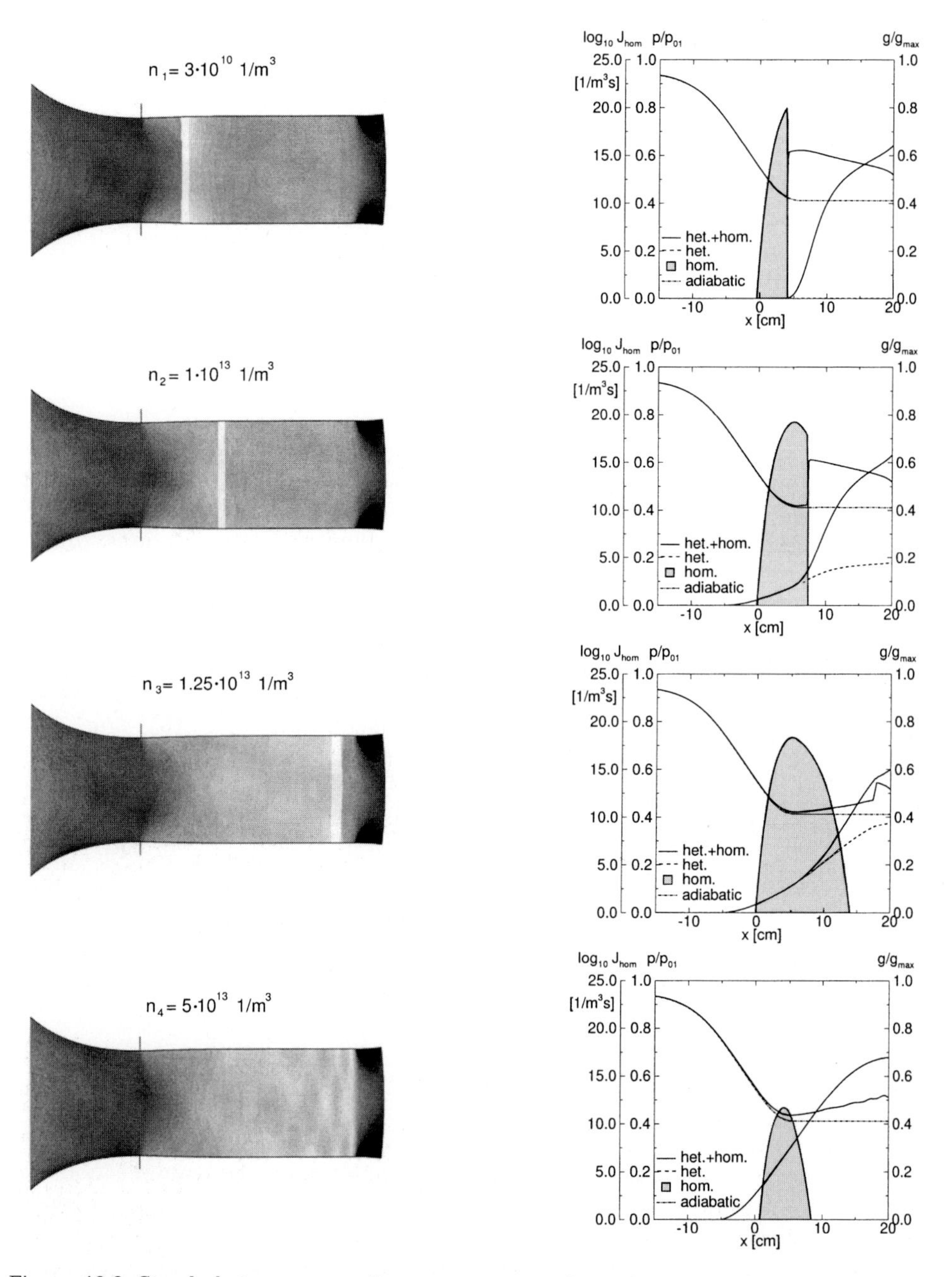

Figure 19.8: Steady heterogeneous/homogeneous condensation in nozzle A1 for different particle densities n, particle radius $r_p = 10^{-7}$ m, contact angle $\Theta = 30°$, reservoir conditions $T_{01} = 295$ K, $p_{01} = 1.0$ bar, $\Phi_0 = 35\%$, steam/carrier gas mixture; left: numerical simulated schlieren pictures, flow from left to right; right: pressure ratio p/p_{01}, condensate mass fraction g/g_{max} and nucleation rate J along the nozzle axis.

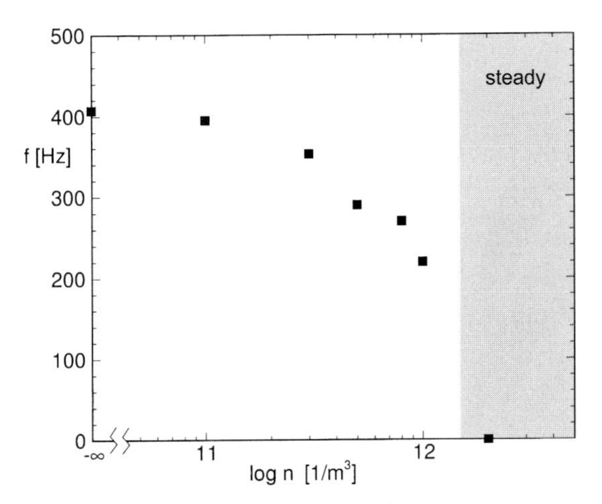

Figure 19.9: Unsteady heterogeneous/homogeneous condensation in nozzle A1, frequency dependence on particle density n, particle radius $r_p = 10^{-7}$ m, contact angle $\Theta = 30°$, reservoir conditions $T_{01} = 295$ K, $p_{01} = 1.0$ bar, $\Phi_0 = 75\%$, steam/carrier gas mixture.

19.5 Steady Flow in Low Pressure Steam Turbine Cascade L

For the investigation of the condensing blade to blade flow, the fifth stage stator blade of an operating turbine has been chosen. In this stage, high losses and hence the primary nucleation has been detected [20]. The flow is assumed to be inviscid. Therefore, the trailing edge flow is modelled using a cusp. The shape of the cusp and the base pressure are determined from the control of the momentum balance according to Denton [21]. Figure 19.10 shows the numerically simulated schlieren pictures for three different reservoir and downstream conditions. The main difference between the cases L1, L2 and L3 is the pressure ratio p_{01}/p_2, the thermodynamic state at the inlet is approximately the same. The increase of the pressure ratio causes a delay downstream of the maximum of the nucleation and therefore of the condensate formation. The effect of this behavior can be seen in Figure 19.10. For case L1, a shock due to condensation appears at the suction side, which coincides with the pressure side shock coming from the trailing edge. Case L2 and L3 are characterized by a continuous pressure increase (indicated by the bright area) due to condensation, which is weaker and delayed further downstream for case L3 (highest pressure ratio). Figure 19.11 compares the static pressure distribution at the blade surface for the different inlet and outlet conditions. Obviously, high accurate inviscid Euler simulations are not able to reproduce stronger shock boundary layer interactions with local separation bubbles (Figure 19.11, left). However, good agreement with experiments is observed for weak shocks (Figure 19.11, middle, right). More details including the nucleation and condensate formation can be found in [10, 22].

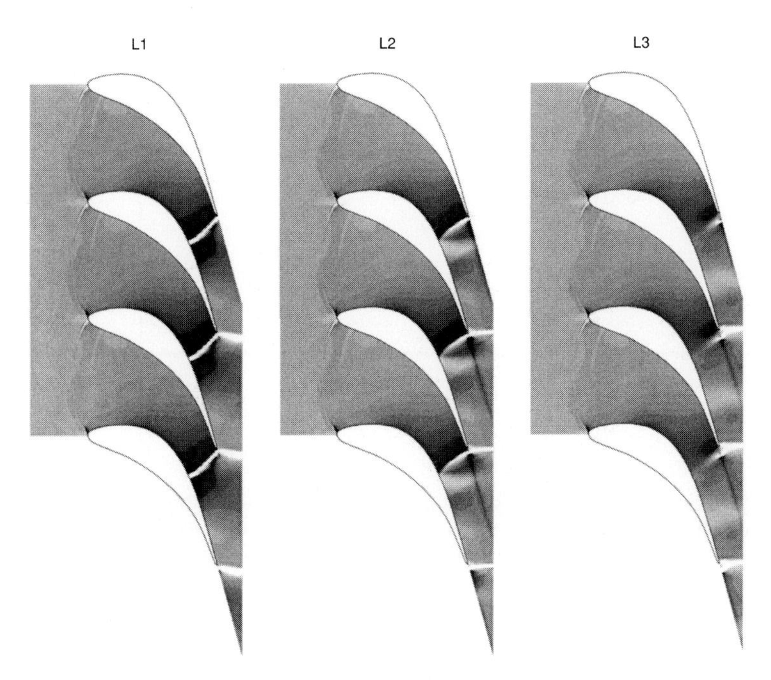

Figure 19.10: Steady condensing flow in the LP steam turbine cascade L for three different reservoir and downstream conditions; numerically simulated schlieren picture, flow from left to right; left: case L1 – T_{01}=354 K, p_{01}=0.403 bar, p_2=0.163 bar; middle: case L2 – T_{01}=354 K, p_{01}=0.409 bar, p_2=0.194 bar; right: case L3 – T_{01}=357.5 K, p_{01}=0.417 bar, p_2=0.206 bar.

19.6 Rotor/Stator Interaction in Linear Cascade CA7.5

All unsteady effects discussed in the previous chapters are caused by self-excited oscillations due to the interaction of the compressibility of the fluid and the high intensity of the latent heat release from nonequilibrium condensation of the steam. However, in multistage turbomachinery, e. g. in steam turbines, an additional forced excitation mechanism exists by the wake generated unsteadiness. Therefore, we here deal with the forced wake generated unsteadiness in axial flow turbines, especially in the two-phase flow regime. Since the nucleation process is extremely sensitive to changes in the static temperature, the condensate formation and hence the surface pressure distribution are strongly influenced. For comparison with the self-excited oscillation dynamics in the linear cascade A1A (see Section 19.3.1.3), we investigate a comparable simple cascade geometry. The rotor blades are formed by circular arcs, the thickness parameter d/c=0.075, and the stagger angle is reduced to zero, i.e. β_s=90°. For the same reason, the fluid is a steam/carrier gas mixture. Figure 19.12 shows the steady condensing flow through the rotor. Due to the moderate

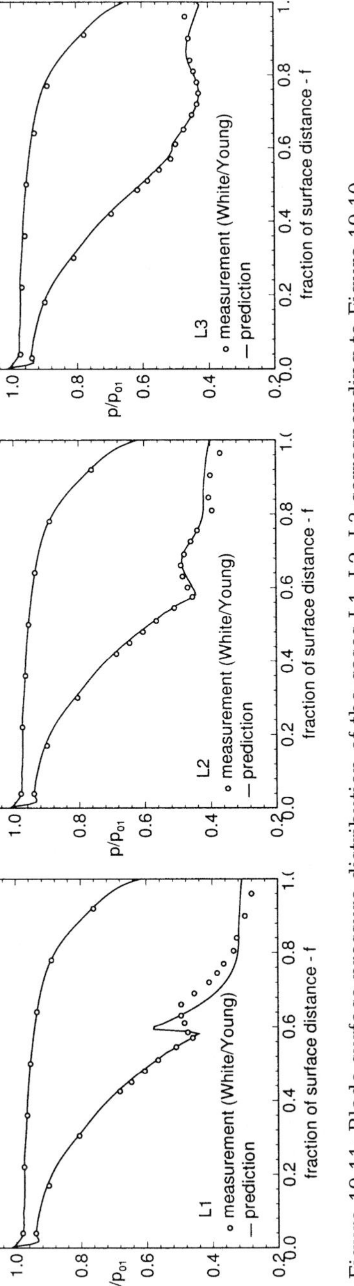

Figure 19.11: Blade surface pressure distribution of the cases L1, L2, L3 corresponding to Figure 19.10.

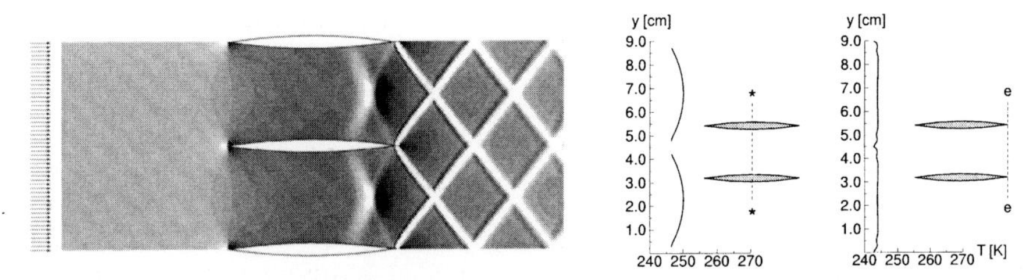

Figure 19.12: Steady condensing flow in the cascade CA7.5 without stator/rotor interaction, steam/carrier gas mixture; inlet conditions $T_1 = 275$ K, $p_1 = 0.76$ bar, $T_1 - T_s(p_{v,1}) = -6.6$ K, $M_{f,1} = 0.65$, $g_{max} = 9.84$ g/kg, averaged downstream conditions $T_2 = 247$ K, $p_2 = 0.37$ bar, $M_{f,2} = 1.24$; left: numerically simulated schlieren picture (flow from left to right) and inlet velocity distribution; right: temperature distribution at the plane indicated by $\cdot - \cdot$, and at the exit plane e – e.

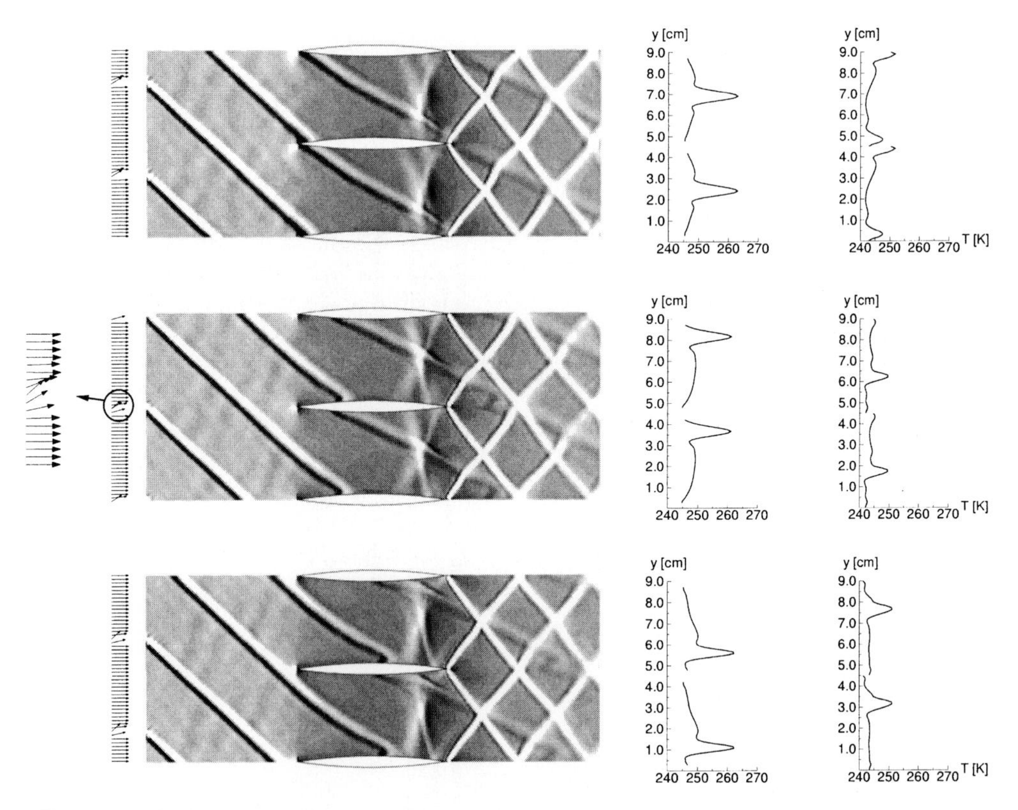

Figure 19.13: One cycle of the condensing flow in the rotating cascade CA7.5 with interaction of upstream (stator) generated wakes, steam/carrier mixture; inlet conditions $p_1 = 0.76$ bar, $T_1 = 275$ K (outside the wake), amplitude $\Delta(c_1)_{max}/c_1 = 0.5$, maximum static temperature variation $\Delta T_1 = 25$ K, wake width $b/p = 0.1$, averaged downstream conditions $T_2 = 247$ K, $p_2 = 0.38$ bar, $M_{f,2} = 1.23$, frequency $f = 4.44$ kHz; left: numerically simulated schlieren pictures (flow from left to right) and inlet velocity distribution; right: temperature distribution at the plane indicated by $\cdot - \cdot$, and at the exit plane e – e.

steam mass content of g_{max}=9.84 g/kg, the condensation process is subcritical, i.e. the flow remains supersonic near the blade surfaces and only a very weak normal shock develops near the centerline. Superimposing the moving temperature wake, the flow becomes unsteady with intense sound waves ahead of the rotor and oblique waves entering and passing through the blade channels (Figure 19.13). It is interesting to note that the axial gradient of the main flow decreases the amplitude of the temperature wake about 50% [23] (see Figure 19.13, right). However, the condensation zone is significantly disturbed. From the intense interaction with the nucleation rate J and with the wetness fraction g/g_{max}, we conclude that the dispersed droplet spectrum can not accurately be determined if the rotor/stator interaction is not taken into account. More details, especially concerning the modelling of the rotor/stator interaction, are given in [22].

19.7 Interaction of Vortex Shedding and Condensation in VKI-1 Turbine Cascade

Another unsteady effect in turbine cascade flows is caused by vortex shedding. In our case, the interaction of the unsteady flow field due to vortex shedding with the nonequilibrium condensation is considered in particular. Therefore, the condensing unsteady, viscous, turbulent flow in the VKI-1 cascade [24–26] is investigated. The reservoir and downstream conditions chosen are given in the caption of Figure 19.14. Figure 19.14 (top) shows the Mach number contours for the steady flow (left – numerical dissipation artificially increased) and the flow with vortex shedding (right). The formation of vortices changes the wake flow field significantly. Furthermore, it causes a slight oscillation of the pressure side shock. Both effects induce a time-dependent change of the nucleation and droplet growth. Thus, the formation of condensate (Figure 19.14, bottom) and the droplet spectrum are altered significantly compared to the steady flow.

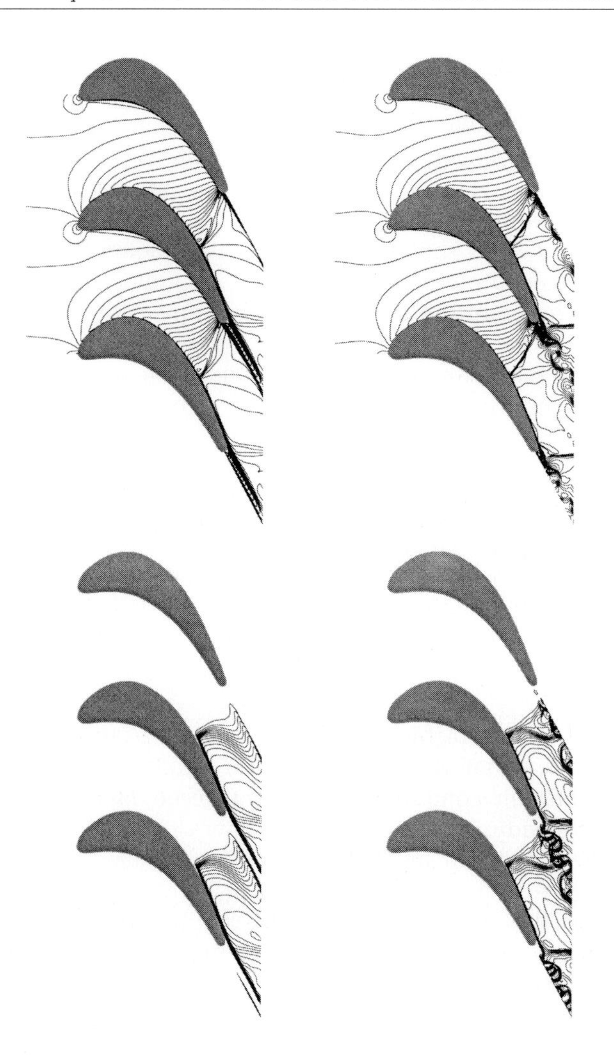

Figure 19.14: Condensing, viscous, turbulent flow in VKI-1 cascade, pure steam, reservoir conditions $T_{01}=357.5$ K, $p_{01}=0.417$ bar, downstream condition $p_2=0.206$ bar, isentropic exit Mach number $M_{2,\text{is}}=1.08$, Reynolds number based on exit velocity $Re_2=5\cdot10^5$, inlet flow angle $\beta_1=30^\circ$; top – left: Mach number contours – steady flow; top – right: Mach number contours – unsteady flow; bottom – left: condensate mass contours – steady flow; bottom – right: condensate mass contours – unsteady flow.

Acknowledgements

The authors would like to express their gratitude to the Deutsche Forschungsgemeinschaft for supporting our research by the contracts Zi 18/35-1,2 and Schn 352/13-3,4,5.

References

[1] J. Frenkel: *Kinetic Theory of Liquids.* Oxford University, Oxford, 1946.

[2] G. Gyarmathy: *Two-Phase Steam Flow in Turbines and Separators.* In: M.J. Moore, C.H. Sieverding (Eds.): *Hemisphere.* Chapter 3, 1976, pp. 57–82.

[3] J.B. Young: *The Condensation and Evaporation of Liquid Droplets in a Pure Vapour at Arbitrary Knudsen Number.* Int. J. Heat Mass Transfer **34** (1991) 1649–1661.

[4] J.B. Young: *The Condensation and Evaporation of Liquid Droplets at Arbitrary Knudsen Number in the Presence of an Inert Gas.* Int. J. Heat Mass Transfer **36** (1993) 2941–2956.

[5] F. Peters, K.A.J. Meyer: *Measurement and Interpretation of Growth of Monodispersed Water Droplets Suspended in Pure Vapour.* Int. J. Heat Mass Transfer **38** (1995) 3285–3293.

[6] S. Kotake, I.I. Glass: *Flows with nucleation and condensation.* Prog. Aerospace Sci. **19** (1981) 129–196.

[7] S. Kotake, I.I. Glass: *Condensation of water vapour in rarefaction waves: II. Heterogenous nucleation.* AIAA Journal **15** (1977) No. 2.

[8] S. Kotake, I.I. Glass: *Condensation of water vapour on heterogenous nuclei in a shock tube.* UTIAS Report No. 207, CN ISSN 0082-5255, April, 1976.

[9] U.C. Goldberg: *Toward a Pointwise Turbulence Model for Wall-Bounded and Free Shear Flows.* J. Fluids Eng. **116** (1994) 72–76.

[10] G.H. Schnerr, M. Heiler: *Two-Phase Flow Instabilities in Channels and Turbine Cascades.* In: M. Hafez, K. Oshima (Eds.): Computational Fluid Dynamics Review. John Wiley & Sons, New York, London, 1998.

[11] G.H. Schnerr, S. Adam, K. Lanzenberger, R. Schulz: *Multiphase Flows: Condensation and Cavitation Problems.* In: M. Hafez, K. Oshima (Eds.): Computational Fluid Dynamics Review 1. John Wiley & Sons, New York, London, 1995.

[12] G. Mundinger: *Numerische Simulation instationärer Lavaldüsenströmungen mit Energiezufuhr durch homogene Kondensation.* Ph. D. Thesis, Fakultät für Maschinenbau, Universität Karlsruhe, Germany, 1994.

[13] S. Adam, G. H. Schnerr: *Instabilities and Bifurcation of Nonequilibrium Two-Phase Flows.* J. Fluid Mech. October, 1997.

[14] G.H. Schnerr, S. Adam: *Visualization of Unsteady Gas/Vapor Expansion Flows.* In: Journal of Thermal Science, Science Press Beijing, New York, 1997, pp. 171–180.

[15] S. Adam: *Numerische und experimentelle Untersuchung instationärer Düsenströmungen mit Energiezufuhr durch homogene Kondensation.* Ph. D. Thesis, Fakultät für Maschinenbau, Universität Karlsruhe, Germany, 1996.

[16] J. Zierep, G.H. Schnerr, S. Adam: *Fluiddynamische Instabilitäten und Oszillationen durch Kondensationsverzug und Phasenumwandlung im Nichtgleichgewicht.* Zwei-

ter Zwischenbericht zum DFG-Schwerpunktprogramm „Transiente Vorgänge in mehrphasigen Systemen mit einer oder mehreren Komponenten" (Zi 18/35-2/Schn 325/13-3), März, 1995.

[17] G.H. Schnerr, S. Adam, M. Heiler: *Unsteady Condensing Flows in Channels and Cascades.* In: Aerodynamics of Turbomachinery, IMechE Seminar Publication 1996-21 Supplementum, The Institution of Mechanical Engineers, Bury St Edmunds and London.

[18] G.H. Schnerr, M. Heiler, S. Adam: *Steady and Unsteady Two-Phase Flows in Turbine Cascades.* Proc. Japanese-German Symposium on Multiphase Flows, Tokyo, September 25–27, 1997, Eds.: Takamoto Saito, Ulrich Müller, pp. 117–131.

[19] D. Barschdorff: *Verlauf der Zustandsgrößen und gasdynamische Zusammenhänge bei der spontanen Kondensation reinen Wasserdampfs in Lavaldüsen.* Forsch. Ing.-Wes. **37**(5) (1971) 146–157.

[20] A.J. White, J.B. Young, P.T. Walters: *Experimental Validation of Condensing Flow Theory for a Stationary Cascade of Steam Turbine Blades.* Phil. Trans. R. Soc. London A **354** (1996) 59–88.

[21] J.D. Denton, L. Xu: *The Trailing Edge Loss of Transonic Turbine Blades.* J. Turbomachinery **112** (1990) 277–285.

[22] G.H. Schnerr, M. Heiler, G. Winkler: *Steady and Unsteady Condensate Formation in Turbomachinery – Blade to Blade Flow and Rotor/Stator Interaction.* Proc. Third Int. Conf. on Multiphase Flow, Lyon, France, June 8–12, 1998.

[23] D.E. van Zante, J.J. Adamczyk, A.J. Strazisar, T.H. Akiishi: *Wake Recovery Performance Benefit in a High-Speed Axial Compressor.* ASME97-GT-535, 1997.

[24] R. Kiock, F. Lehthauf, N.C. Baines, C.H. Sieverding: *The Transonic Flow through a Plane Turbine Cascade as Measured in Four European Wind Tunnels.* Transactions of the ASME, Journal of Engineering for Gas Turbines and Power **108** (1986) 277–284.

[25] C.H. Sieverding, H. Heinemann: *The Influence of Boundary Layer State on Vortex Shedding From Flat Plates and Turbine Cascades.* Transactions of the ASME, Journal of Turbomachinery **112** (1990) 181–187.

[26] A. Arnone, R.C. Swanson: *A Navier-Stokes Solver for Turbomachinery Applications.* Transactions of the ASME, Journal of Turbomachinery **115** (1993) 305–313.

20 Numerical Simulation of Thermal and Mechanical Non-Equilibrium Effects in Critical and Near-Critical Flows of Hot-Water at Subcooled and Saturated Conditions

Achim Dittmann, Jörg Huhn and Michael Wein *

Abstract

A two-fluid two-phase flow model has been developed to examine non-equilibrium effects in critical and near-critical two-phase flows of initially subcooled or saturated hot water through pipes and nozzles. The six-equation model consists of the phasic conservation equations for mass and momentum, the liquid thermal energy, and a transport equation for the bubble number density. To solve for the unknown variables, a semi-implicit finite difference method is utilized. The closure of the set of differential equations is accomplished by thermodynamic relationships and additional constitutive equations describing momentum transport, and interphase heat and mass transfer, which account for different flow regimes. For flashing systems, wall and bulk nucleation models were included to describe the initial state of delayed vapor generation. In this way, thermal non-equilibrium is considered to be the consequence of excessive energy states required to activate nucleation sites, of restricted interfacial area and limited heat transfer between the phases.

The numerical results obtained by this model for two-phase flows with various inlet conditions agree well with experimental data for flow rate, pressure and void fraction distribution.

* Technische Universität Dresden, Institut für Technische Thermodynamik, Mommsenstr. 13, D-01062 Dresden, Germany

20.1 Introduction

Two-phase flows are of much interest in the power and process industries. The accurate prediction of two-phase transients and critical discharge is essential for the thermal hydraulic sizing of safety devices and for the analysis of hypothetical ruptures of vessels or pipes.

Critical or choked flow corresponds to fluid dynamical maximum on the flow rate in pipes and nozzles for a given upstream fluid state. When lowering the downstream pressure, the flow rate increases until a maximum value is reached. The flow rate will then remain constant independent of further reduction of the downstream pressure. In contrast to single-phase flow, where the critical flow velocity is equal to the sonic velocity, there is no such generally accepted relationship for two-phase flows, since the assumptions of isentropic change of state and fluid homogeneousness are not valid.

In the homogeneous equilibrium model [1], the two-phase mixture is regarded as a single-phase fluid incorporating equal phase velocities and temperatures. This model shows reasonable results for two-phase flows undergoing relatively slow processes, such as the flow through long pipes. By way of contrast, it fails in describing flows where there is insufficient time to reach equilibrium, e.g. blow-down through short tubes and nozzles.

A large number of models for critical two-phase flow have been suggested to account for mechanical and thermal non-equilibrium. Elias and Lellouche [2] presented a comprehensive review on common methods aimed at modelling critical two-phase flow. They found that only models based on the conservation and balance equations gave reliable predictions of critical mass fluxes.

The purpose of this report is to present recent results of our current research on critical and near-critical two-phase flows. A two-fluid model developed for high mass flux two-phase flows through differently shaped flow ducts, which comprises both hydrodynamic and thermal non-equilibrium effects will be described in the following sections. The model contains more information in the basic formulation regarding nucleation and bubble transport and features improvements in the constitutive relationships. Furthermore, it will be shown that the utilized method provides modelling capabilities needed for further development of interphase transfer relationships.

20.2 Hydrodynamic Two-Phase Flow Model

The basic structure of the two-fluid two-phase flow model is outlined in Figure 20.1.

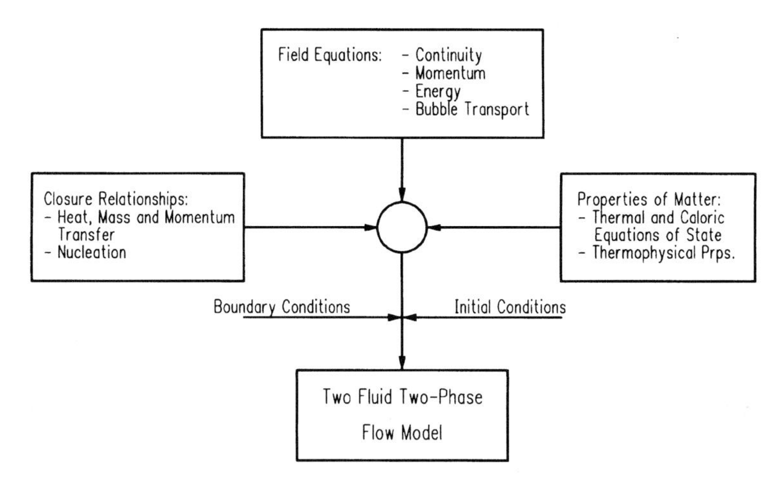

Figure 20.1: Basic structure of the two-fluid two-phase flow model.

The construction starts from the field equations, which consist of the continuity, the momentum and the energy equation of each phase. In addition, a vapor bubble transport equation will be used. To accomplish closure of the resulting set of equations, supplementary relationships are needed. These comprise thermal and caloric equations of state, and thermophysical data, respectively, as well as closure relations regarding heat, mass and momentum transfer. Furthermore, correlations describing nucleation phenomena are included for initially single-phase flows undergoing a phase change process. Herein, the focus is on evaporation.

The development of the two-fluid model is mainly motivated by the deficiencies and uncertainties in the description of hydrodynamic non-equilibrium effects in the drift-flux model of Wein and Huhn [3, 4] preceding this work, where different phase velocities were implied by semi-empirical and algebraic slip correlations. As the convective heat transfer, and with it the mass transfer, is also affected by hydrodynamic non-equilibrium in terms of the relative velocity, the degree of information on thermal non-equilibrium is extended by the two-fluid approach as well.

In the strict sense, single-phase flow in a variable area duct is already three-dimensional. This applies more strongly for multi-phase flow with non-uniform phase-distribution. Consequently the quasi one-dimensional treatment employed here is an approximation, where all flow properties are assumed uniform across any cross-section of the flow, and are functions of the axial coordinate only for steady-state flow. Within the framework of the quasi one-dimensional two-fluid model, multi-dimensional effects such as bulk and wall nucleation, or heat transfer at differently shaped vapor quantities with complex phase interactions as in the bubbly-slug flow, are introduced by means of appropriate volume-related source terms.

Although experimental observations reported by Zimmer et al. [5] and Fincke [6] have indicated multi-dimensional effects in choked flow such as sheet cavitation, these phenomena have not yet gained a solid theoretical explanation and are beyond the scope of this work.

20.2.1 Field equations of the two-fluid model

The one-dimensional formulation of the two-fluid two-phase flow model used in this work is obtained by time or statistical averaging of the phase-separated local, instantaneous conservation equations as explained in depth by Ishii [7]. For a one-component liquid/vapor system, this leads to two continuity, two momentum, and two energy equations [8].

In the following, equal pressures are assumed for both vapor and liquid phase, except at nucleation. Furthermore, the vapor phase is assumed to be saturated with respect to the local pressure, whereas the liquid phase may pass through subcooled, saturated, or superheated states, respectively. As the focus herein is directed to two-phase flows with small vapor quantities, the vapor at saturation assumption seems quite reasonable. Consistently, there is no large deviation from thermal equilibrium for the vapor phase to be expected. At higher void fractions, e.g. annular flow, the vapor bulk temperature may depart from saturation temperature, since the thermal conductivity of the vapor is much smaller than that of the liquid phase.

With the thermal state of the vapor varying only with pressure, the vapor energy equation can be dropped. For the liquid phase energy equation, changes of potential energy as well as viscous and turbulent dissipation terms are neglected to obtain the thermal formulation of the liquid energy balance equation. To account for variation in the bubble number density, a bubble transport equation as proposed by Kocamustafaogullari and Ishii [9] is used in addition to the phasic balance equations. Thus, the basic set of equations utilized to describe wall-adiabatic, quasi one-dimensional two-phase flow consists of six partial differential equations:

- Vapor Mass Equation

$$\frac{\partial(\alpha_v \rho_v)}{\partial t} + \frac{1}{A}\frac{\partial}{\partial z}(\alpha_v \rho_v v_v A) = \Gamma_v \,, \tag{1}$$

- Liquid Mass Equation

$$\frac{\partial(\alpha_l \rho_l)}{\partial t} + \frac{1}{A}\frac{\partial}{\partial z}(\alpha_l \rho_l v_l A) = -\Gamma_v \,, \tag{2}$$

- Vapor Momentum Equation

$$a_\mathrm{v}\rho_\mathrm{v}\frac{\partial v_\mathrm{v}}{\partial t}+a_\mathrm{v}\rho_\mathrm{v}v_\mathrm{v}\frac{\partial v_\mathrm{v}}{\partial z}=-a_\mathrm{v}\frac{\partial p}{\partial z}-a_\mathrm{v}\rho_\mathrm{v}g\sin\theta-f_\mathrm{wv}+\Gamma_\mathrm{v}(v^*-v_\mathrm{v})-f^*\,, \tag{3}$$

- Liquid Momentum Equation

$$a_\mathrm{l}\rho_\mathrm{l}\frac{\partial v_\mathrm{l}}{\partial t}+a_\mathrm{l}\rho_\mathrm{l}v_\mathrm{l}\frac{\partial v_\mathrm{l}}{\partial z}=-a_\mathrm{l}\frac{\partial p}{\partial z}-a_\mathrm{l}\rho_\mathrm{l}g\sin\theta-f_\mathrm{wl}-\Gamma_\mathrm{v}(v^*-v_\mathrm{l})+f^*\,, \tag{4}$$

- Liquid Thermal Energy Equation

$$\frac{\partial}{\partial t}(a_\mathrm{l}\rho_\mathrm{l}e_\mathrm{l})+\frac{1}{A}\frac{\partial}{\partial z}(a_\mathrm{l}\rho_\mathrm{l}e_\mathrm{l}v_\mathrm{l}A)=-\frac{p}{A}\frac{\partial}{\partial z}(a_\mathrm{l}v_\mathrm{l}A)-p\frac{\partial a_\mathrm{l}}{\partial t}-\Gamma_\mathrm{v}h_\mathrm{l}^\mathrm{x}-A^*q^*\,, \tag{5}$$

- Bubble Transport Equation

$$\frac{\partial N_\mathrm{b}}{\partial t}+\frac{1}{A}\frac{\partial}{\partial z}(N_\mathrm{b}v_\mathrm{v}A)=\phi_\mathrm{wn}+\phi_\mathrm{bn}\,. \tag{6}$$

The six primary dependent variables in the set of field equations (Equations (1)–(6)) are the static pressure p, the void fraction a_v, the velocities of the vapor v_v and the liquid phase v_l, the specific internal energy of the liquid e_l, and the bubble number density N_b. The independent variables are the time t and the axial coordinate z. Further, the secondary dependent variables are the phasic densities ρ_v and ρ_l, the liquid temperature T_l, and the saturation temperature T^s associated with the interfacial and vapor properties. Note that the momentum equations (Equations (3) and (4)), the liquid thermal energy equation (Equation (5)), and the bubble transport equation (Equation (6)) are written in nonconservation form.

The liquid volume fraction a_l is related to the void fraction (vapor volume fraction) by:

$$a_\mathrm{v}=1-a_\mathrm{l}\,. \tag{7}$$

The term Γ_v on the right hand side of the vapor continuity equation (Equation (1)) describes the volumetric rate of mass exchange due to phase change, which represents a source term for evaporation, and a sink term for condensation, respectively. For overall continuity, this requires Γ_v to appear with a negative sign in Equation (2).

The force terms on the right-hand sides of the momentum equations (Equations (3) and (4)) represent – from left to right – the pressure gradient, the force of gravity with θ the angle of the duct inclination, the wall friction, the momentum transfer due to mass exchange, and the volumetric force f^* due to interfacial friction and virtual mass.

In the liquid energy equation (Equation (5)), the terms $-\Gamma_\mathrm{v}h_\mathrm{l}^\mathrm{x}$ and $-A^*q^*$ constitute the volumetric heat transfer rates from the vapor to the liquid phase due to mass exchange at the vapor/liquid interface (evaporation or condensation), and the thermal energy exchange due to the bulk temperature potential between the phases. The enthalpy associated with the interface mass transfer in Equation (5) are defined as

$$h_l^x = \begin{cases} h_l & \text{for vaporization}\,, \\ h_l^s & \text{for condensation}\,. \end{cases} \tag{8}$$

The source terms ϕ_{wn} and ϕ_{bn} in the bubble transport equation (Equation (6)) represent the volumetric wall and bulk nucleation, respectively.

20.2.2 Thermal and caloric equations of state, thermophysical properties

The two-fluid model presented embodies three independent thermodynamic state variables, namely the static pressure p, the void fraction α_v, and the liquid temperature $T_l = f(p, e_l)$. All remaining thermal, caloric, and thermophysical variables as well as derivatives are determined in terms of the independent state variables. In order to obtain short program execution times, fast formulations for thermodynamic properties and supplementary equations are necessary. For that reason, approximations according to the new industrial standard for water and steam IAPWS-IF97 [10, 11] have been implemented into the program.

20.2.3 Constitutive models

In order to obtain closure of the basic set of equations (Equations (1) to (6)), additional relationships describing momentum, heat and mass transfer are needed. In addition, an approach to formulate delayed bubble generation and flashing of initially subcooled flows has to be specified. Within this section, the constitutive models for wall and interphase friction, virtual mass forces, bubble nucleation as well as interfacial area, heat and mass transfer are presented.

An important aspect of utilizing correlations originally developed for steady-state flow conditions is the assumption that these relationships can be applied to transient flow conditions in a quasi-steady way. However, since we are interested in the steady flowfield variables, the time-marching technique used herein is simply a means to achieve this ends and therefore steady-state correlations should be applicable.

Since all of the above mentioned models are closely tied with the internal two-phase flow structures, geometrical shape of the phases and their distribution within the flow duct, a flow regime map has to be defined.

20.2.3.1 Flow regime map

A simple flow regime map proposed by Blinkov et al. [12] has been embodied in the computer code making use of the void fraction as the main criterion for defining transitions between the flow regimes.

Bubbly flow is supposed to exist for $\alpha \leq 0.3$. At void fractions $0.3 < \alpha < 0.8$, the bubbles agglomerate to form bubble slugs and partially coalesce to form larger portions of vapor taking the idealized shape of cylindrical plugs (Taylor bubbles). For $\alpha \geq 0.8$, dispersed droplet flow develops due to bubble disintegration and droplet entrainment from the lateral surface of the plugs. Similar flow maps can be found in Dobran [13], Richter [14], and Schwellnus and Shoukri [15]. Stratification effects were not considered, since we concentrate on high mass-flux flows.

Sample calculations with small changes in the flow regime limits of void fraction of up to ± 0.05 did not reveal significant influence on the results. However, no systematic sensitivity analysis considering the assumed flow regime transitions was conducted.

20.2.3.2 Wall friction

To account for wall friction in *single-phase flow*, the well-known correlations for the friction coefficient of Blasius for smooth tubes, and Colebrook-White for rough tubes have been utilized. Thus, the wall friction force per unit volume for pure liquid flow is obtained by:

$$f_{\mathrm{wl}} = 2\, c_{\mathrm{fl}} \frac{\rho_{\mathrm{l}} v_{\mathrm{l}}^2}{D_{\mathrm{h}}} \,. \tag{9}$$

For *two-phase flow*, a friction multiplier approach according to Friedel [16] is applied to yield the frictional force per unit volume exerted by the duct wall on the liquid as:

$$f_{\mathrm{wl}} = f_{\mathrm{wl},2\phi} = 2\, c_{\mathrm{f,lo}} R \frac{G^2}{\rho_{\mathrm{l}} D_{\mathrm{h}}} \,, \tag{10}$$

where the two-phase multiplier R is defined as the ratio of the frictional pressure drops of two-phase flows and single-phase flows, both taken at the total mass flux of the mixture G:

$$R = \left. \frac{(\Delta p/\Delta z)_{2\phi}}{(\Delta p/\Delta z)_{1\phi}} \right|_G \,. \tag{11}$$

The single-phase frictional coefficient $c_{\mathrm{f,lo}}$ in Equation (10) is calculated using the liquid-only Reynolds number:

$$Re_{\mathrm{lo}} = \frac{G D_{\mathrm{h}}}{\mu_{\mathrm{l}}} \,, \tag{12}$$

where G is the total mass flux of the two-phase mixture given as:

$$G = \alpha_v \rho_v v_v + \alpha_l \rho_l v_l \,. \tag{13}$$

This seems to be a reasonable approximation for convergent ducts where the pressure gradient is expected to keep the boundary layers thin. When applied to divergent sections, where the boundary layer displacement in two-phase flow may significantly differ from the single-phase flow, greater uncertainties arise. However, pressure drop due to wall friction plays a minor role for high mass flux two-phase flow through relatively short pipes or nozzles. In that case, the pressure variation is dominated by fluid acceleration, or deceleration, respectively. On the other hand, the inclusion of wall friction was found to be important for correct prediction of two-phase flow through long pipes.

The vapor phase is expected to have almost no contact with the wall. Thus, the wall friction can be neglected:

$$f_{wv} \approx 0. \tag{14}$$

Pressure drop due to wall friction turned out to be of minor importance for flows of high mass flux through relatively short pipes or nozzles, where pressure variations are dominated by fluid acceleration, or deceleration, respectively. On the other hand, the inclusion of wall friction was found to be important for correct prediction of two-phase flow through long pipes.

20.2.3.3 Interphase momentum transfer

The interfacial volumetric force f^* in the momentum equations (Equations (3) and (4)) comprises the drag force and the virtual (or apparent) mass force due to relative acceleration between the phases:

$$f^* = f^*_d + f^*_{vm} \,. \tag{15}$$

Both terms are described in the following two subsections. The momentum transfer due to mass exchange is described in a separate subsection. The additional reaction force due to bubble growth is not considered in the present study.

Interphase friction: The interfacial friction force per unit volume for *bubbly flow*, $\alpha_v \leq 0.3$, is expressed in terms of the drag coefficient and the interfacial area density [14]:

$$f^*_d = f^*_{d,b} = \frac{C'_{d,b}}{2} \frac{3}{4R_b} \alpha_v (1-\alpha_v)^3 \rho_l (v_v - v_l)|v_v - v_l| \,. \tag{16}$$

A modified drag coefficient as suggested by Rowe and Henwood [17] is used to account for bubble interference:

$$C'_{d,b} = C_{d,b}(1-\alpha_v)^{-2n} \,. \tag{17}$$

The exponent n in the above equation is approximately $n = 2.35$ for $Re_b > 1000$ as found by Wallis [1]. The drag coefficient for a single bubble $C_{d,b}$ depends on the bubble Reynolds number:

$$Re_b = \frac{(1 - a_v)\rho_l |v_v - v_l| 2 R_b}{\mu_l} \tag{18}$$

and is obtained by

$$C_{d,b} = \begin{cases} 24\, Re_b^{-1}(1 + 0.15\, Re_b^{0.687}) & Re_b < 1 \cdot 10^3 \\ 0.44 & Re_b \geq 1 \cdot 10^3 \end{cases} \tag{19}$$

as derived by Wallis [1]. One should mention that the assumption of spherical shape for the bubbles is only a rough estimate. In reality, the flow conditions are far more complex due to bubble deformation and internal circulation. Further, for bubbles closely following one after another, a reduction in drag should become apparent.

In the *bubbly-slug flow* regime, $0.3 < a_v < 0.8$, it is suggested to model the total interfacial friction as composed of two terms: the part coming from drag exerted on the bubbles within the slugs, and the part, which stems from interfacial friction at the lateral area of the plugs (cylinder-like Taylor bubbles). Thus, the interfacial friction can be correlated as:

$$f_d^* = \frac{a_{v,bs}}{a_v} f_{d,bs}^* + \frac{a_{v,p}}{a_v} f_{d,p}^* \,, \tag{20}$$

where $f_{d,bs}^*$ is determined from Equation (16) at $a_{v,bs}$. To evaluate the interfacial friction of the plugs $f_{d,p}^*$, a friction coefficient method is applied:

$$f_{d,p}^* = \frac{2\, c_{f,p}^*}{D_h} a_{v,p}^{1/2} \rho_v (v_v - v_l)|v_v - v_l| \,, \tag{21}$$

where the annular-type interfacial friction coefficient c_f^* is given by Richter [14] as:

$$c_{f,p}^* = 0.005 \left[1 + 75\left(1 - a_{v,p}\right)\right]. \tag{22}$$

The total void fraction in the bubbly-slug flow regime is the sum of the portions due to the plugs and the not yet agglomerated bubbles left in the liquid film and slug region:

$$a_v = a_{v,p} + a_{v,bs} \,. \tag{23}$$

The method used to predict the void fractions $a_{v,p}$ and $a_{v,bs}$ is presented in a subsequent section on interfacial area.

It is recognized that the proposed relation (Equation (20)) implies a certain degree of uncertainty, as the relative velocity of the bubbles within the slugs may be significantly different from the velocity of the Taylor-bubbles, due to differently pronounced mechanical coupling between vapor and surrounding liquid. However, Equation (20) provides the desired smooth transition at the void fraction boundaries $\alpha_{v,b} = 0.3$ (bubbly to bubbly-slug), and $\alpha_{v,a} = 0.8$ (annular flow). Further, this formulation was found to give much better results when compared to the interpolation procedures carried out by Richter [14] and Schwellnus and Shoukri [15].

At present, a drag coefficient relationship analogous to Equation (16) is used for the *dispersed droplet flow* regime $\alpha_v > 0.8$. However, none of the presented calculation results passes the assumed void fraction transition boundary of $\alpha_{v,a} = 0.8$.

Virtual mass force: The total effective mass of an accelerated body moving through a surrounding fluid consists of the mass of itself plus a virtual mass (also called apparent mass) that arises from the inertial properties of the fluid in the immediate vicinity of the body. Formulations of the virtual mass term introduce a coefficient C_{vm}, which describes the volume of displaced fluid that contributes to the effective mass. Fundamental contributions to virtual mass effects can be found in Prandtl et al. [18], Zuber and Findlay [19], Zuber [20], Drew et al. [21], Cook and Harlow [22].

From the above explanations, the volumetric virtual mass force, or inertial drag force per unit volume, can be expressed by the product of displaced mass and the total substantial derivative of the relative velocity between vapor and liquid as:

$$f^*_{vm} = C_{vm}\alpha_l\alpha_v\rho_m\left[\frac{\partial(v_v - v_l)}{\partial t} + v_l\frac{\partial v_v}{\partial z} - v_v\frac{\partial v_l}{\partial z}\right], \tag{24}$$

where the displaced fluid is represented by the homogeneous mixture density ρ_m defined by:

$$\rho_m = \alpha_v\rho_v + \alpha_l\rho_l\,. \tag{25}$$

Proper inclusion of the virtual mass effect was found to be important [21], especially in the bubbly flow regime [15]. When omitted, vapor bubbles can gain unreasonably large accelerations through the liquid phase. The resulting pressure oscillations largely affect the numerical stability.

For the *bubbly flow*, $0 < \alpha_v \leq 0.3$, and the *dispersed droplet flow*, $0.8 \leq \alpha_v < 1$, the virtual mass coefficient is given as:

$$C_{vm,b} = C_{vm,d} = 0.5\,. \tag{26}$$

Equation (26) is consistent with findings of Zuber [20] and Wijngaarden [23], who have shown that for bubbly and dispersed droplet flow $C_{vm} \approx 0.5$, whereas for separated flow $C_{vm} \to 0$. For this reason, the virtual mass effect was fully neglected by Richter [14] and Schwellnus and Shoukri [15] for void fractions $\alpha_v > 0.3$.

In order to find a description of the virtual mass coefficient for the *bubbly-slug flow*, a phenomenological approach based on the following observations is proposed: at void fractions of $\alpha_{v,b} = 0.3$, phase separation due to bubble coalescence is assumed to set in, which should result in lowering of the virtual mass coefficient. On the other hand, at higher void fractions, the virtual mass coefficient should rise again as droplets entrained from the liquid annulus and from the liquid released by the slugs are formed.

Promising results were obtained by the following formulation for the transitional regime between bubbly and dispersed droplet flow:

$$C_{vm,bs} = \frac{\alpha_{v,bs}}{\alpha_v} C_{vm,b} + \frac{\alpha_{v,p}}{1 - \alpha_v} C_{vm,p} \,, \tag{27}$$

where the virtual mass coefficient for the Taylor bubbles is estimated to be $C_{vm,p} = 0.125$. The determination of the void fraction portions due to bubbles in the slugs $\alpha_{v,bs}$, and due to the plugs $\alpha_{v,p}$, respectively, are presented in a subsequent section on interfacial area.

The virtual mass coefficient according to Drew et al. [21] does not regard to the flow regime:

$$C_{vm} = \begin{cases} 0.5\,(1 + 2\alpha_v)(1 - \alpha_v)^{-1}, & \text{for } 0 < \alpha_v \leq 0.5 \\ 0.5\,(3 - 2\alpha_v)\alpha_v^{-1}, & \text{for } 0.5 < \alpha_v < 1 \,. \end{cases} \tag{28}$$

A comparison between the virtual mass coefficients obtained from Equations (26) to (28) as functions of the void fraction only is given in Figure 20.2. Note

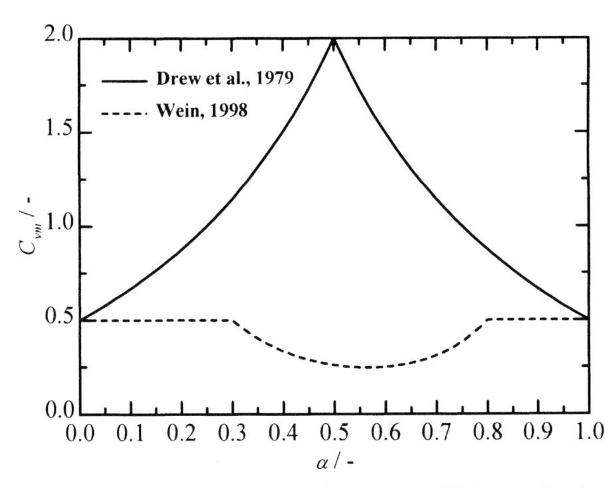

Figure 20.2: Comparison between the virtual mass coefficients obtained from Drew et al. [21] and the present formulation.

that the suggested correlation accounts for the phase separation effect when entering the bubbly-slug flow regime depicted by a decrease of the virtual mass coefficient. When approaching the droplet flow regime, where phase dispersion due to bubble disintegration and entrainment effects occurs according to the flow regime assumptions, the virtual mass coefficient increases again to reach its value for dispersed flow.

Momentum transfer due to mass exchange: The volumetric force associated with the change in velocity when mass is transferred across the liquid/vapor interface due to evaporation or condensation is incorporated in the model using the following definition for the interfacial velocity

$$v^* = (1 - \eta)\, v_{\mathrm{V}} + \eta v_{\mathrm{l}} \,, \tag{29}$$

where η is a distribution parameter. Thus, the terms associated with mass exchange on the right-hand sides of the momentum equations of vapor and liquid (Equations (3) and (4)) can be expressed as:

$$\Gamma_{\mathrm{V}}(v^* - v_{\mathrm{V}}) = -\eta \Gamma_{\mathrm{V}}(v_{\mathrm{V}} - v_{\mathrm{l}}) \,, \tag{30}$$

$$\Gamma_{\mathrm{V}}(v^* - v_{\mathrm{l}}) = -(1 - \eta)\Gamma_{\mathrm{V}}(v_{\mathrm{V}} - v_{\mathrm{l}}) \,. \tag{31}$$

In the case of evaporation, the evaporated liquid is changing its velocity from v_{l} to v_{V}, whereas condensing vapor ends at v_{l}. The distribution parameter η weighs the portion of the force terms mentioned above. Thus, the total force is exerted on the vapor phase only for $\eta = 1$, or on the liquid phase only for $\eta = 0$, respectively.

For reversible flow, Wallis [1] has shown that the parameter η reads:

$$\eta = 0.5 \,. \tag{32}$$

This value of η is also used in the present study.

20.2.3.4 Nucleation model

In flows with transition from an initially single-phase to a two-phase state (flashing flows), wall and heterogeneous bulk nucleation mechanisms are taken into account in order to provide information about the number of instantaneously generated vapor bubbles. The source terms on the right-hand side of the bubble transport (Equation (6)) represent these mechanisms.

Wall nucleation: Wall nucleation is assumed to be a cyclical process of bubble growth ending with bubble departure and a subsequent dwell period until the next bubble is nucleated [24]. The main quantities and relations used for the application of their model are repeated here for convenience. A least square analysis of experimental data reported in literature resulted in the following dimensional expression for the nucleation frequency per site:

$$\nu_{ns} = \frac{10^4}{s \cdot K^3}(T_l - T_v)^3 . \tag{33}$$

The nucleation site density is related to the bubble departure radius and the cavity size as:

$$N_{ns} = 0.25 \cdot 10^{-7} \frac{R_{dep}^2}{R_{cs}^4} . \tag{34}$$

The cavity size R_{cs} of an active site was assumed to be equal to the critical radius of a spherical vapor bubble:

$$R_{cs} = \frac{2\sigma T_v}{\rho_v \Delta h_{lv}(T_l - T_v)} . \tag{35}$$

By balancing drag and surface tension forces acting on a single bubble assumed to grow entirely within the viscous sublayer, the departure radius was obtained as:

$$R_{dep} = 0.58 \left[\left(\frac{\sigma R_{cs}}{\rho_l} \right)^{0.5} \left(\frac{\mu_l}{\tau_w} \right)^{0.7} \left(\frac{\rho_l}{\mu_l} \right)^{0.3} \right]^{5/7} , \tag{36}$$

where the wall shear stress τ_w is calculated by the Blasius relation. Amongst other effects such as non-sphericity and bubble interference, the influence of contact angle has been neglected in this correlation.

Combining of Equations (33) and (34) gives the perimeter-averaged bubble generation rate from active nucleation sites as:

$$\phi_{wn} = \frac{\nu_{ns} N_{ns} \xi}{A} \tag{37}$$

with ξ the duct perimeter and A the cross-sectional area of the channel [9]. It is presumed that the frequency ν_{ns} is uniform around the channel perimeter. Wall nucleation is assumed to start at the location where the local pressure drops to the saturation pressure according to the liquid temperature. It then continues in the bubbly flow up to the transition to the next flow regime.

Bulk nucleation: For critical flashing flows through pipes, heterogeneous nucleation initiated by foreign particles and dissolved gases suspended in the bulk fluid is assumed, where the bubble number density at flashing inception is calculated as a function of the pipe's length-to-diameter ratio using:

$$\frac{N_b}{m^{-3}} = \begin{cases} \exp\left[24 + 2\ln\left(\frac{L}{D_h}\right)\right], & L/D_h \le 10 \\ \exp\left[35 - 2.8\ln\left(\frac{L}{D_h}\right)\right], & L/D_h > 10 , \end{cases} \tag{38}$$

as found by Dagan et al. [25]. They presumed a minimum liquid superheat of 3 K to determine the location of flashing inception. In this study, the latter was substituted by the superheat criterion in Equation (35). The bulk nucleation rate is taken to be zero everywhere except for the mesh cell corresponding to the flashing plane. At this cell, Equation (38) is used to calculate the bulk nucleation rate approximately as:

$$\phi_{bn} \simeq N_b \frac{v_l}{\Delta z} , \tag{39}$$

where Δz is the axial step-size.

20.2.3.5 Interfacial area

The constitutive equations for the interfacial area per unit volume depend on assumptions for the geometrical shape and the phase distribution.

In the *bubbly flow* regime, $0 < \alpha_v \leq 0.3$, the interfacial area per unit volume for spherical shaped vapor bubbles is given by:

$$A^* = A_b^* = \frac{3\alpha_v}{R_b} = (36\,\pi\,N_b)^{1/3} \alpha_v^{2/3} . \tag{40}$$

In the *bubbly-slug flow*, $0.3 < \alpha_v < 0.8$, it is assumed that some spherical bubbles continue to grow, while others coalesce to form cylindrical plugs (Taylor-like bubbles). Thus, the total interfacial area per unit volume in this flow regime is the sum of the surface portions of bubbles and plugs:

$$A^* = A_{bs}^* + A_p^* , \tag{41}$$

$$A_p^* = \frac{4\alpha_{v,p}^{2/3}}{\alpha_{v,a}^{1/6} D_h} \tag{42}$$

is the interfacial area density of the plugs. Note that at $\alpha_p = \alpha_a$, Equation (42) yields the relationship for the interfacial area of annular flow. The plug portion of void fraction is given by:

$$\alpha_{v,p} = \frac{1}{1 - \alpha_{v,b}} \left[\alpha_v - \alpha_{v,b} \left(1 - \frac{(\alpha_v - \alpha_{v,b})(1 - \alpha_{v,a})}{\alpha_{v,a} - \alpha_{v,b}} \right) \right] . \tag{43}$$

Thus, the bubble slugs portion of void fraction can be determined from:

$$\alpha_{v,bs} = \alpha_v - \alpha_{v,p} . \tag{44}$$

The interfacial area density of the bubble slugs is obtained from Equation (40) as:

$$A_{bs}^* = A_b^*(\alpha_{v,bs}) . \tag{45}$$

For *dispersed droplet flow*, $\alpha_v \geq 0.8$, the interfacial area density is determined from:

$$A^* = A_d^* = \frac{3(1-\alpha_v)}{R_d} , \tag{46}$$

where the droplet radius is calculated using:

$$R_d = \frac{\sigma We}{2\rho_v(v_v - v_l)^2} . \tag{47}$$

The Weber number in Equation (47) is assumed to be constant at $We = 5$ [12].

20.2.3.6 Interphase heat transfer

The heat flux from the liquid to the vapor/liquid interface, which is assumed to be at saturation, can be expressed as:

$$q^* = h^*(T_l - T^s) , \tag{48}$$

where h^* is the interfacial heat transfer coefficient to be modelled according to the flow regime assumptions.

For the *bubbly flow* regime, $\alpha_v \leq 0.3$, Blinkov et al. [12] applied a relationship for the thermal phase of bubble growth in superheated liquid developed by Labuntzov et al. [26] to yield the interfacial heat transfer coefficient as:

$$h = h_b = \frac{12}{\pi} Ja \left[1 + \frac{1}{2}\left(\frac{\pi}{6\,Ja}\right)^{2/3} + \frac{\pi}{6\,Ja}\right] \frac{k_l}{2\,R_b} , \tag{49}$$

where the Jakob number is defined as:

$$Ja = \frac{\rho_l c_{pl}(T_l - T_v)}{\rho_v \Delta h_{lv}} . \tag{50}$$

Equation (49) is further used in the *bubbly-slug flow* regime, $0.3 < \alpha_v < 0.8$, for the yield the heat transfer of the bubbles left in the slug. The heat transfer coefficient for the plugs is correlated according to:

$$h_p^* = 0.0073\,\rho_l v_l c_{pl} . \tag{51}$$

The net heat transfer rate is the sum of the heat transfer rates for the bubbles and the plugs, which yields for the heat transfer coefficient according to Blinkov et al. [12] as:

$$h^* = h_{bs}^* = \frac{h_b^* A_{bs}^* + h_p^* A_p^*}{A^*} . \tag{52}$$

In the *dispersed droplet flow* regime,

$$h_\mathrm{d}^* = \frac{k_\mathrm{l} Nu}{2\,R_\mathrm{d}}\,, \tag{53}$$

where the droplet Nusselt number is assumed to be constant $Nu = 16$ [27].

To account for vapor recondensation phenomena due to pressure recovery in two-phase flows through divergent ducts, the heat transfer into the liquid phase was evaluated according to Hughes et al. [28].

20.2.3.7 Interphase mass transfer

The rate of phase change is due to latent heat transfer at the interface, and is given by a function of interfacial area density and net heat flux to the interface [29]:

$$\Gamma_\mathrm{v} = \frac{q^* A^*}{h_\mathrm{v}^\mathrm{s} - h_\mathrm{l}^\mathrm{s}}\,, \tag{54}$$

where h_v^s and h_l^s are the specific saturation of vapor and liquid, respectively. Note that $\Gamma_\mathrm{v} > 0$ for vaporization, and $\Gamma_\mathrm{v} < 0$ for condensation, respectively.

20.3 Numerical Method

In order to solve the nonlinear system of partial differential equations (Equations (1) to (6)), a semi-implicit finite difference method is applied [8, 30, 31]. The basic idea of this method is to replace the original set of differential equations by a system of finite-difference equations partially implicit with respect to time. As can be seen later, the degree of implicitness chosen is such that the field equations applied to an individual mesh cell can be reduced to a single difference equation in terms of the pressure alone.

A staggered spatial mesh is imposed on the geometry in question, where thermodynamic variables are fixed at cell centers and velocities are evaluated at cell boundaries (Figure 20.3). Continuous grid refinement was applied to en-

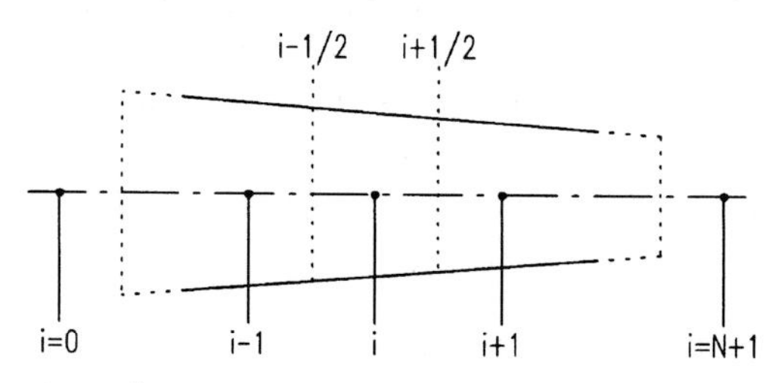

Figure 20.3: Nodalization scheme.

hance spatial resolution at regions where strong pressure and bubble number density gradients are expected, e.g. in the vicinity of a nozzle throat. To provide necessary relationships among variables at cell centers and cell faces, a weighted donor cell technique has been used [30]. This approach represents a compromise between the very stable but first-order accurate full donor cell differencing, and the more accurate but less stable central difference method. The overall stability criterion for the time step size of the semi-implicit method is given by the material Courant limit as:

$$\Delta t \leq \min\left[\frac{\Delta z_i}{|v_{v,i}|}, \frac{\Delta z_i}{|v_{l,i}|}\right], \quad i = 0 \dots N+1 . \tag{55}$$

In the finite-difference formulations of the mass, thermal energy and bubble transport equations, only the respective time-dependent variables, the pressure, the phasic velocities, and the source terms are calculated at the new time level t^{n+1}. All other variables involved, including the convection terms, are calculated explicitly at the old time level t^n. To give an example, the finite-difference form of the vapor mass equation (Equation (1)) reads:

$$(a_v\rho_v)_i^{n+1} - (a_v\rho_v)_i^n + \frac{1}{A_i}\frac{\Delta t}{\Delta z_i}\left\{(a^n\rho_v^n v_v^{n+1} A)_{i+1/2} - (a^n\rho_v v_v^{n+1} A)_{i-1/2}\right\} = \Delta t\,\Gamma_{v,i}^{n+1} , \tag{56}$$

where i denotes variables at cell centers, and $i \pm 1/2$ terms at cell faces.

The finite difference mixture momentum equations (Equations (3) and (4)) contain only the time-dependent terms in question and the pressure gradient terms at the forward time $n+1$, and can be expressed symbolically as:

$$v_{v,i+1/2}^{n+1} = -\beta_{v,i+1/2}^n (p_{i+1}^{n+1} - p_i^{n+1}) + \gamma_{v,i+1/2}^n , \tag{57}$$

$$v_{l,i+1/2}^{n+1} = -\beta_{l,i+1/2}^n (p_{i+1}^{n+1} - p_i^{n+1}) + \gamma_{l,i+1/2}^n , \tag{58}$$

where the coefficients $\beta_{v,i+1/2}^n$, $\gamma_{v,i+1/2}^n$, $\beta_{l,i+1/2}^n$, and $\gamma_{l,i+1/2}^n$ comprise convection and force terms at time t^n. Note that Equations (57) and (58) represent simple linear relations between the phasic velocities and the pressure.

The finite difference equations forms originating from the remaining balance equations (Equations (2), (5) and (6)) along with relationships between cell centers and faces plus the constitutive equations represent a nonlinear algebraic system of equations. Herein, a Newton-iteration method is applied, where all equations have to be linearized around the latest iterate values of the unknowns. Linearization of Equation (56) yields:

$$\Delta p_i^k\left\{\left[\alpha_v\frac{d\rho_v}{dp}\right]_i^k+\frac{\Delta t}{A_i\Delta z_i}\left[(\alpha_v\rho_v\beta_v A)_{i+1/2}^n+(\alpha_v\rho_v\beta_v A)_{i-1/2}^n\right]-\Delta t\left[\frac{\partial\Gamma_v}{\partial p}\right]_i^k\right\}$$

$$+\Delta\alpha_i^k\left\{\rho_{vi}^k-\Delta t\left[\frac{\partial\Gamma_v}{\partial\alpha_v}\right]_i^k\right\}+\Delta T_{l,i}^k\Delta t\left[\frac{\partial\Gamma_v}{\partial T_l}\right]_i^k$$

$$-\frac{1}{A_i}\frac{\Delta t}{\Delta z_i}\left[\Delta p_{i+1}^k(\alpha_v\rho_v\beta_v A)_{i+1/2}^n+\Delta p_{i-1}^k(\alpha_v\rho_v\beta_v A)_{i-1/2}^n\right]$$

$$=-\left\{\left(\alpha_v^k\rho_v^k-\alpha_v^n\rho_v^n\right)_i-\Delta t\Gamma_{v,i}^k+\frac{1}{A_i}\frac{\Delta t}{\Delta z_i}\left[(\alpha_v^n\rho_v^n v_v^{n+1}A)_{i+1/2}-(\alpha_v^n\rho_v^n v_v^{n+1}A)_{i-1/2}\right]\right\}, \tag{59}$$

where k denotes iterative quantities at the forward time $n+1$.

The same procedure has to be carried out for the remaining four equations, namely on the difference equations originating from Equations (2), (5) and (6). The resulting set of equations at each mesh cell expressed in matrix form read:

$$-\boldsymbol{b}_1\Delta p_{i-1}^k+\boldsymbol{M}\boldsymbol{x}_i^k-\boldsymbol{b}_2\Delta p_{i+1}^k=\boldsymbol{c}, \tag{60}$$

where $\boldsymbol{x}_i^k=(\Delta p,\Delta\alpha_v,\Delta T_l,\Delta N_b)_i^k$ is the transposed vector of unknowns at node i. $\boldsymbol{b}_1$ and $\boldsymbol{b}_2$ are vectors containing the factors of the pressure corrections at adjacent mesh cells. The elements of the 4×4 matrix M and the vector $\boldsymbol{c}$ are easily identified by inspection of Equation (59) and its equivalents not listed, where $\boldsymbol{c}$ represents terms on their right-hand sides.

Equations (57) and (58) give the velocities at time t^{n+1} to be appearing in the matrices M and the vector $\boldsymbol{c}$ of Equation (60). With the velocities at hand, the first term in Equation (60) is a function of the pressure alone and can be transformed into a tri-diagonal system of equations, or written for mesh cells $i=1$ to N:

$$L_i\Delta p_{i-1}^k+\Delta p_i^k+U_i\Delta p_{i+1}^k=R_i\,. \tag{61}$$

At the boundary cells located outside the considered geometry have to specified, the following inlet and exit pressure corrections depending on the exit mixture velocity are imposed:

$$\Delta p_0^k=0\,,\ p_0=p_{in}=\text{const.} \tag{62}$$

$$\Delta p_{N+1}^k=\begin{cases}0, & p_0=p_{in}=\text{const. for subsonic flow: } v_{m,N+1/2}<c_s\\ \Delta p_N^k & \text{for sonic, or supersonic flow: } v_{m,N+1/2}\geq c_s\,,\end{cases} \tag{63}$$

where the mixture velocity v_m is defined as

$$v_m = \frac{\alpha_v \rho_v v_v + \alpha_l \rho_l v_l}{\rho_m}, \tag{64}$$

and c_s is the sonic velocity of the two-phase mixture [32].

With the pressure corrections obtained from the tri-diagonal system (Equation (61)), all other corrections can be determined from Equation (60). These corrections will then be added to the results of the preceding iteration step generating new values for p, α_v, T_l, and N_b. The iterative process is repeated until the maximum relative variation of pressure drops below a sufficiently small prescribed limit. The typical number of iterations per time step ranged between two and five at early stages of solution and one or two iterations near steady state.

In addition to the boundary conditions for the pressure, the inlet liquid temperature was assumed to be constant. For subcooled inlet conditions, the void fraction and the bubble number density at $i = 0$ are set equal to zero. In case of a two-phase mixture already present at the inlet, the void fraction at $i = 0$ is calculated by means of a simple energy balance utilizing measured stagnation enthalpies and assuming homogeneous flow. Moreover, an estimated inlet bubble number density has to be specified. All other variables are calculated by linear extrapolation of downstream and upstream results.

20.4 Results

For the purpose of code validation, calculations were conducted for various experiments reported in literature and for own experiments with plain nozzles. Regarding inlet temperature levels, both subcooled and saturated inlet conditions were regarded. Since modelling of flashing flows is a more challenging task due to the need to simulate delayed nucleation additionally, this report concentrates on such flows.

In order to illustrate the changes with respect to results obtained by works preceding this one, we will distinguish the following two-phase flow models:

- HNEM: Homogeneous Non-Equilibrium Model (from [3], where $S = v_v/v_l = 1$),
- DFM: Drift-Flux Model (from [3] with slip correlation of [33]),
- TFM: Two-Fluid Model (this work).

Bolle et al. [34] carried out a set of flashing experiments on a real safety valve and on a model of it, constructed to enable pressure measurements. The shape of the valve model of 26.5 mm ID at the inlet, and 10.43 mm ID at the second

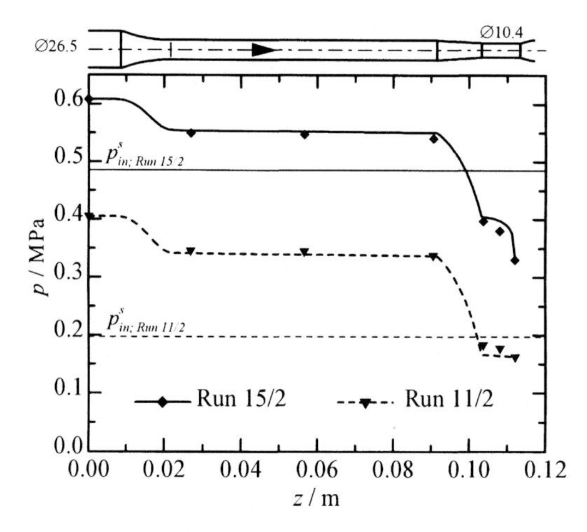

Figure 20.4: Pressure distributions [34], Runs 11/2 and 15/2.

straight pipe section, is shown on top of Figure 20.4, comparing measured and calculated pressure distributions of Runs 11/2 and 15/2. As simulated flashing takes place not before the vicinity of the narrowest cross-section following the second convergent valve section, void fractions are considerably small within the computational domain. Consequently, there is no difference in the axial pressure profiles between the two-fluid and drift-flux in Figure 20.4.

Table 20.1 lists additional experimental data and calculated mass flow rates from the drift-flux and the two-fluid model.

Reports on the experiments conducted at the Brookhaven National Laboratory (BNL) with upwards directed flashing flows through a convergent-divergent nozzle are given by Abuaf et al. [35] and Zimmer et al. [5]. As can be

Table 20.1: Measured and calculated data of the experiments [34].

Run	p_{in}	$T^{\text{s}}_{\text{in}} - T_{\text{l,in}}$	$\dot{m}_{\text{exp}}$	$\dot{m}^{\text{DFM}}_{\text{clc}}$	$\dot{m}^{\text{TFM}}_{\text{clc}}$
	MPa	K	kg/s	kg/s	kg/s
1/2	0.600	141.64	2.70	2.59	2.56
7/2	0.556	35.70	2.21	2.33	2.30
8/2	0.610	39.49	2.35	2.49	2.46
11/2	0.405	24.37	1.75	1.81	1.80
12/2	0.512	33.64	2.12	2.20	2.16
14/2	0.572	37.88	2.27	2.39	2.36
15/2	0.608	8.66	1.65	1.63	1.61
16/2	0.502	2.10	1.31	1.26	1.16
17/2	0.560	6.47	1.61	1.54	1.48
18/2	0.614	10.05	1.77	1.76	1.70

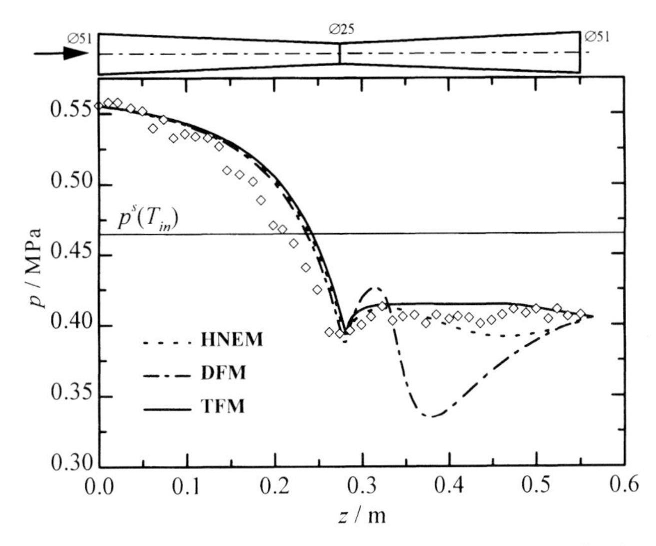

Figure 20.5: Comparison of measured and calculated pressure distributions, Run 309 [35], $p_{\text{in}} = 0.556$ MPa, $\Delta T_{\text{in}} = 6.8$ K, $\dot{m}_{\text{exp}} = 8.94$ kg/s, $\dot{m}_{\text{clc}}^{\text{TFM}} = 8.45$ kg/s.

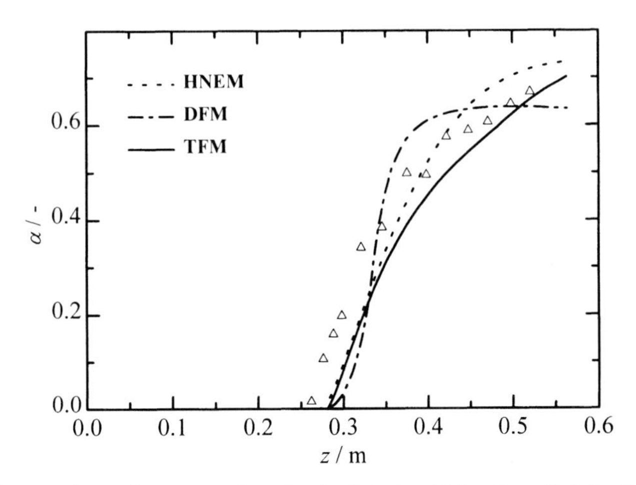

Figure 20.6: Comparison of measured and calculated void fraction distributions, Run 309 [5].

seen on top of Figure 20.5, the nozzle was symmetrically to the throat of 25 mm ID and 51 mm ID at both ends. Detailed pressure and void fraction profiles were measured. The simulated location of flashing inception always occurred in the immediate vicinity of the nozzle throat, where the highest liquid superheatings are attained. This corresponds to experimental observations as reported by Zimmer et al. [5].

Figures 20.5 and 20.6 compare measured and calculated axial pressure and void fraction profiles of the high mass flux test Run 309.

The two-fluid and the homogeneous model show excellent agreement with experimental pressure and void fraction distributions. The almost constant pressure in the diverging section accompanied by a steep rise in void fraction were found to be typical at high mass-fluxes. Here, the rapid vapor generation seems to completely offset the increasing cross-sectional area. The calculated void fraction profiles are slightly displaced to the nozzle exit with respect to measured data. This may be due to pronounced non-uniform radial void fraction distribution at high mass fluxes, which can not be resolved by the models. The examination of the chordal void fraction measurements of Zimmer et al. [5] revealed that most of the vapor phase was in the vicinity of the wall. This might imply that vapor bubbles mainly nucleate at the wall and rather tend to coalesce downstream to build up a vapor annulus than to migrate into the center of the nozzle. However, an improved model must take the effect of non-uniform void distribution into account.

Figure 20.7 describes the calculated phasic velocity conditions for this test run. As expected, nearly equal velocities of vapor and liquid can be found for bubbly flow regime, since the liquid and vapor phase are strongly coupled in this regime from the mechanical viewpoint. Considerable deviation between the phasic velocities appears when the void fraction exceeds the assumed transition boundary from bubbly to bubbly-slug flow at $\alpha_v = 0.3$.

At lower mass fluxes (cf. Figure 20.8), the measured and calculated axial pressure profiles increase within the diverging part of the nozzle and thus have a qualitatively similar appearance as in single-phase flow through a venturi. As can be seen from Figure 20.9, vapor generation short downstream of the throat is overestimated at relatively low mass flow rates. In this region, intermittent bubble appearance followed by sudden bubble collapse in the pressure recovery zone was detected. The large scatter of measured void fraction downstream of the throat reflects this phenomenon. The rapid condensation process taking

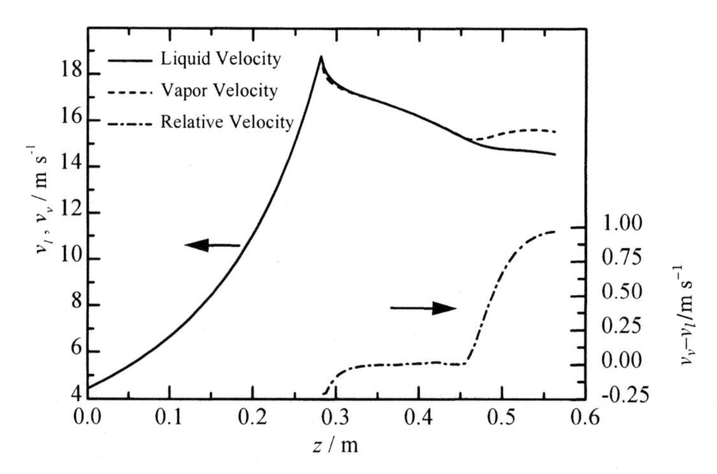

Figure 20.7: Calculated liquid and vapor velocity distributions, Run 309 [35].

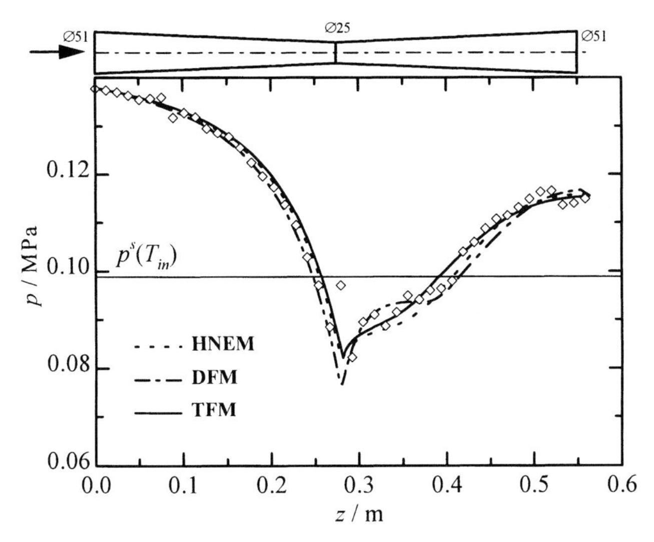

Figure 20.8: Comparison of measured and calculated pressure distributions, Run 780 [5], $p_{\text{in}} = 0.138$ MPa, $\Delta T_{\text{in}} = 9.6$ K, $\dot{m}_{\text{exp}} = 5.13$ kg/s, $\dot{m}_{\text{clc}}^{\text{TFM}} = 4.98$ kg/s.

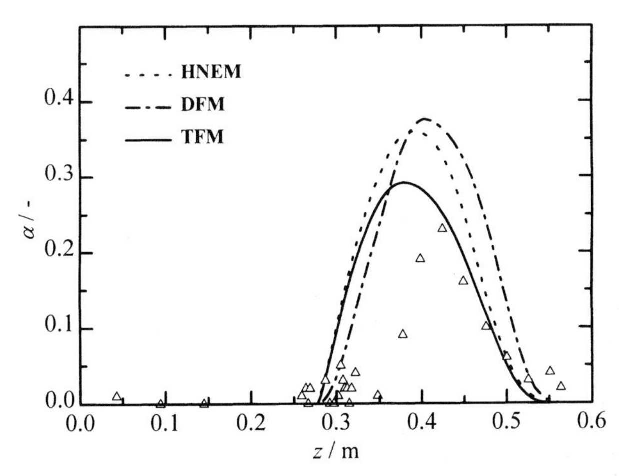

Figure 20.9: Comparison of measured and calculated void fraction distributions, Run 780 [35].

place where the local pressure exceeds the saturation pressure is reproduced quite well by the two-fluid model.

Additional data of considered BNL test runs are listed in Table 20.2.

Further calculations have shown that the predictive quality of the present calculation model depends on mass flux. As outlined above, its use should therefore be restricted to high-mass flux two-phase flows.

Table 20.2: Measured and calculated data of BNL experiments [5, 35].

Run	p_{in}	$T^{\text{s}}_{\text{in}} - T_{\text{l,in}}$	$\dot{m}_{\text{exp}}$	$\dot{m}^{\text{HNEM}}_{\text{clc}}$	$\dot{m}^{\text{DFM}}_{\text{clc}}$	$\dot{m}^{\text{TFM}}_{\text{clc}}$
	MPa	K	kg/s	kg/s	kg/s	kg/s
148	0.305	12.8	7.50	7.95	8.26	7.83
273	0.573	8.4	8.86	9.21	9.45	9.05
288	0.530	4.8	7.25	7.65	7.83	7.52
309	0.556	6.8	8.94	8.56	8.79	8.45
780	0.138	9.6	5.13	5.05	5.33	4.98

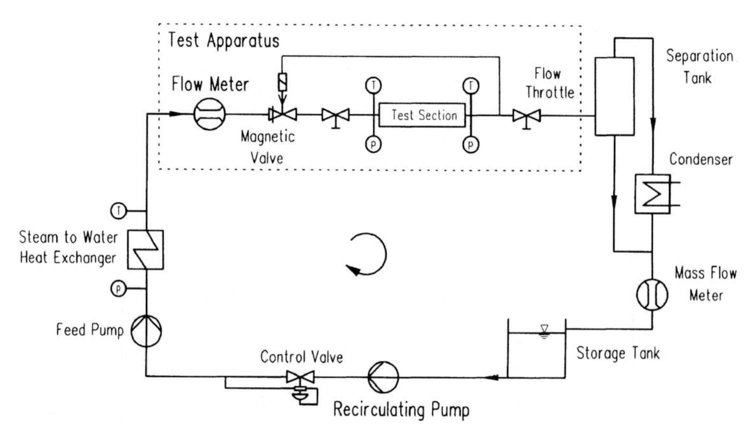

Figure 20.10: Schematic diagram of test facility with test apparatus.

As a part of the present work, a test apparatus was developed in order to study flashing flow in circular and plain convergent-divergent nozzles. The test apparatus was permanently attached to the local district heating test facility as shown in Figure 20.10.

The test fluid, water, is recirculated continuously through the loop at constant upstream operating conditions of pressure, temperature, and flow rate. Simultaneous recordings of flow rate as well as pressures and temperatures at inlet and exit are taken. In addition, wall pressure distributions are measured at interchangeable nozzle elements inserted into the transparent test section as sketched in Figure 20.11.

Flash photographs were taken to examine developing flow patterns, void fraction distribution, and geometric shape of generated vapor. Figure 20.12 is the contour image of such a flash photograph, where the flow direction is from the left to the right. Several different flow regimes are observed as illustrated.

Subcooled water enters the nozzle and is rapidly accelerated in the convergent section. At the sharp edged throat, a liquid jet is formed with a vapor layer at the upper and lower wall. The liquid jet then extends approximately

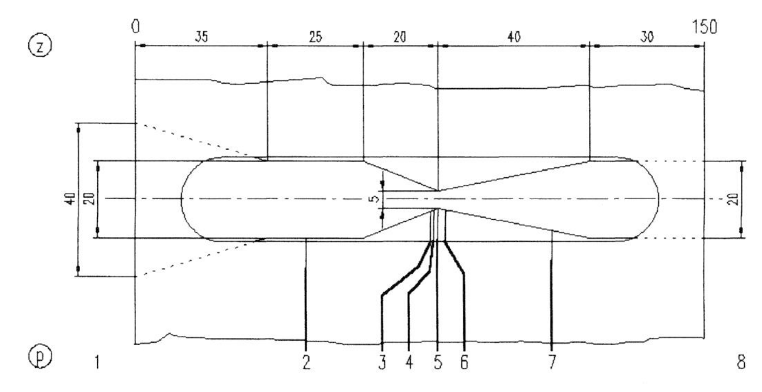

Figure 20.11: Sketch of transparent plain nozzle with wall pressure connections.

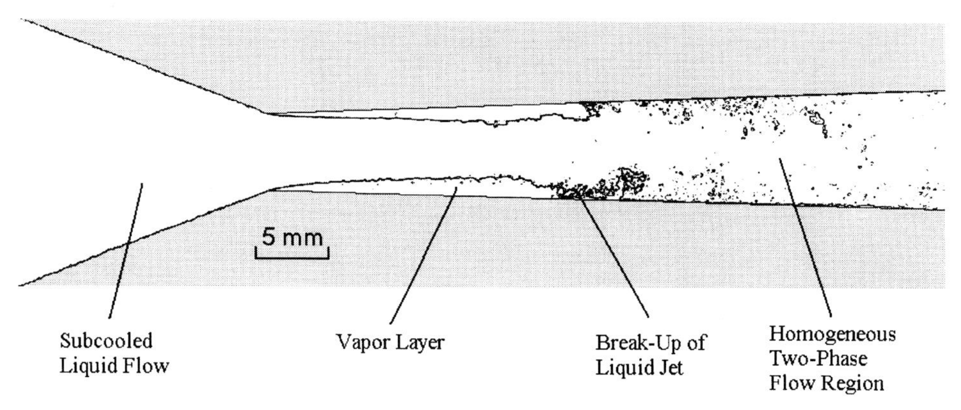

Figure 20.12: Flashing in two-phase flow through plain convergent-divergent nozzle, contour image of Run RB02, $p_{\text{in}} = 3.66$ bar, $\Delta T_{\text{in}} \approx 10$ K, $\dot{m}_{\text{exp}} = 0.452$ kg/s, $\dot{m}_{\text{clc}} = 0.453$ kg/s.

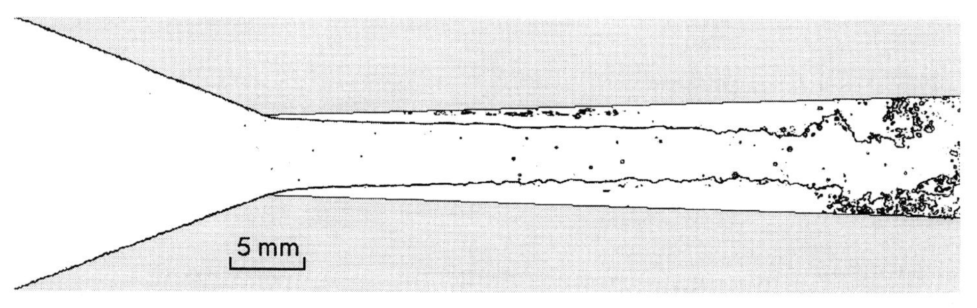

Figure 20.13: Contour image of Run RB03, $p_{\text{in}} = 3.12$ bar, $\Delta T_{\text{in}} \approx 5$ K, $\dot{m}_{\text{exp}} = 0.322$ kg/s, $\dot{m}_{\text{clc}} = 0.310$ kg/s.

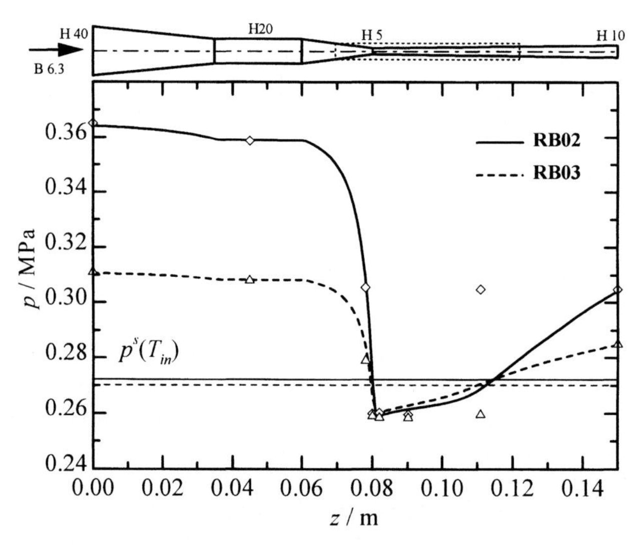

Figure 20.14: Comparison of measured and calculated pressure distributions, Runs RB02 and 03.

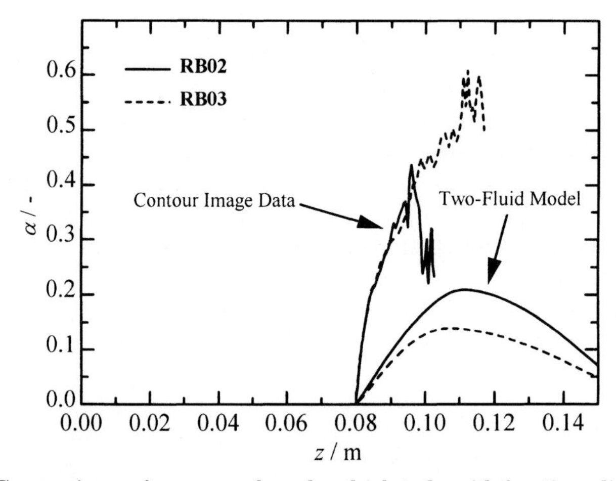

Figure 20.15: Comparison of measured and calculated void fraction distributions, Runs RB02 and 03.

four throat heights downstream before it breaks up, and homogeneous two-phase flow appears.

At lower flow rates and at a lower degree of inlet subcooling, the liquid jet extends much further downstream and spherical bubble appear within the liquid core (Figure 20.13).

These phenomena are quite similar to those observed by Abuaf et al. [35] and Zimmer et al. [5], as described above. Though mass flow rates are predicted quite accurately, the two-fluid model is not capable of reproducing the steep rise of void fraction due to detachment-induced cavitation at the throat. Figures 20.14 and 20.15 compare experimental and calculated pressure and void fraction distributions.

20.5 Conclusion

A quasi one-dimensional two-fluid model for the analysis of thermal and hydrodynamic non-equilibrium effects in hot-water two-phase flow has been developed. It is based on the solution of six time-dependent balance equations by means of a semi-implicit finite-difference method in conjunction with a Newton-iteration technique.

The model includes distributed nucleation processes for flashing flows, bubble growth and coalescence as well as variations of the bubble number density. It simulates three flow regimes: bubbly, bubbly slug and dispersed droplet flow. Two-phase flow of both saturated and subcooled inlet conditions can be simulated.

A novel formulation for the description of the virtual mass effect based on observations and extensive numerical studies is proposed. The relation has been extensively tested against available experimental data found in literature and data from own experiments showed excellent agreement with measured pressure distributions and mass flow rates.

Thus, the model is expected to be an efficient tool for further improvement of correlations needed to describe the initial vapor generation as well as interphase momentum, heat and mass transfer in two-phase critical and near-critical two-phase flow.

Further development is needed to simulate detachment-induced cavitation phenomena at convex-formed walls and condensation shocks in divergent ducts more accurately.

List of Symbols

A cross-sectional area, area
c specific heat capacity, coefficient
C coefficient
D diameter
e specific internal energy
f volumetric force
g acceleration due to gravity
G total mass flux

h specific enthalpy, heat transfer coefficient
i mesh cell index
Ja Jakob number
k thermal conductivity
L pipe or nozzle length
$\dot{m}$ mass flow rate
N number density, number
Nu Nusselt number
p pressure
q heat flux
R radius
S slip ratio
R two-phase multiplier
Re Reynolds number
t time
T temperature
v velocity
We Weber number
x vapor flow quality
z axial coordinate

Greek Symbols

α volume fraction
Γ volumetric rate of phase change
η distribution parameter
θ angle of inclination
μ dynamic viscosity
ν frequency
ξ perimeter
ρ density
σ surface tension
τ shear stress
ϕ volumetric nucleation rate

Subscripts

a annular
b bubble, bubbly
bn bulk nucleation
bs bubble slugs
clc calculated
cs cavity size
dep departure
in inlet
l liquid
lo liquid only
m mixture
ns nucleation site
p isobaric, plug
r relative

d	droplet, drag	s	sonic
ex	exit	v	vapor, vaporization
exp	experimental	vm	virtual mass
f	friction	wn	wall nuclation
h	hydraulic	1ϕ	single-phase
i	cell index	2ϕ	two-phase

Superscripts

k	iteration count	s	saturation
n	time step	$*$	interfacial property

References

[1] G. Wallis: *One-Dimensional Two-Phase Flow.* McGraw-Hill, New York, 1969.

[2] E. Elias, G.S. Lellouche: *Two-phase critical flow.* Int. J. Multiphase Flow **20** (1994) 91–168.

[3] M. Wein, J. Huhn: *Numerical Simulation of Non-Equilibrium Hot-Water Two-Phase Flows.* In: Convective Flow and Pool Boiling Conference, Kloster Irsee, Germany, 1997.

[4] M. Wein, J. Huhn: *Numerical Simulation of Non-Equilibrium Hot-Water Two-Phase Flows.* Submitted for publication to Taylor and Francis, Philadelphia, USA, 1998.

[5] G.A. Zimmer, B.J.C. Wu, W.J. Leonhardt, N. Abuaf, O.C. Jones, Jr.: *Pressure and Void Distributions in a Converging-Diverging Nozzle with Nonequilibrium Water Vapor Generation.* Tech. Rep. BNL-NUREG-26003, Brookhaven National Laboratory, Informal Report, 1979.

[6] J.R. Fincke: *The Correlation of Nonequilibrium Effects in Choked Nozzle Flow with Subcooled Upstream Conditions.* In: ANS Small Break Specialists Mtg., California, 1981, pp. 4–30.

[7] M. Ishii: *Thermo-Fluid Dynamic Theory of Two-Phase Flow.* Eyrolles, Paris, France, 1974.

[8] J.A. Trapp, R.A. Riemke: *A Nearly-Implicit Hydrodynamic Numerical Scheme for Two-Phase Flows.* J. Computational Physics **66** (1986) 62–82.

[9] G. Kocamustafaogullari, M. Ishii: *Interfacial Area and Nucleation Site Density in Boiling Systems.* Int. J. Heat Mass Transfer **26**(9) (1983) 1377–1387.

[10] W. Wagner, J.R. Cooper, A. Dittmann, J. Kijima, H.-J. Kretzschmar, A. Kruse, R. Mares, K. Oguchi, H. Sato, I. Stücker, O. Sifner, Y. Takaishi, I. Tanishita, J. Trübenbach, T. Willkommen: *The IAPWS Industrial Formulation 1997 for the Thermodynamic Properties of Water and Steam.* Zur Veröffentlichung eingereicht beim ASME Journal of Engineering for Gas Turbines and Power, 1998.

[11] W. Wagner, A. Kruse: *Zustandsgrößen von Wasser und Wasserdampf. Der Industrie-Standard IAPWS-IF97 für die thermodynamischen Zustandsgrößen und ergänzende Gleichungen für andere Eigenschaften.* Europäische Ausgabe deutsch/englisch edn., Springer-Verlag, 1998.

[12] V.N. Blinkov, O.C. Jones, Jr., B.I. Nigmatulin: *Nucleation and Flashing in Nozzles – 2, Comparison with Experiments Using a Five Equation Model for Vapour Void Development.* Int. J. Multiphase Flow **19**(6) (1993) 965–986.

[13] F. Dobran: *Nonequilibrium Modeling of Two-Phase Critical Flows in Tubes.* Trans. ASME, J. Heat Transfer **109** (1987) 731–738.

[14] H.J. Richter: *Separated two-phase flow model: Application to critical two-phase flow.* Topical Report EPRI NP-1800, Electric Power Research Institute, 1981.

[15] C.F. Schwellnus, M. Shoukri: *A Two-Fluid Model for Non-Equilibrium Two-Phase Critical Discharge.* The Canadian Journal of Chemical Engineering **69** (1991) 188–197.

[16] L. Friedel: *Improved Friction Pressure Drop Correlations for Horizontal and Vertical Two-Phase Pipe Flow.* 3R intern. **18**(7) (1979) 485–491.

[17] P.N. Rowe, C.A. Henwood: *Drag Forces in a Hydraulic Model of a Fluidized Bed-I.* Trans. Inst. of Chem. Eng. **39** (1961) 43–54.

[18] L. Prandtl, K. Oswatitsch, K. Wieghardt: *Führer durch die Strömungslehre.* Chap. 6.7, 9th edn., Verlag Vieweg, 1990.

[19] N. Zuber, J.A. Findlay: *The Effects of Non-Uniform Flow and Concentration Distributions and the Effect of the Local Relative Velocity on the Average Volumetric Concentration in Two-Phase Flow.* Tech. rep., General Electric Company, 1964.

[20] N. Zuber: *On the Dispersed Two-Phase Flow in the Laminar Flow Regime.* Chemical Engineering Science **19** (1964) 897–917.

[21] D.A. Drew, L.Y. Cheng, R.T. Lahey, Jr.: *The Analysis of Virtual Mass Effects in Two-Phase Flow.* Int. J. Multiphase Flow **5** (1979) 233–242.

[22] T.L. Cook, F.H. Harlow: *Virtual Mass in Multiphase Flow.* Int. J. Multiphase Flow **10**(6) (1984) 691–696.

[23] L. van Wijngaarden: *Hydrodynamic Interaction Between Gas Bubbles in Liquid.* J. Fluid Mech. **77**(1) (1976) 27–44.

[24] T.S. Shin, O.C. Jones: *Nucleation and Flashing in Nozzles – 1. A Distributed Nucleation Model.* Int. J. Multiphase Flow **19**(6) (1993) 943–964.

[25] R. Dagan, E. Elias, E. Wacholder, S. Olek: *A Two-Fluid Model for Critical Flashing Flows in Pipes.* Int. J. Multiphase Flow **19**(1) (1993) 15–25.

[26] D.A. Labuntzov, B.A. Kolchugin, V.S. Golovin, E.A. Zakharova, L.N. Vladimirova: *High Speed Camera Investigation of Bubble Growth for Saturated Boiling in a Wide Range of Pressure Variations.* Thermophysics of High Temperature **2**(2) (1964) 446–453.

[27] C.W. Solbrig, J.H. McFaden, R.W. Lyczkowski, E.D. Hughes: *Heat Transfer and Friction Correlations Required to Describe Steam-Water Behavior in Nuclear Safety Studies.* AIChE Symp. Ser. **74**(174) (1978) 105–109.

[28] E.D. Hughes, M.P. Paulsen, L.J. Agee: *A Drift-Flux Model of Two-Phase Flow for RETRAN.* Nuclear Technology **54** (1981) 410.

[29] P. Saha: *A Nonequilibrium Vapor Generation Model for Flashing Flows.* Trans. ASME, J. Heat Transfer **106** (1984) 198–203.

[30] D.R. Liles, W.H. Reed: *A Semi-Implicit Method for Two-Phase Fluid Dynamics.* J. Computational Physics **26** (1978) 390–407.

[31] EPRI: *Review and Application of the TRAC-PD2 Computer Code.* Report EPRI NP-2826, Electric Power Research Institute, 1983.

[32] L. Cheng, R.T. Lahey, Jr., D.A. Drew: *The Effect of Virtual Mass on the Prediction of Critical Flow.* In: Transient Two-Phase Flow, Springer, 1983, pp. 323–340.

[33] R. Huq, J.L. Loth: *Analytical Two-Phase Flow Void Prediction Method.* J. Thermophysics and Heat Transfer **6**(1) (1992) 139–144.

[34] L. Bolle, P. Downar-Zapolski, J. Franco, J.M. Seynhaeve: *Flashing Water Flow through a Safety Valve.* J. Loss Prev. Process Ind. **8**(2) (1995) 111–126.

[35] N. Abuaf, O.C. Jones, Jr., B.J.C. Wu: *A Study of Nonequilibrium Flashing of Water in a Converging-Diverging Nozzle, Vol. 1 – Experimental.* Tech. Rep. NUREG/CR-1864 and BNL-NUREG-51317, U.S. Nuclear Regulatory Commission, 1981.

21 Numerical Studies of Flow in Fuel Injector Nozzles – Interaction of Separation and Cavitation

Günter H. Schnerr, Claas Vortmann and Jürgen Sauer *

Abstract

The flow through injector nozzles is important because it affects the spray and the atomization process. If cavitation occurs inside the nozzle, the numerical modelling is quite complicated. The first part of this report deals with the numerical investigation of single-phase flow through an injector nozzle. Experimental data are compared with laminar and turbulent flow calculations. The results are discussed by assuming that cavitation occurs if the pressure drops beyond the vapor pressure. The influence of the throat inlet radius is investigated. Preliminary studies of cavitating flow were performed by using a "law of state"-cavitation model.

21.1 Introduction

The study of internal flow through injector nozzles is very important in order to optimize the combustion process in engines and to reduce emissions, although the relation between the injector design and combustion process is quite complicated [1]. Experiments showed that, depending on the injection pressure, cavitation is present leading to very complicated flow structures, as can be seen from Figure 21.1. The picture is taken from an experimental study of Roosen et al. [2], the fluid is Decalin, the injection pressure is $p_{inj} = 30$ bar, the pressure at the nozzle outlet is $p_{outlet} = 1$ bar. At the nozzle throat inlet, a cavity forms (sheet cavitation). It becomes unstable with increasing cavity length, breaks off and cavitation clouds travel downstream. Experimental studies of Chaves et al. [3] showed that

* Universität Karlsruhe (TH), Fachgebiet Strömungsmaschinen, Kaiserstr. 12, D-76128 Karlsruhe, Germany

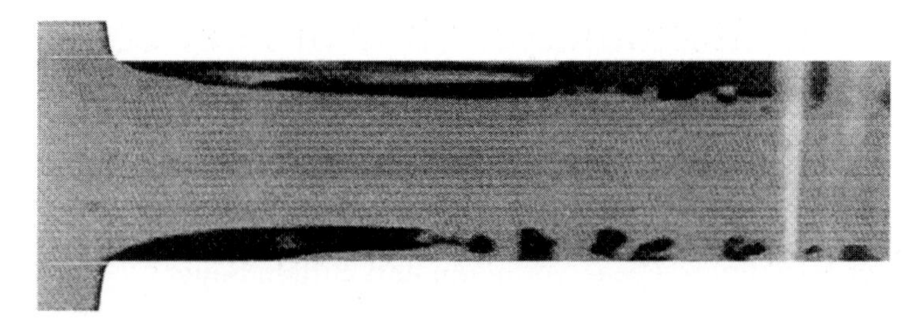

Figure 21.1: Experimental visualisation of cavitating flow through an injector nozzle investigated by Roosen et al. [2]; fluid Decalin, flow from left to right, inlet pressure = 30 bar, outlet pressure = 1 bar; dark areas denote zones where cavitation occurs.

the presence of cavitation can significantly alter the characteristics of the nozzle flow, such as the spray angle, etc. To study this flow numerically, the cavitation model must be capable of modelling the sheet cavitation and the travelling cloud cavitation. The development of such a model can base on a combination of a model for sheet cavitation [4–7] with a cloud cavitation model [7], or to use a cavitation model that does not explicitly distinguish between these cavitation types, as proposed by Delannoy and Kueny [8], Schmidt [9], or Chen and Heister [10]. In the present study, the model of Delannoy and Kueny [8] will be used, a summary of the numerical approach and the description of the model will be given in the following sections. The first part of the present study discusses in detail the single-phase flow through an injector nozzle, especially the effects of the throat inlet radius on pressure distribution and flow structure are addressed. The second part is dedicated to the investigation of cavitating flows in a duct with a triangular obstacle using two-phase flow analysis.

21.2 Numerical Scheme and Cavitation Model

The CFD-Software[1] that is applied in this study solves the two-dimensional continuity equation for a compressible fluid and the Navier-Stokes equations in cartesian coordinates:

$$\frac{\partial \rho}{\partial t} + \frac{\partial (\rho v_i)}{\partial x_i} = 0\,, \quad i = 1,2 \tag{1}$$

$$\rho \frac{\mathrm{D} v_i}{\mathrm{D} t} = -\frac{\partial p}{\partial x_i} + \frac{\partial \tau_{ij}}{\partial x_j}\,, \quad i,j = 1,2\,, \tag{2}$$

[1] The original version of the CFD-Software was developed by Matthias Krömer, Institute for Ship Building at the Technical University Hamburg-Harburg.

where the viscous part of the stress tensor for a Newtonian fluid τ_{ij} is defined as:

$$\tau_{ij} = \mu\left(\frac{\partial v_i}{\partial x_j} + \frac{\partial v_j}{\partial x_i}\right), \quad i,j = 1,2. \tag{3}$$

The compressible term of the viscous stress tensor is not yet implemented in the code. This term is only non-zero, if turbulent, compressible effects occur. For the modelling of the turbulent flow, the k-ω-model of Wilcox [11] is used. Hence, it is necessary to solve the transport equation for the turbulent, kinetic energy $k = 1/2\,\overline{v_i' v_i'}$ and the equation for the rate of dissipation energy ω· The momentum and turbulence equations are seen as special cases of the general transport equation ($\Phi = v_j$ for momentum conservation, $\Phi = k$ respectively $\Phi = \omega$ for the turbulent equations).

$$\underbrace{\rho\frac{\partial \Phi}{\partial t}}_{\text{unsteady}} + \underbrace{\rho v_i \frac{\partial \Phi}{\partial x_i}}_{\text{convective}} = \underbrace{\rho v_i \frac{\partial \Gamma \Phi}{\partial x_i}}_{\text{diffusive}} + \underbrace{q}_{\text{Source}}, \quad i = 1,2. \tag{4}$$

The integral form of this transport equation is solved by using an implicit Finite-Volume-Method and non-orthogonal structured, single-block grids with a collocated arrangement of variables. For time discretization, the implicit Euler method is applied. With the deferred correction approach, an explicit second order accurate central differencing scheme (CDS) is combined with an implicit first order downwind differencing scheme (UDS) for the convective and diffusive terms. At solid walls, the boundary conditions for the turbulent quantities are specified using wall functions, which rely on a logarithmic velocity profile.

By dividing the solution domain into a finite number of control volumes and applying the integral form of Equation (4) to each control volume, an algebraic system of equations is obtained. Each equation has the form:

$$A_P \Phi + \sum_l A_l \Phi_l = Q_p^{\Phi} \quad l = E, W, N, S, \tag{5}$$

where A_P is the coefficient for the value of Φ of the control volume centered around node P, and A_l are the coefficients of the neighboring control volumes: east, west, north, south. The linear equation system is sequentially solved by SIP (Strongly Implicit Procedure) for each variable Φ. The coupling of pressure, velocity and density is achieved by a SIMPLE-algorithm that is also able to treat compressible flow. For further information on the above described methods see Ferziger and Peric [12].

Since the cavitating flow observed in the experiment (Figure 21.1) shows unsteady effects, it is necessary to choose a cavitation model that is able to predict unsteady phenomena. The "law of state"-model presented first by Delannoy and Kueny [8] has produced good results for unsteady cavitation, which agreed to some extent with experimental observations. Schmidt [9] achieved qualitative good results with a similar model for flows in injector nozzles. In the

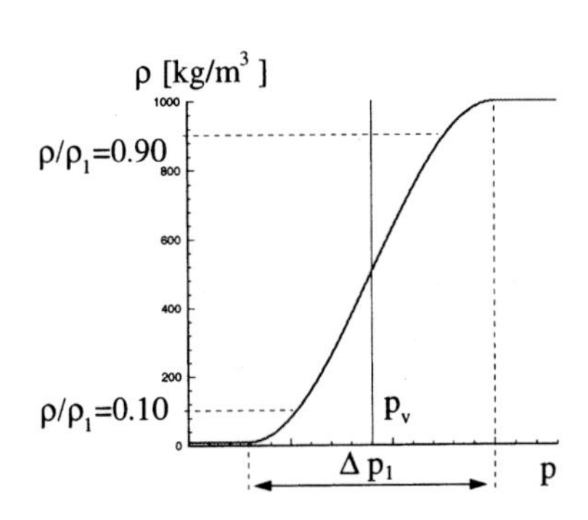

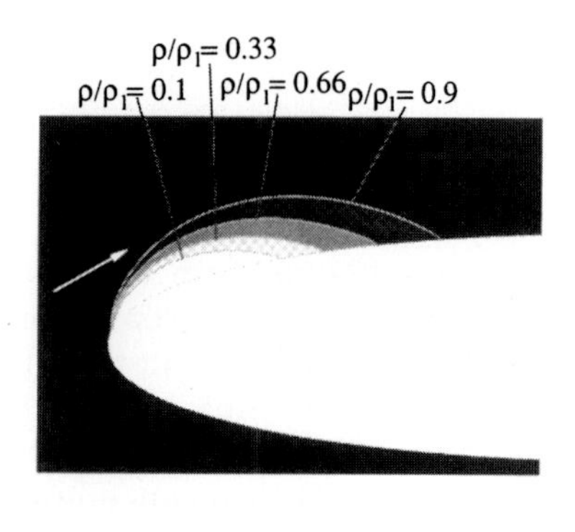

Figure 21.2: Cavitation model using an equation of state; left: equation of state for the mixture; right: iso-density-lines for the law of state, the tip of a NACA 0012 profile is used as an illustrative example.

law of state-method, a homogeneous fluid with varying density is considered. The density ρ is connected with the pressure by the law of state (Figure 21.2, left), which can be divided in three different zones: vapor with density ρ_v, liquid with the density ρ_l and an intermediate region (mixture) following a sine law. The intermediate medium was introduced in order to avoid a vanishing sonic speed a ($a^2 = dp/d\rho$), which would be non-realistic and numerically unstable. The structure of cavitation areas that are obtained by such a cavitation model are expressed in the fact that the phase boundary is not marked by a sharp contour, but by several iso-density-lines. If the pressure interval Δp_l of the intermediate area (Figure 21.2, left) is increased, the boundary interface in Figure 21.2, right, would become wider.

21.3 Single-Phase Flow through an Injector Nozzle

In order to verify the numerical scheme, a benchmark test on single-phase flow through an injection nozzle was performed. The investigated geometry was taken to be the same as Roosen et al. [2] used for their experiments, a sketch of the geometry is given in Figure 21.3. In this experiment, the injection pressure was set equal to $p_{inj} = 80$ bar and the outlet pressure to $p_{outlet} = 26$ bar. Using this pressure set-up, cavitation effects were not observed in the experiments and, therefore, this case can be used for comparison with single-phase numerics. Since the injection pressure is specified in the plenum located upstream of the nozzle inlet, this pressure could not be used as an inlet boundary condition. Instead, a boundary condition with a prescribed mass flow rate was used. The flow rate was determined by averaging a measured spanwise velocity pro-

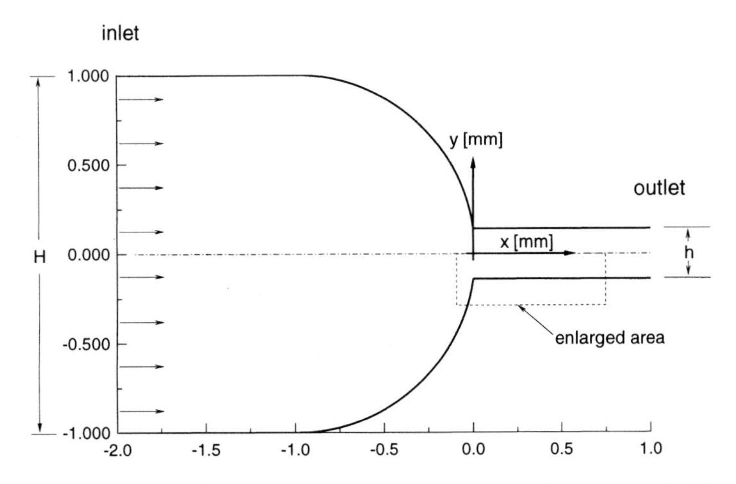

Figure 21.3: Geometry of the injection nozzle investigated by Roosen et al. [2].

file obtained in the nozzle throat and lead to a Reynolds number of $Re = 21600$. Since the flow structure (laminar or turbulent) is not known a priori, numerical simulation under the assumption of turbulent and laminar flow were performed and compared with experimental data. For symmetry reasons, only a half of the nozzle was calculated.

21.3.1 Laminar single-phase nozzle flow

Figure 21.4 presents the predicted flow structure if laminar flow is assumed. One cycle of the periodic vortex shedding is shown, from top to bottom. The time increment is $T/4$. As can be seen from the figure, vortices develop at the sharp edge of the throat inlet and travel downstream. The shedding frequency is $f = 1/T = 138.8$ kHz.

21.3.2 Turbulent single-phase nozzle flow

The turbulent calculation of the nozzle flow was performed using a k-ω turbulence model. The turbulence level at the nozzle inlet was set equal to $Tu_{\mathrm{inl}} = 0.8\%$. As can be seen from Figure 21.5, the flow structure differs completely from the previously presented laminar results. The main flow is steady, with a large recirculation zone near the throat inlet.

Comparing Figures 21.4 and 21.5, it can be seen that the modelling seriously affects the flow structure. This is also evident from Figure 21.6, which shows the wall friction coefficient for three different times $t_1 = t_{\mathrm{ref}}$, $t_2 = t_{\mathrm{ref}} + T/4$, $t_3 = t_{\mathrm{ref}} + T/2$, and for turbulent flow. For the laminar flow case, c_{f} is negative indicating reverse flow from the throat outlet towards the throat inlet, which was

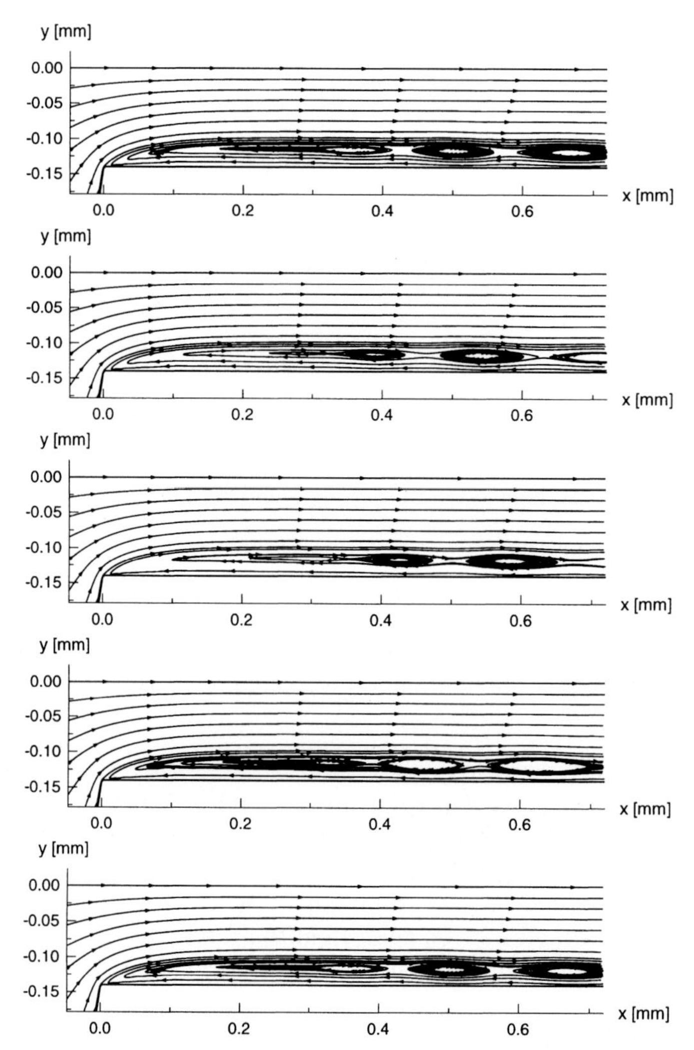

Figure 21.4: Unsteady laminar single-phase flow through the sharp edged throat of the injector nozzle: streamlines, $Re = 21\,600$, one cycle of the periodic vortex shedding is shown.

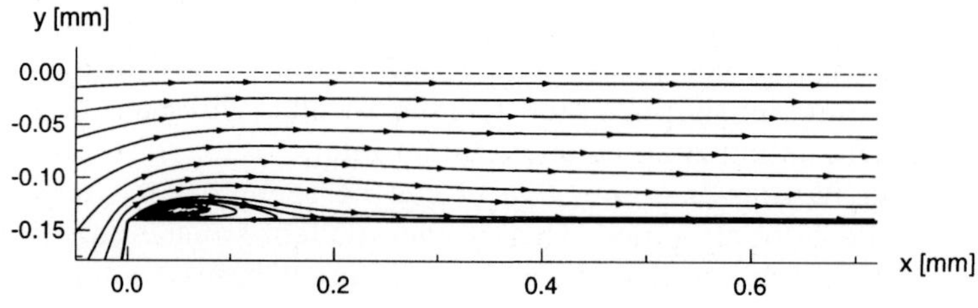

Figure 21.5: Turbulent single-phase flow through the sharp edged throat of the injector nozzle: streamlines, $Re = 21600$.

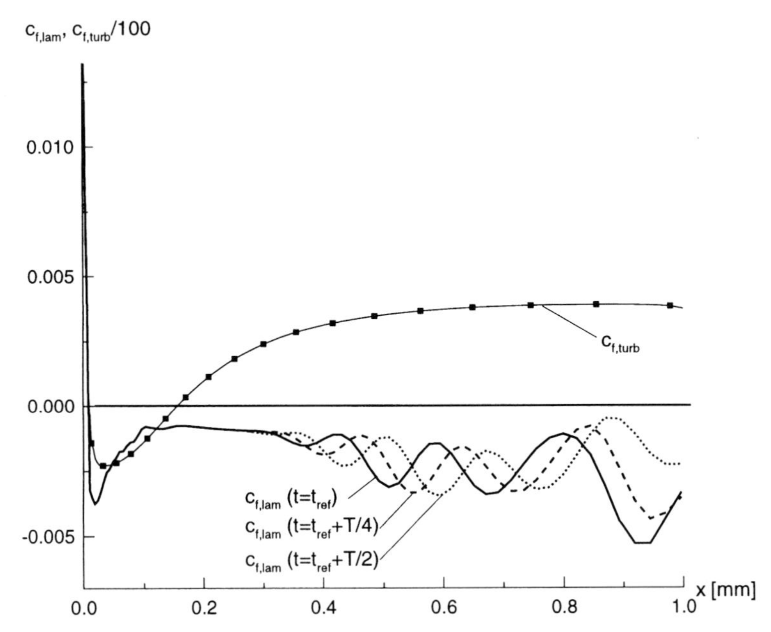

Figure 21.6: Wall friction coefficient for unsteady laminar and turbulent flow case, $Re=21600$.

not observed in the experiments. Furthermore, comparison of numerically predicted spanwise velocity profiles in the throat with experimental data shows good agreement with the turbulent calculations, therefore the flow can be assumed to be turbulent.

21.4 Flow through an Injector Nozzle Including Cavitation Effects

The first part of this section deals with single-phase flow through the injection nozzle, which was investigated experimentally by Roosen et al. [2]. The effects of a throat inlet radius variation are discussed with respect to the flow structure, pressure distribution and occurrence of cavitation. For the discussion of cavitation, the water temperature is assumed to be constant at $T_\infty=25\,^{\circ}\mathrm{C}$, corresponding to a vapor pressure of $p_v=p_{sat}(T_\infty)=3170$ Pa. The second part is dedicated to preliminary results for cavitating flow in a channel using two-phase flow analysis.

21.4.1 Predicting cavitation effects based on single-phase flow analysis

Experimental studies of cavitating flows in injector nozzles are of great interest, since cavitation can significantly change the spray characteristics such as the spray angle of the nozzle [3], and therefore affects the combustion process and the overall engine performance. In the experimental work, the effects of a pressure variation (injection pressure and outlet pressure) and the influence of the nozzle geometry (diameter, length) on cavitation inception are studied in great detail, whereas the information described as how to the throat inlet radius affects cavitation is rather poor. The experimental study of a throat inlet radius variation is very difficult due to the small sizes of injection nozzles, and therefore the effects of a radius variation can be studied more efficiently by numerical simulation. The geometry under consideration is presented in Figure 21.7, the enlarged areas show a selection of investigated throat inlet radii. For discretization of the inlet wall curvature, a minimum of 8 grid points is used. The radius variation was performed with respect to $r_1 = 14$ µm. The largest radius investigated is $r_0 = 16 r_1 = 224$ µm, the smallest $r_5 = r_{1/16} = 0.875$ µm. The Reynolds number based on the throat height h and the inlet velocity is kept constant at a value of $Re = 21600$, the outlet pressure is set equal to $p_{\text{outlet}} = 6$ bar.

Figure 21.8 presents the σ-distribution at the nozzle wall for different throat inlet radii r, where the cavitation number is defined as:

$$\sigma = \frac{p - p_v}{\frac{1}{2}\rho U_\infty^2} \,. \tag{6}$$

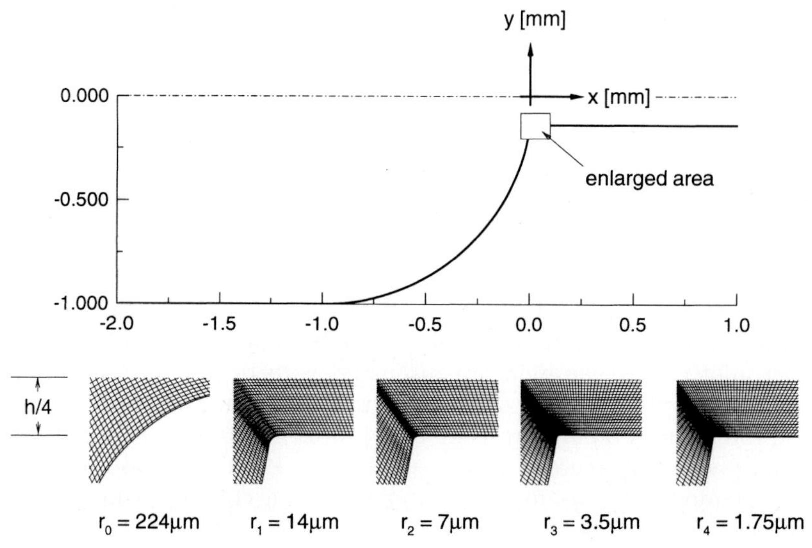

Figure 21.7: Investigated nozzle geometry, the different inlet radii investigated are shown in detail.

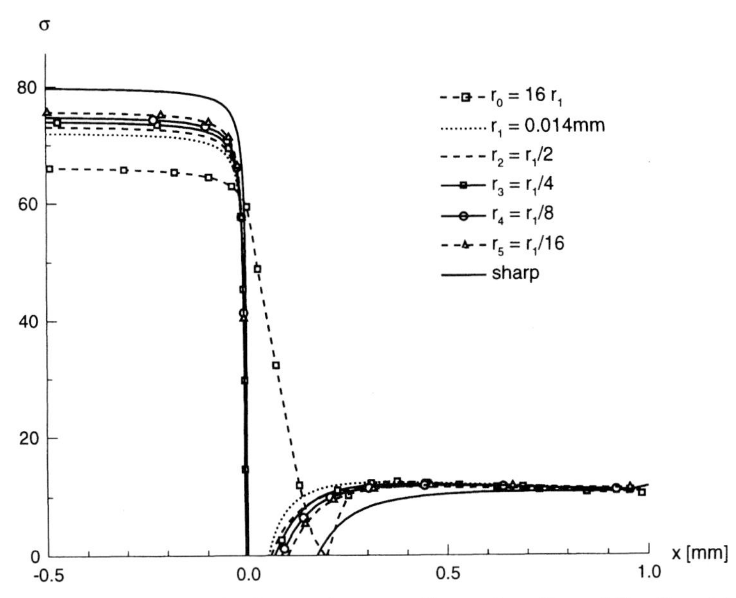

Figure 21.8: σ-Distribution at the lower throat wall as a function of the throat inlet radius r, $Re=21\,600$.

For all radii investigated, except for r_0, the minimum pressure is below the vapor pressure p_v and, consequently, under the assumption that the liquid cannot sustain any pressure lower than the vapor pressure, cavitation occurs. The radius r_0 was found in an iterative procedure by continuously increasing the inlet radius until the minimum pressure no longer drops below the vapor pressure. This investigation shows that for the investigated pressure set-up cavitation can be avoided if the inlet radius is sufficiently large.

From Figure 21.8, it can be seen that the region, where the vapor pressure is reached, continuously thickens when reducing the inlet radius, which is also evident from Figure 21.9. For the remainder of this section, the area, where the pressure is lower than the vapor pressure, will be referred to as the cavitation

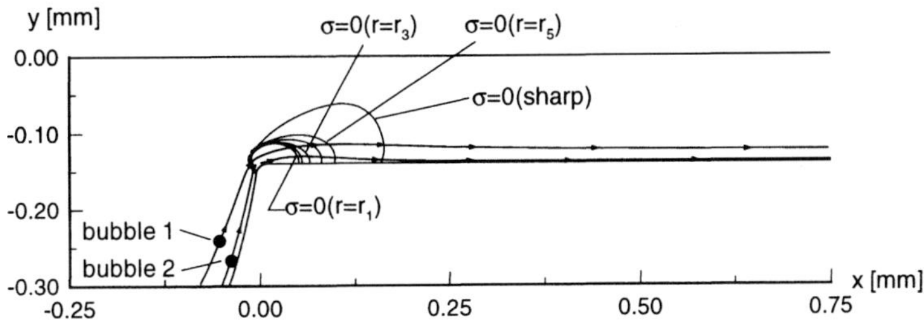

Figure 21.9: Cavitation regions (identified as $\sigma=0$ contour lines) as a function of the throat inlet radius r, $Re=21\,600$.

region. To discuss the effects of the σ-distribution, cavitation will be regarded as a heterogeneous growth process of pre-existing nuclei when the pressure falls below the vapor pressure, which is a widely used approach in several studies such as Kubota et al. [13], Schnerr et al. [7]. Starting from a nuclei (gas bubble or particle), vapor bubbles develop and grow or collapse with a speed depending on the surrounding pressure p_∞, temperature T_∞ and inertia field. A measure for the time interval, during which a bubble can grow, is the transit time Δt needed to cross the region $\sigma<0$. Consider two vapor bubbles moving on streamlines as depicted in Figure 21.9.

Assuming that the bubbles are convected with the bulk flow, the velocity of bubble 1 is much greater than that of bubble 2, which moves close to the wall, where viscous effects decrease the bulk velocity. The typical transit time for bubble 1 is on the order of $\Delta t \sim 10^{-7}$ s and much shorter compared to the transit times of bubble 2, which are $\Delta t_1 = 7.8 \cdot 10^{-4}$ s, $\Delta t_2 = 8.7 \cdot 10^{-4}$ s, $\Delta t_3 = 1.04 \cdot 10^{-3}$ s, $\Delta t_4 = 1.3 \cdot 10^{-3}$ s, $\Delta t_5 = 1.6 \cdot 10^{-3}$ s, $\Delta t_s = 2.6 \cdot 10^{-4}$ s, as depending on the throat inlet radius. The bubbles enter the cavitation region with the critical radius $r_c = 2\sigma_t/(p_v - p_{cav})$, σ_t is the surface tension. The pressure in the cavitation region is set equal to a constant at $p = p_{cav}$, the temperature is set as $T_\infty = 25\,^\circ$C, and the bubble grows under those conditions for the time interval of Δt seconds. Since the pressure in the cavitation region is not given, it is varied in the range of 700 to 1500 Pa to investigate its effects on the bubble growth. The growth problem is modelled analogous to the vapor bubble growth in a uniformly superheated liquid. The superheat is given as $\Delta T = T_\infty - T_{sat}(p_{cav})$,

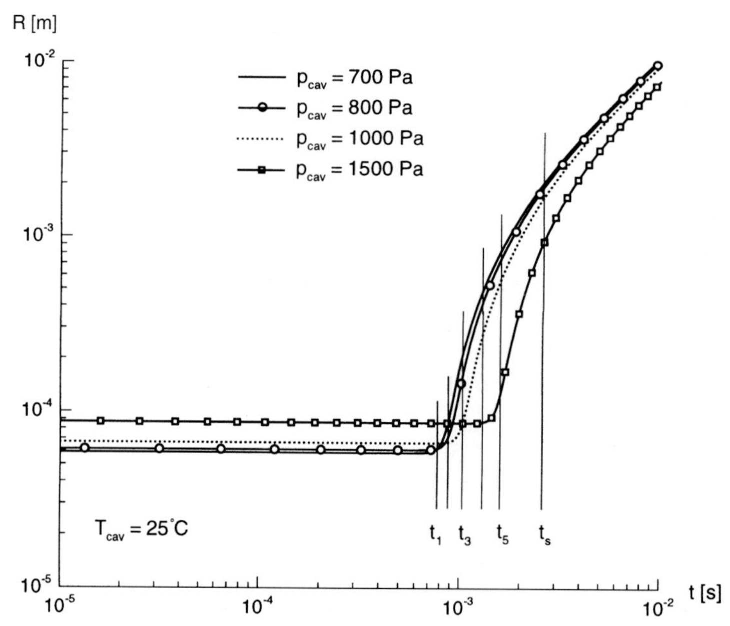

Figure 21.10: Spherical vapor bubble growth depending on the pressure in the cavitation region p_{cav}.

where T_{sat} is the saturation temperature. The solution of the bubble growth is governed by the Rayleigh-Plesset equation combined with the energy equation, details of the numerical scheme can be found in Lee and Merte [14].

Figure 21.10 presents the predicted vapor bubble growth for different pressures p_{cav}. Significant bubble growth is observed for transit times larger 10^{-3} s. Due to the short transit time, the growth of bubble 1 while passing the region $\sigma<0$ can be neglected, whereas for bubble 2, the transit time is on the order of 10^{-3} s and growth effects can be expected. The intersection points of the vertical lines with the curves $R(t,p_{cav})$ yield the final radius that bubble 2 reaches under those conditions. Especially for the lower pressures, the final radius is very sensitive to the transit time and consequently, it is sensitive to a variation of the throat inlet radius r. In the case of $p_{cav}=700$ Pa, the final radii differ by more than one order of magnitude leading to the conclusion that in the case of $r=r_1$, no cavitation will be observed, whereas in the case of $r=r_5$, strong interaction of the growing bubbles with the bulk flow is to be expected.

Besides the pressure distribution, the throat inlet radius also strongly affects the flow structure, leading to an attached flow for large radii, to separated flows with a recirculation zone for the case $r=r_5$ or the sharp-edged inlet. Figure 21.11 shows the wall friction coefficient c_f as a function of the throat inlet radius r.

The change in the flow structure, especially the appearance of a recirculation zone, does significantly affect the cavitation dynamics when discussed with respect to transit time and bubble growth. A nuclei caught in the recircu-

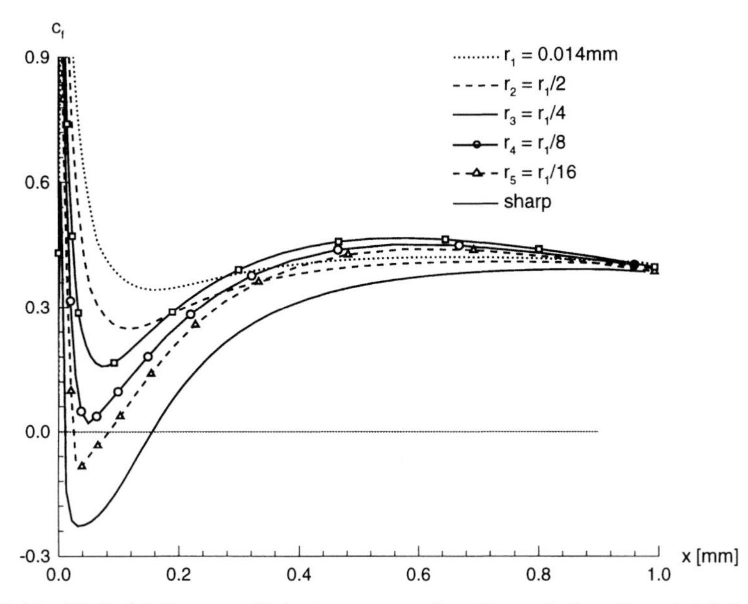

Figure 21.11: Wall friction coefficient c_f as a function of the throat inlet radius r, $Re=21\,600$.

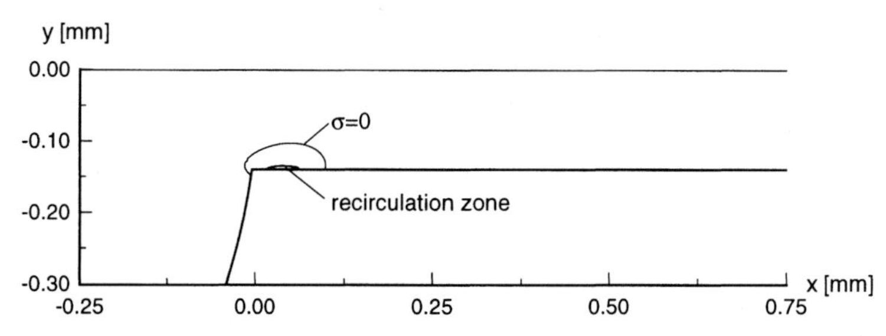

Figure 21.12: Extension of recirculation zone and cavitation region for $r = r_1/16$, $Re = 21\,600$.

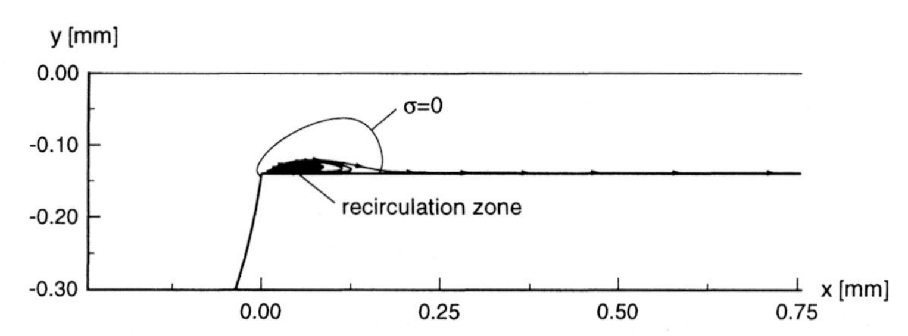

Figure 21.13: Extension of recirculation zone and cavitation region for the sharp edge, $Re = 21\,600$.

lation zone will remain there, consequently the growth time is not limited. In the cases where separation occurs (resulting in a recirculation zone), which are shown in Figures 21.12 and 21.13, respectively, the recirculation zone overlaps with the cavitation region. Therefore, the nuclei caught in the recirculation zone can grow significantly, resulting in strong interactions with the bulk flow. If this interaction would be taken into account, the resulting extent of the recirculation area will grow and reach approximately twice the dimension that is shown in this work.

21.4.2 Predicting cavitation effects based on two-phase flow analysis

The numerical simulation of cavitating flow is performed using the cavitation model of Delannoy and Kueny [8], as described in section 21.2. The investigated duct geometry is shown in Figure 21.14, which is adopted from the experimental study of Lush and Peters [15]. This geometry was found to be more suitable for preliminary studies of two-phase flows than the previous used injector nozzle, since detailed information about cavity length, volume, shedding frequency, etc., exist. Furthermore, the Lush geometry does also have a sharp edge and therefore resembles to some extent to the nozzle geometry. The liquid density is set to $\rho_l = 1000\ \mathrm{kg/m^3}$, the density of the vapor to $\rho_v = 1\ \mathrm{kg/m^3}$.

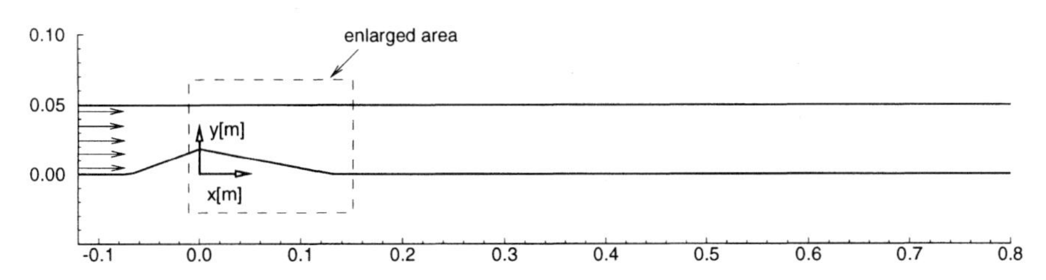

Figure 21.14: Geometry of the duct with a triangular obstacle.

The flow direction is from left to right and the fluid is assumed to be inviscid. This assumption was found to be reasonable because the steady single-phase flow calculations of the turbulent and the inviscid case do not differ significantly. The inlet velocity is set equal to $U_{inl}=13.5$ m/s and the outlet pressure to $p_{outlet}=1$ bar.

Figure 21.15 shows one cycle of the periodically cavitating flow. Cavitation clouds, identified as regions of low densities, develop at the edge of the triangular obstacle due to the low pressure. The cavity grows, breaks off and moves downstream where it finally collapses. The shedding frequency is $f=55$ Hz, the maximum attached cavity length was $l=20$ mm. In the experiments, a typical shedding frequency of the order $f=100$ Hz and typical cavity lengths of 50 mm have been observed, leading to a Strouhal number in the range of 0.28–0.35 [16]. In order to quantitatively match the experiments, in terms of shedding frequency, cavity length or volumes, further numerical investigations have to be performed to see how the model parameters affect the quantities stated above.

21.5 Conclusions

The first part of this report deals with the numerical analysis of single-phase flow through an injection nozzle. Comparison of experiments with numerical data showed that the investigated nozzle flows are turbulent. For the discussion with respect to the occurrence of cavitation, it is assumed that the fluid cannot sustain any pressure lower than the vapor pressure, and hence cavitation occurs whenever the vapor pressure is reached. The calculations showed that the pressure distributions and the flow structure are very sensitive to a variation of the throat inlet radius. The region, where the vapor pressure is reached, does significantly thicken when the inlet radius is decreased. If the radius is smaller than a critical threshold value, the flow structure changes from attached flow to separated flow with a recirculation zone near the throat inlet. For the case investigated, this critical radius is $r_c=h/32$, where h is the height of the nozzle throat. Since the region, where the pressure is below the vapor

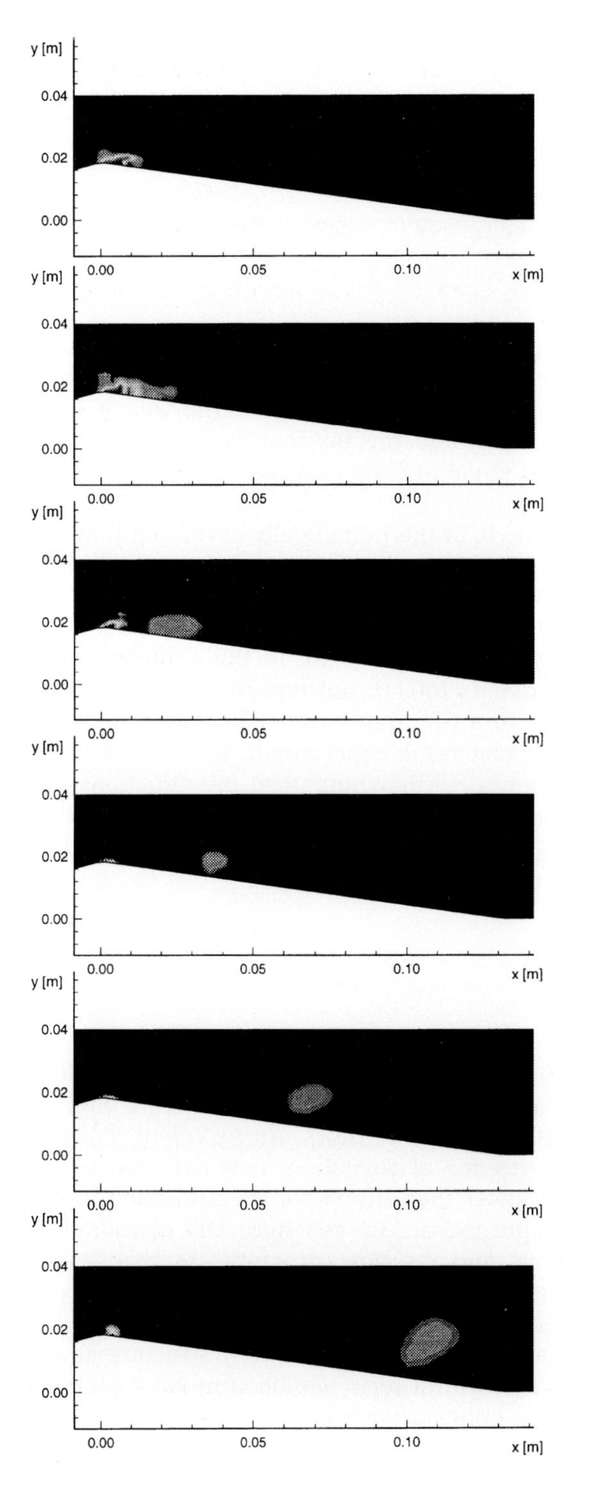

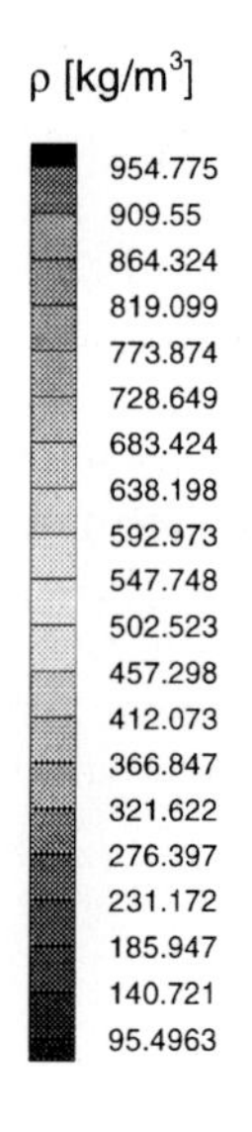

Figure 21.15: Unsteady cavitating flow in a duct with a triangular obstacle, density distribution.

pressure, does overlap with the recirculation zone, nuclei caught in the recirculation zone can grow significantly, and strong interaction with the bulk flow is to be expected. It has also been shown that cavitation can be avoided if the inlet radius is chosen to be sufficiently large.

Preliminary studies of cavitating flow in a duct showed that the employed two-phase analysis is capable to reproduce flow structures comparable to those observed in the experiments. The cavity grows, breaks off and moves downstream where it finally collapses. In order to quantitatively match the experiments, in terms of discharge coefficient, cavity length or volumes, further numerical investigations have to be performed.

Acknowledgements

The funding support from the Deutsche Forschungsgemeinschaft by the contract Schn 352/14-1 is gratefully acknowledged. Furthermore, the authors would like to thank M. Peric and M. Krömer for committing the single-phase version of the numerical code to us.

References

[1] D.T. Montgomery, M. Chan, T. Chang, P.V. Farrel, R.D. Reitz: *Effect of injector nozzle hole size and number on spray characteristics and the performance of a heavy duty d.i. diesel engine.* SAE Paper 962002, 1996.

[2] P. Roosen, S. Kluitmann, K.-F. Knoche: *Untersuchung und Modellierung des transienten Verhaltens von Kavitationserscheinungen bei ein- und mehrkomponentigen Kraftstoffen in schnell durchströmten Düsen.* Final report, Aachen, 1995.

[3] H. Chaves, M. Knapp, A. Kubitzek, F. Obermeier, T. Schneider: *Experimental study of cavitation in the nozzle hole of Diesel injectors using transparent nozzles.* SAE 950290, 1995.

[4] G.H. Schnerr, K. Lanzenberger: *A chimera grid scheme for simulation of cavitating flow.* In: L.C. Wrobel, B. Sarler, C.A. Brebbia (Eds.): Computational Modelling of Free and Moving Boundary Problems III. Computational Mechanics Publications, Southampton, 1995, pp. 75–82.

[5] G.H. Schnerr, K. Lanzenberger: *Vapor/Liquid Interfaces in Cavitating Flows, Cavitation and Multiphase Flow Forum.* In: J. Katz, Y. Matsumoto (Eds.): ASME, FED-Vol. 210, New York, 1995, pp. 17–22.

[6] G.H. Schnerr, K. Lanzenberger, C. Spengler: *Modelling of Boundary Conditions for Closed Sheet Cavities.* 3rd International Symposium on Aerothermodynamics of Internal Flow, Beijing, China, 1996.

[7] G.H. Schnerr, S. Adam, K. Lanzenberger, R. Schulz: *Multiphase Flows: Condensation and Cavitation Problems.* In: M. Hafez, K. Oshima (Eds.): Computational Fluid Dynamics REVIEW. John Wiley & Sons, Ltd., 1995.
[8] Y. Delannoy, J.L. Kueny: *Two-phase flow approach in unsteady cavitation modelling.* In: O. Fumya (Ed.): Cavitation and Multiphase Flow Forum. ASME FED-Vol. 98, New York, 1990, pp. 153–158.
[9] D.P. Schmidt: *Cavitation in Diesel Fuel Injector Nozzles.* Ph. D. thesis at the University of Wisconsin-Madison, USA, 1997.
[10] Y. Chen, S.D. Heister: *Modelling cavitation flow in diesel injectors.* Atomisation and Sprays **6** (1996) 709–726.
[11] D.C. Wilcox: *Turbulence Modelling for CFD.* DCW Industries Inc., La Cañada, California, 1993.
[12] J.H. Ferziger, M. Peric: *Computational methods for fluid dynamics.* Springer Verlag, 1996.
[13] A. Kubota, H. Kato, H. Yamaguchi: *A New Modelling of Cavitating Flow: A Numerical Study of Cavitation on a Hydrofoil Section.* J. Fluid Mech. **240** (1992) 59–96.
[14] H.S. Lee, H. Merte: *Spherical vapor bubble growth in uniformly superheated liquids.* Int. J. Heat Mass Transfer **39**(12) (1996) 2427–2447.
[15] P.A. Lush, P.I. Peters: *Visualisation of the cavitating flow in a venturi-type duct using high speed cine photography.* IAHR, Amsterdam, 1982.
[16] R.A. Furness: *Studies of the mechanics of fixed cavities in a two-dimensional convergent-divergent nozzle.* I. Mech. E. Conf. on Cavitation, Heriot Watt University, 1974.

22 Numerical Simulation of Cavitation Phenomena in Accelerated Liquids

Christian Dickopp and Josef Ballmann *

Abstract

With the goal to investigate the development of cavitation in highly accelerated liquids and its transient behavior, the liquid flow together with single, gas- or vapor-filled cavitation bubbles is simulated numerically. For this purpose, an explicit finite element method has been developed that approximates the solution of the Navier-Stokes equations for compressible fluid flow of both phases on an optionally moving, hybrid grid in two space dimensions by linear shape functions. This method has been extended by a multiblock technique that allows the application of the code to single vapor bubbles, which arise where the pressure of a liquid flow undergoes the vapor pressure. The wall of a single bubble is treated as a contact discontinuity that is fitted during the simulation using the appropiate jump conditions there. Its changing shape is also followed by a moving line of the grid. With the help of one-dimensional model equations, bounds for the time step are derived to ensure a stable behavior of the numerical solution. With the background that the used finite element method is a multi-dimensional approach without any dimensional splitting like used in most of the Godunov-type methods, the idea behind ENO-schemes for reconstruction in Godunov-type methods has been transferred to the finite element method in order to improve the implemented flux-corrected-transport-limiter (FCT) for the control of the order of spatial approximation. Different examinations have shown that this ENO-version of the FCT-limiter preserves better the symmetric properties in the case of a symmetric bubble collapse than the classical FCT-strategy, and it reduces also the dependence of the numerical approximation on the discretization.

* RWTH Aachen, Lehr- und Forschungsgebiet für Mechanik, Templergraben 64, D-52062 Aachen, Germany

22.1 Introduction

Cavitation phenomena arise in many technical applications of flowing liquids, especially in case of high acceleration and flow around sharp corners of solid walls: The pressure may locally undergo the vapor pressure of the liquid causing cavitation bubbles. These bubbles affect the flow in their environment by oscillations or even by collapses.

For example, bubbles occur in fuel injector channels behind inlet corners [1]. The existence of these bubbles can either disturb the flow through the nozzle or improve the quality of the spray and the vaporization behind the injector's exit in the combustion chamber [2, 3]. So, one goal of the project is to analyse the injector flow in cooperation with experimentally working research groups.

The flow simulation through the injector nozzle is divided into two steps: First, a transient liquid flow is calculated starting with given initial conditions in the whole flow field and boundary conditions along the injector walls, in the preinjection chamber's cross-section and at the entrance of the combustion chamber. The regions where the liquid pressure undergoes the vapor pressure are determined. At these locations, single bubbles are inserted into the flowfield in such a manner that for each bubble, a separate calculation block is created and the data at the block boundary are exchanged between the blocks. So, the second step of the simulation consists in a multiblock application of the basic finite element method for transient one- and two-phase flows.

The underlying numerical method for the simulation of the transient behavior of single bubbles has been developed originally for the problem of cavitation damaging, which is another important technical problem. Its mechanism is still not clarified satisfactorily. To contribute to its understanding, the collapse of a bubble near a solid surface is simulated in a fully coupled manner together with the dynamical response of the elastic solid using the numerically approximated solution of the elastodynamical equations according to [4]

The paper is organized as follows: In Sections 22.2 and 22.3, the mathematical model and the implemented numerical method are described. Linear and nonlinear stability of the finite element discretization for the different kinds of the treated one- or two-phase flows are considered in Section 22.4, and Section 22.5 presents numerical results for the starting process of the flow through an injector and the initial phase of a bubble collapse near the surface of an elastic solid.

22.2 Mathematical Model

22.2.1 Equations of motion

Independent variables of the regarded flows are two space coordinates and time, either using cartesian coordinates x and y for planar problems or cylindrical coordinates r, z and ϕ for problems with rotational symmetry.

The mathematical formulation and the numerical method should be applicable to single phase or two-phase flows. In both cases, the fluids are modelled as compressible, viscous and heat conductive gas and liquid. The computational grid is optionally moving with the time-dependent velocity $\boldsymbol{v}^{\text{grid}}$ or fixed in space.

Under these assumptions, the governing system of Navier-Stokes equations can be written in the following conservative form for the case of rotational symmetry with respect to the z-axis:

$$\boldsymbol{U}_{,t}(r,z,t) + \boldsymbol{F}_r(r,z,t)_{,r} + \boldsymbol{F}_z(r,z,t)_{,z} + \frac{1}{r}\boldsymbol{Q}(r,z,t) = \boldsymbol{0}\,. \tag{1}$$

$\boldsymbol{U}$ denotes the vector of conservative variables and $\boldsymbol{F}_r$ $\boldsymbol{F}_z$ are the flux vectors, while

$$\boldsymbol{Q} = \begin{pmatrix} \rho(v_r - v_r^{\text{grid}}) \\ \rho v_r(v_r - v_r^{\text{grid}}) - \tau_{rr} - \tau_{\phi\phi} \\ \rho v_z(v_r - v_r^{\text{grid}}) - \tau_{rz} \\ \rho e(v_r - v_r^{\text{grid}}) + v_r(p - \tau_{rr}) - v_z\tau_{rz} + q_r \end{pmatrix} \tag{2}$$

represents source terms resulting from the transformation into cylindrical coordinates. In the case of planar flows, the source terms vanish (i.e. $\boldsymbol{Q}=\boldsymbol{0}$) and the coordinates are renamed: $x = r$, $y = z$. To close the system of equations, material laws are needed depending on the fluid model. For the heat flux $\boldsymbol{q}$, Fourier's law is used, whereas the viscous stress tensor is described following Stokes' ansatz that includes also source terms in the case of cylindrical coordinates. The coefficients of heat conductivity λ and of viscosity μ are alternatively determined by models of transport phenomena following Sutherland or Chapman and Enskog.

As in many similar projects, the behavior of a compressible liquid is described by Tait's equation of state in a barotropic way:

$$\frac{p+B}{p_0+B} = \left(\frac{\rho}{\rho_0}\right)^n \quad \Leftrightarrow \quad p = (p_0+B)\left(\frac{\rho}{\rho_0}\right)^n - B = \frac{p_0+B}{\rho_0^n}\, v^{-n} - B \tag{3}$$

with the following values of the constants for pure water:

$$p_0 = 7025\,\text{Pa}, \quad B = 3.01\times10^8\,\text{Pa}, \quad \rho_0 = 998{,}2\,\frac{\text{kg}}{\text{m}^3}, \quad n = 7.15\,. \tag{4}$$

The equations for the internal energy u and for the speed of sound can be derived from the thermal equation of state by thermodynamical relations [5].

22.2.2 Conditions at phase boundaries

The wall of a single bubble as a phase boundary is treated as a contact discontinuity, which is fitted during the simulation using the appropriate conditions there. The needed relations at such a liquid-gas or liquid-vapor interface consist of the jump conditions for a viscous contact discontinuity and allow to include the effects of surface tension with the coefficient σ and of the mass transport caused by evaporation.

A part of the numerical results presented later concerns the model problem for cavitation damaging, i.e. the collapse of a single bubble near an elastic solid wall.

In this case, the liquid-solid interface is another contact discontinuity of the problem. There, we neglect heat conduction so that the jump condition reduces to the dynamic condition [5–7]:

$$-p + \boldsymbol{n}_D \tau \boldsymbol{n}_D = \boldsymbol{n}_D \sigma_S \boldsymbol{n}_D \,, \quad \boldsymbol{n}_D \tau \boldsymbol{t}_D = \boldsymbol{n}_D \sigma_S \boldsymbol{t}_D \tag{5}$$

and the viscous contact condition

$$\boldsymbol{v} = \boldsymbol{v}_S \,. \tag{6}$$

Here σ_S stands for the stress tensor in the solid, whereas τ denotes again the viscous stress tensor in the fluid. In case of neglected viscosity, the Navier-Stokes equations reduce to the Euler equations, and the jump condition reduces to the conditions of continuous pressure and continuous velocity normal to the interface.

22.3 Numerical Method

22.3.1 The weighted residual formulation

According to the concept in [8, 9 and 10] for single fluid flows, the finite element flow solver for the two fluid problem is based on a weak formulation of the Navier-Stokes equations in conservation form, i.e. Equation (1). It is obtained by the orthogonal projection of the residual equation for an approximate solution $\overline{\boldsymbol{u}}(r, z, t)$

$$l(\overline{\boldsymbol{u}}(r, z, t)) = \varepsilon \quad \forall (r, z) \in \Omega \tag{7}$$

with the functional I for the governing equations

$$I(\boldsymbol{U}(r,z,t)) = \boldsymbol{U}_{,t}(r,z,t) + \boldsymbol{F}_r(r,z,t)_{,r} + \boldsymbol{F}_z(r,z,t)_{,z} + \frac{1}{r}\boldsymbol{Q}(r,z,t)\,, \tag{8}$$

and a residual vector ε onto a set of weighting functions w_m, $m \in M$:

$$\langle I(\overline{\boldsymbol{u}}(r,z,t)), w_m\rangle = \langle \varepsilon, w_m\rangle = 0 \qquad \forall (r,z) \in \Omega, \quad m \in M\,. \tag{9}$$

Here $\langle , \rangle$ denotes the scalar product that is defined for two continuous functions $a, b \in C^0(\Omega)$ by:

$$\langle a, b\rangle := \int_\Omega ab\,\mathrm{d}\Omega\,. \tag{10}$$

On each element e of the grid, shape functions $\Phi^{(e)}_{N(e)}$ are assumed, which interpolate between the unknown values $u_{N(e)}$ of the conservative variables at the nodes $N(e)$ of e:

$$\overline{u} = \sum_{\text{all elements } e} \; \sum_{\text{all nodes } N(e)} \boldsymbol{u}_{N(e)} \Phi^{(e)}_{N(e)} = \sum_N \Phi_N \boldsymbol{u}_N\,, \tag{11}$$

where $\boldsymbol{u}_N$ *depends on the time t.*

For flow problems, piecewise linear or even piecewise constant (if viscosity and heat flow are neglected) shape functions are sufficient.

The governing equations are discretized in conservation form that allows the capturing of discontinuities. As an example, the weighted residual formulation is now derived for the energy equation:

$$(\rho e)_{,t} + \left(\rho e(v_{x^i} - v^{\text{grid}}_{x^i}) + (p\delta_{ij} - \tau_{ij})v_{x^j} + q_i\right)_{,x^i} + \frac{Q_e}{r} = 0\,, \tag{12}$$

($x^1 := r$, $x^2 := z$) with the source term $\frac{Q_e}{r}$. The functional for the energy equation consists of two parts: The integrals over the domain Ω resulting from the scalar product of the noted energy equation with a weight function w_m and the boundary integral term incorporating the transition condition for the energy at a gas-liquid interface:

$$\begin{aligned}
&\int_\Omega \Phi_N w_m\,\mathrm{d}\Omega\,(\rho e)_{N,t} + \int_\Omega (\Phi_N)_{,x^i} w_m\,\mathrm{d}\Omega\big((\rho e)_N(v_{x^i N} - v^{\text{grid}}_{x^i N})\\
&+ (p_N\delta_{ij} - (\tau_{ij})_N)v_{x^j N} + (q_i)_N\big) + \int_\Omega \frac{\Phi_N}{r} w_m\,\mathrm{d}\Omega\,(Q_e)_N\\
&+ \int_{\Gamma_{GL}} \Phi^*_N w^*_m\,\mathrm{d}s\left(|[(p_N E - \tau_N)\boldsymbol{v}_N + \boldsymbol{q}_N]|\boldsymbol{n}_D + (\sigma\left(\frac{1}{R_1} + \frac{1}{R_2}\right)_N E\boldsymbol{v}_N)\boldsymbol{n}_D\right) = 0\,.
\end{aligned} \tag{13}$$

Here, w_m^* und Φ_N^* indicate the restrictions of the weight functions or the shape functions to a boundary.

After an assembling process over all elements of the grid, the weak form of the Navier-Stokes equations can be written as

$$\boldsymbol{MU}_{,t} = \boldsymbol{RS}(\boldsymbol{U}) \tag{14}$$

with the consistent mass matrix M. Its components are

$$M_{Nm} = \int_{\Omega} \Phi_N w_m \, \mathrm{d}\Omega \, . \tag{15}$$

The right hand side $RS(\boldsymbol{U})$ consists of integral terms for the fluxes, transition conditions and boundary conditions as presented for the energy equation.

22.3.2 Petrov-Galerkin discretization and time integration

For time integration, an explicit two-step Taylor method is used that can be implemented in a predictor-corrector strategy:

$$\begin{aligned}
&\text{Predictor step: } M\boldsymbol{U}^{n+1/2} = M\boldsymbol{U}^n + \frac{1}{2}(\Delta t)M\boldsymbol{U}^n_{,t} = M\boldsymbol{U}^n + \frac{1}{2}(\Delta t)RS_{\mathrm{p}}(\boldsymbol{U}^n)\\
&\text{Corrector step: } M\boldsymbol{U}^{n+1} = M\boldsymbol{U}^n + (\Delta t)M\boldsymbol{U}^{n+1/2}_{,t} = M\boldsymbol{U}^n + (\Delta t)RS_k(\boldsymbol{U}^{n+1/2}) \, .
\end{aligned} \tag{16}$$

Because this scheme with second order of approximation is an explicit one, the time step is restricted by a stability criterion being derived in a separate section.

If the shape and weight functions are chosen as piecewise linear functions L_N in the sense of a classical Galerkin method, then in both time steps the full consistent mass matrix M with the entries

$$M_{ij} = \int_{\Omega} L_i L_j \, \mathrm{d}\Omega \tag{17}$$

has to be inverted, which needs a high effort. This was the motivation for the development of so-called collocation methods which are obtained by use of locally constant functions P_e (= 1 on the element e; = 0 otherwise) or P_N (= 1 on elements, which contain the node N; = 0 otherwise) as weight functions in the predictor step.

In the case of the weight functions P_e, the approximations have the following form for the two time integration steps:

$$\begin{aligned}
&\text{Predictor step:} \quad w_m = P_e\,, && \text{Corrector step:} \quad w_m = L_m\,,\\
&\boldsymbol{U}^n = L_N \boldsymbol{U}_N^n\,, && \boldsymbol{U}^n = L_N \boldsymbol{U}_N^n\,,\\
&\boldsymbol{U}^{n+1/2} = P_e \boldsymbol{U}_e^{n+1/2}\,, && \boldsymbol{U}^{n+1} = L_N \boldsymbol{U}_N^{n+1}\,,\\
&\boldsymbol{F}_r^{an} = L_N (\boldsymbol{F}_r^{an})_N\,, && \boldsymbol{F}_r^{n+1/2} = P_e (\boldsymbol{F}_r^{n+1/2})_e\,,\\
&\boldsymbol{F}_z^{an} = L_N (\boldsymbol{F}_z^{an})_N\,, && \boldsymbol{F}_z^{n+1/2} = P_e (\boldsymbol{F}_z^{n+1/2})_e\,,\\
&\boldsymbol{Q}^{an} = L_N (\boldsymbol{Q}^{an})_N\,, && \boldsymbol{Q}^{n+1/2} = P_e (\boldsymbol{Q}^{n+1/2})_e\,,\\
&\boldsymbol{R}^{an} = L_N (\boldsymbol{R}^{an})_N\,, && \boldsymbol{R}^{n+1/2} = P_e (\boldsymbol{R}^{n+1/2})_e\,.
\end{aligned} \tag{18}$$

The consequence of the collocation is that the matrix in the predictor step can be inverted in a direct way without any solver and so the computational effort becomes reduced.

If the collocation is applied for the nodes of the grid, i.e. the functions P_N (N is the number of a node) are the weight functions, the following discretization is obtained:

$$\begin{aligned}
&\text{Predictor step:} \quad w_m = P_m\,, && \text{Corrector step:} \quad w_m = L_m\,,\\
&\boldsymbol{U}^n = L_N \boldsymbol{U}_N^n\,, && \boldsymbol{U}^n = L_N \boldsymbol{U}_N^n\,,\\
&\boldsymbol{U}^{n+1/2} = L_N \boldsymbol{U}_N^{n+1/2}\,, && \boldsymbol{U}^{n+1} = L_N \boldsymbol{U}_N^{n+1}\,,\\
&\boldsymbol{F}_r^{an} = L_N (\boldsymbol{F}_r^{an})_N\,, && \boldsymbol{F}_r^{n+1/2} = L_N (\boldsymbol{F}_r^{n+1/2})_N\,,\\
&\boldsymbol{F}_z^{an} = L_N (\boldsymbol{F}_z^{an})_N\,, && \boldsymbol{F}_z^{n+1/2} = L_N (\boldsymbol{F}_z^{n+1/2})_N\,,\\
&\boldsymbol{Q}^{an} = L_N (\boldsymbol{Q}^{an})_N\,, && \boldsymbol{Q}^{n+1/2} = L_N (\boldsymbol{Q}^{n+1/2})_N\,,\\
&\boldsymbol{R}^{an} = L_N (\boldsymbol{R}^{an})_N\,, && \boldsymbol{R}^{n+1/2} = L_N (\boldsymbol{R}^{n+1/2})_N\,.
\end{aligned} \tag{19}$$

Both described collocation methods have been implemented and applied to standard test cases whereby an important difference concerning the reproduction of the speeds of informations was found: The method with the collocation points in the nodes is able to reproduce the true physical wave velocities, whereas solutions calculated by the other version of collocation exhibit wrong speeds of pro-

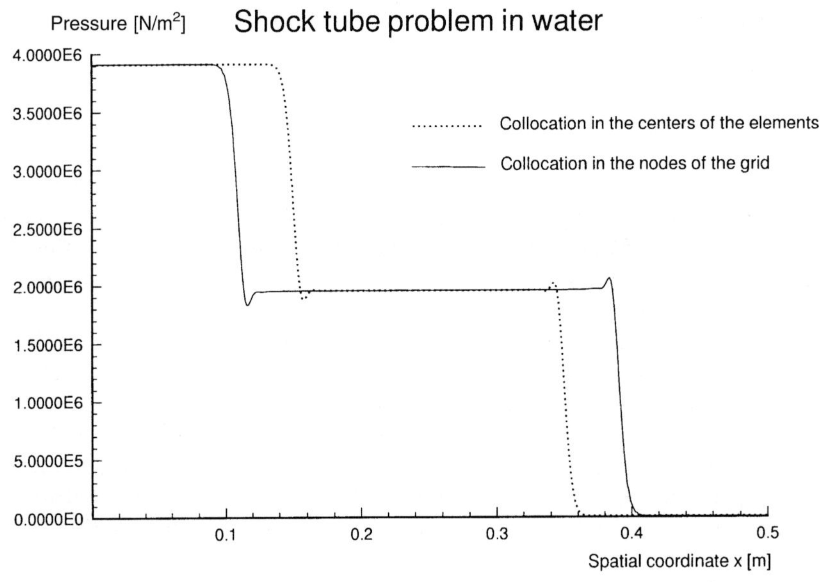

Figure 22.1: Pressure for a shock tube problem in water after $9.77 \cdot 10^{-5}$ s calculated by two methods of collocation.

pagation. This behavior is independent of the chosen CFL-number within the range of stability discussed in the next section. As an example, Figure 22.1 shows results for a shock tube problem in water obtained with the two collocation methods. The initial conditions were for both test runs two regions of constant pressure, a pressure jump at $x = 0.25$ and the fluid at rest everywhere. Both simulations used the same equidistant grid with the distance $\Delta x = 10^{-3}$ m. Because the speed of sound for water at the regarded densities is about 1476.5 m/s and has only little changes during the simulations, it is expected that the pressure waves propagate a distance of $1476.5 \text{ m/s} \times 9.77 \cdot 10^{-5}\,\text{s} \approx 0.144\,\text{m}$ in the physical time interval from $t = 0$ until $t = 9.77 \cdot 10^{-5}$ s. So, the pressure waves should have the positions $x_1 \approx 0.106$ m and $x_2 \approx 0.394$ m. Only the method with the collocation in the nodes calculates these wave positions correctly.

So, all other results presented in this paper have been calculated by the scheme that uses the collocation at the grid nodes in the predictor step.

22.4 Stability Analysis

22.4.1 Linear aspects of the stability

Because the time integration method is an explicit one, the choice of the time step is restricted by a stability criterion. The nonlinearity of the governing system of equations requires a control of the spatial order of approximation to stabilize nonlinear phenomena as shocks within the solution. This difficult topic is treated in the next section.

In smooth parts of the solution, a local linearisation of the system is possible, and the application of the classical Lax-Richtmyer theory seems to be reasonable. The fundamental theorem says that a method converges if and only if it is consistent and stable. For the used two-step Taylor method and the linear shape functions, consistency is obvious. Analysis of stability, which means to find a bound for the norm of the operator of the method, is more difficult, and many concepts of stability have been derived, which differ in the damping properties, which are required for the operator.

The von Neumann approach, which may be interpreted as a kind of Fourier analysis, is often used to study methods for conservation laws because it is close to the situation within a simulation: An error function with wave length of the grid size and its multiples is supposed with an exponential behavior in time:

$$u_j^n = q^n e^{ij\Delta x \phi} , \tag{20}$$

$i = \sqrt{-1}$, and inserted in the method. The demand $|G_j| \leq 1$ for the amplification factor

$$G_j := \frac{u_j^{n+1}}{u_j^n} = q \tag{21}$$

means that the error will not be amplified by the method and will lead to the wanted bound for the time step Δt.

With the goal to investigate first a one-dimensional analogue to the described finite element discretization for two-phase problems, the model equation

$$u_{,t} + \lambda u_{,x} = 0 \tag{22}$$

is regarded. After some algebra, the amplification factor Equation (21) can be evaluated for all phase angles $\psi \in [0, 2\pi]$ and dependent on the Courant number

$$\eta = \frac{\lambda \Delta t}{\Delta x} . \tag{23}$$

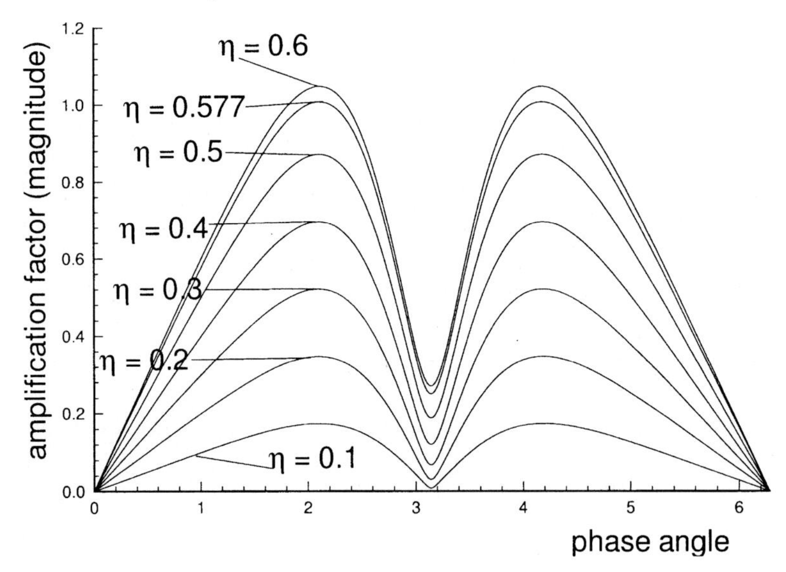

Figure 22.2: Error amplification dependent on the Courant number according to the von Neumann stability analysis.

If the quantity $|q|$ is plotted (Figure 22.2), it can be recognized that the condition $|q| \leq 1$ is fulfilled only if

$$\eta \leq \frac{1}{\sqrt{3}} \approx 0.577 \, . \tag{24}$$

To analyse the finite element discretization for two-phase problems, further extensions of the model equation are needed. As a motivation, the following scalar equation is derived for the temporal change of the sum of the static pressure p and the flux of momentum $\rho \mathbf{v}^2$ is derived by combining the conservation laws (for simplicity without diffusion) and the general equation of state:

$$\mathrm{d}p = c^2 \mathrm{d}\rho \, , \tag{25}$$

where c denotes the local speed of sound. Only one space dimension is considered.

$$(p + \rho \mathbf{v}^2)_{,tt} = (c^2 + \mathbf{v}^2)(p + \rho \mathbf{v}^2)_{,xx} \, . \tag{26}$$

This form is similar to the wave equation:

$$u_{,tt} = a^2 u_{,xx} \, , \tag{27}$$

which describes waves propagating with the speeds a and $-a$ like the system of first order PDEs:

$$u_{,t} + a u_{,x} = 0 \,, \quad u_{,t} - a u_{,x} = 0 \,. \tag{28}$$

To this point, the results already derived can be applied for the stability of solutions.

The same calculations are executed for conservation laws with additional source terms containing $\frac{1}{x}$ and terms at a phase boundary. After some algebra, the following criterion for linear stability of a symmetric two-phase flow based on an one-dimensional, linear and scalar model equation is obtained:

$$(2+\sqrt{4})\eta = \frac{4\lambda \Delta t}{\Delta x} \leq \frac{1}{\sqrt{3}} \quad \leftrightarrow \quad \Delta t \leq \frac{1}{4\sqrt{3}} \frac{\Delta x}{\lambda} \approx 0.144 \frac{\Delta x}{\lambda} \tag{29}$$

with $\lambda = c + |u|$. If the grid is moving with the local velocity $\boldsymbol{v}^{\mathrm{grid}}$, an additional convection term arises in the equations of motion. By an analogous derivation as presented, it can be shown that in this case the velocity must be modified to $\lambda = c + |u - v^{\mathrm{grid}}|$. In case of viscous flow, the bound for the Courant number is modified:

$$4\eta \leq \sqrt{\frac{1}{3} + \frac{1}{Re^2}} - \frac{1}{Re} \,. \tag{30}$$

With the argument that many flow phenomena can be regarded locally as one-dimensional problems, the derived conditions for the time step are also demanded for two-dimensional problems.

22.4.2 Nonlinear aspects of the stability

Because the governing system of equations is nonlinear, the analysis presented in the previous section is not yet sufficient to ensure the convergence of the method. The nonlinearity allows the development and propagation of discontinuous parts of the solution where the assumptions for linearisation fail. In those cases, a nonlinear concept of stability is required.

One of the early concepts was the flux-corrected-transport (FCT) formulation originally presented by Boris and Book [8] mainly for finite difference methods, which was transferred to finite element methods by Löhner et al. [9]. Being based on a local comparison between a high order solution u^{h} and a low order solution u^{l}, it is very similar to TVD methods introduced by Harten [11].

The FCT-strategy can be understood as generating a solution that fulfills the TVD-property for average values over cells or elements. Goodman and LeVeque have proved that a multi-dimensional method, which fulfills a multi-dimensional extension of the TVD-property, has at most the spatial order 1 [12]. Because the dissipation of general first order methods is too high, this result says that the TVD-concept is not applicable for multi-dimensional methods like the used finite element approach or multi-dimensional reconstructions for Godunov-type methods. This theorem gave the motivation to introduce the ENO-concept as a weak form of the TVD-property.

In this context, the reconstruction solution for a Godunov-type method is usually specified by a local selection process between different possible interpolations between the discrete solutions in the environment of the regarded node. For this purpose, all interpolations of certain types are evaluated and valued by a defined measure for the smoothness. Then the smoothest solution is selected.

Within the project presented here, these ideas were transferred to the control of the spatial order for the used finite element method, with the goal to improve the FCT-concept as is explained in the following.

Both presented collocation methods lead in the corrector step to a system of equations with the form:

$$M^{(k)}\Delta \boldsymbol{U}^{n+1} = \Delta t RS(\boldsymbol{U}^{n+1/2}) \tag{31}$$

for the increments $\Delta \boldsymbol{U}^{n+1}$ of the conservative variables U. $M^{(k)}$ denotes the consistent mass matrix, which creates the second order of this approximation in space. To obtain a first order method, this matrix is replaced by the lumped mass matrix M_{L} with the entries

$$(M_{\mathrm{L}})_{ij} = \begin{pmatrix} \sum_l M_{il}^{(k)} = \sum_l M_{li}^{(k)} & \text{for } i = j, \\ 0 & \text{for } i \neq j \end{pmatrix} . \tag{32}$$

For stability reasons, an additional viscosity term with the coefficient c_{d} is incorporated into the first order scheme. So it takes the form:

$$M_{\mathrm{L}}\Delta \boldsymbol{U}^{n+1} = \Delta t RS(\boldsymbol{U}^{n+1/2}) + c_{\mathrm{d}}(M^{(k)} - M_{\mathrm{L}})\boldsymbol{U}^{n} . \tag{33}$$

In the following, $\Delta \boldsymbol{U}^{\mathrm{l}}$ and $\Delta \boldsymbol{U}^{\mathrm{h}}$ (h: high, l: low) denote the solutions for the increments of the conservation variables from the first order and the second order method. The task to control the spatial order of approximation can be expressed as the search for a limiter $0 \leq \Theta_{\mathrm{e}} \leq 1$ on each element e that combines the both solutions in a linear way:

$$\Delta \boldsymbol{U} = \Delta \boldsymbol{U}^{\mathrm{l}} + \sum_{e} \Theta_e(\Delta \boldsymbol{U}_e^{\mathrm{h}} - \Delta \boldsymbol{U}_e^{\mathrm{l}}) . \tag{34}$$

The simplest method to determinate the limiters Θ_{e} is to detect pressure oscillations by the evaluation of the local curvature of the static pressure p and to execute an averaging process over elements. As the comparison of the limiters for a shock tube problem in gas shows (Figure 22.3), this limiter is able to avoid oscillations, but particularly the contact discontinuity is smeared significantly.

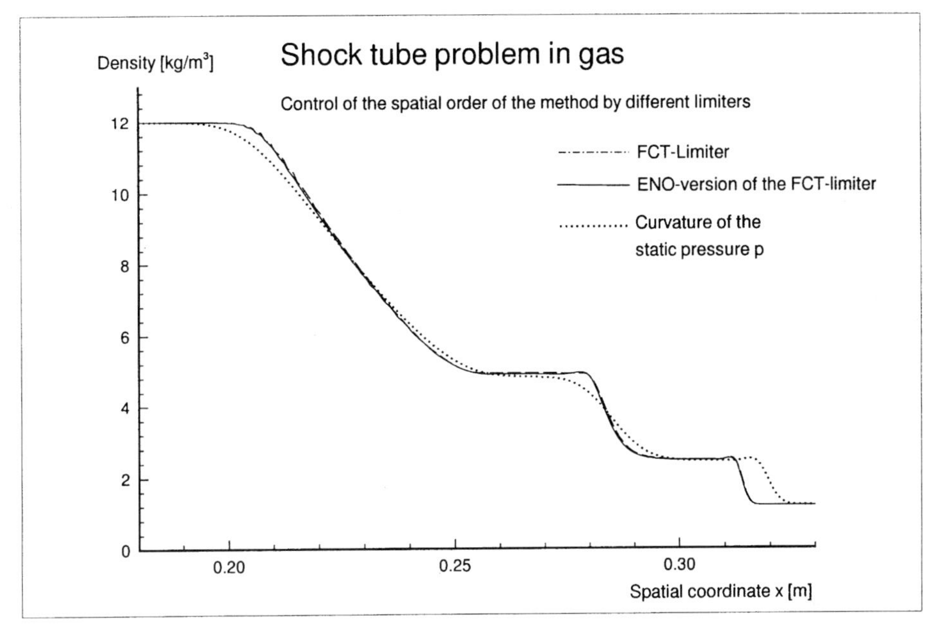

Figure 22.3: Density for a shock tube problem in gas calculated using different limiters to control the spatial order of approximation.

The FCT-limiter following Löhner et al. [9] determines the values in such a way that the combined solution does not over- or undershoot the first order solution in the environment of each node. This approach ensures preserving of monotonicity. The interested reader is referred to [9] or [5]. For the following, it is important to keep in mind that the limiter values $0 \leq \Theta_e \leq 1$ on each element e are calculated as the minimum of limiters $\phi_{N(e)}$ at the vertices $N(e)$ of the element, which are determined themselves in such a way that the combined solution does not exceed the first order solution at the neighbouring nodes.

Based on the truncation error for the one-dimensional Lax-Wendroff method, which is similar to the used finite element formulation, Rick [10] suggested for the diffusion coefficient c_d the following function of the local Courant number η:

$$c_\mathrm{d} = 3\eta\,(1-\eta)\,. \tag{35}$$

In fact, tests have shown that this choice leads to good results for one-dimensional problems as for the shock tube in Figure 22.3, but difficulties arise for real two-dimensional problems as the collapse of a spherical bubble (Figure 22.4).

These problems gave the motivation for us to improve the FCT-limitation and the choice of the diffusion coefficient by following the idea behind ENO-reconstructions within Godunov-type methods.

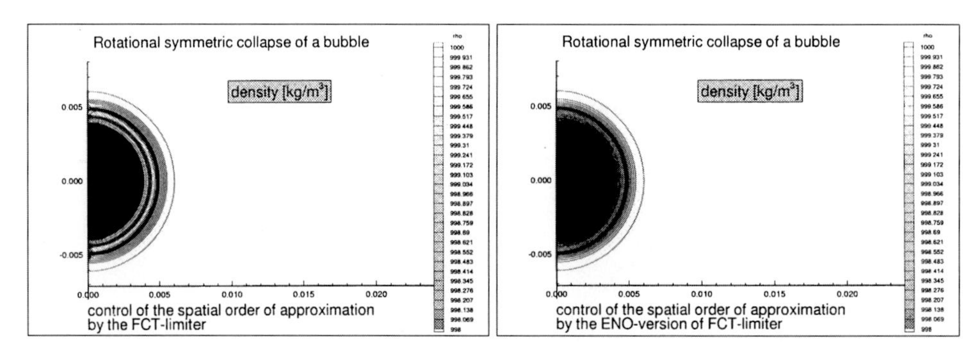

Figure 22.4: Comparison of two limiters for the collapse of a spherical bubble.

Concerning the diffusion coefficient, we take the function of Rick only as an initial guess and then try to improve it by scanning the whole possible interval $[0, 1]$ with steps of about 0.01 with the intention that the first order solution becomes smoother. For this purpose, a local measure of smoothness of the solution is needed that is evaluated for every choice of c_d and at every grid node N. If we use within Godunov-type methods linear polynomials of the form:

$$p_r(x,y) = \sum_{l=0}^{1} \sum_{i+j=l} a_{ij}^r (x-x_0)^i (x-y_0)^j \tag{36}$$

for the reconstruction around the node with the coordinates (x_0, y_0) in a grid with triangular cells a measure of smoothness,

$$\sigma_r = \sum_{i+j=1} |a_{ij}^r| \, , \tag{37}$$

is introduced (see for example [13], p. 246). For a quadrangular cell, only little modifications are needed.

In the case of a finite element method, the shape functions play a similar role as the reconstruction polynomials but are specified before the method starts. So, a similar measure for the smoothness can be used:

form of the element e	shape function	measure σ_e
triangle	$a + b\xi + c\eta$	$\|b\| + \|c\|$
quadrangle	$a + b\xi + c\eta + d\xi\eta$	$\|b\| + \|c\| + \|d\|$. (38)

The coefficients a, b, c and d are determined by the values q_i, $i = 1, 2, 3, 4$ of a regarded quantity q at the nodes:

triangle $a = q_1$, $b = q_2 - q_1$, $c = q_3 - q_1$, $d = 0$

quadrangle $a = q_1$, $b = q_2 - q_1$, $c = q_3 - q_1$, $d = q_4 - q_3 - q_2 + q_1$. (39)

To obtain a maesurement quantity for the nodes, a summation over all elements to which a regarded node belongs is done:

$$\tilde{\sigma}_N = \sum_{\text{elements } e \text{ to which } N \text{ belongs}} \sigma_e \,. \tag{40}$$

Then, at every node N of the grid, the coefficient c_d is selected such that it leads to the smallest value of $\tilde{\sigma}_N$.

The limiters Θ_N, which are now referred to the grid nodes N, are determined by a similar selection process with the difference that not the whole possible interval $[0, 1]$ is scanned but the interval $[min_{\text{neighbornodes } i}\phi_i,$ $max_{\text{neighbornodes } i}\phi_i]$ where ϕ_N represents the already mentioned limiters at the nodes within the FCT-algorithm. We call this improvement of the FCT-strategy by the two selection processes the ENO-version of FCT. The ENO-version of FCT leads to significant better results than the original FCT-limiter for a really two-dimensional problem like the spherical bubble collapse (Figure 22.4). For one-dimensional problems, e.g. shock tube problems (Figure 22.3), both limiters yield almost identical results.

22.5 Examples for Numerical Results

The presented numerical method has been applied to two technical problems involving cavitation:

One important technical example of cavitation in highly accelerated liquids is the phenomenon of cavitation damaging of solid surfaces that is caused by the collapses of bubbles generated upstream in the flow. With the motivation to investigate the still unclarified mechanism of such damaging, the numerical method has been applied to the model problem of a single bubble that collapses near an elastic solid surface. The dynamical behavior of the solid has been also simulated solving numerically the elastodynamical equations by a Godunov-type method similar to Zwas' approach [4]. Both numerical codes are coupled at the fluid-solid interface in a time-accurate way by a predictor-corrector strategy. The following Figures 22.5 and 22.6 show time steps of this simulation with an initially spherical bubble of radius 4 mm. Initially, all three

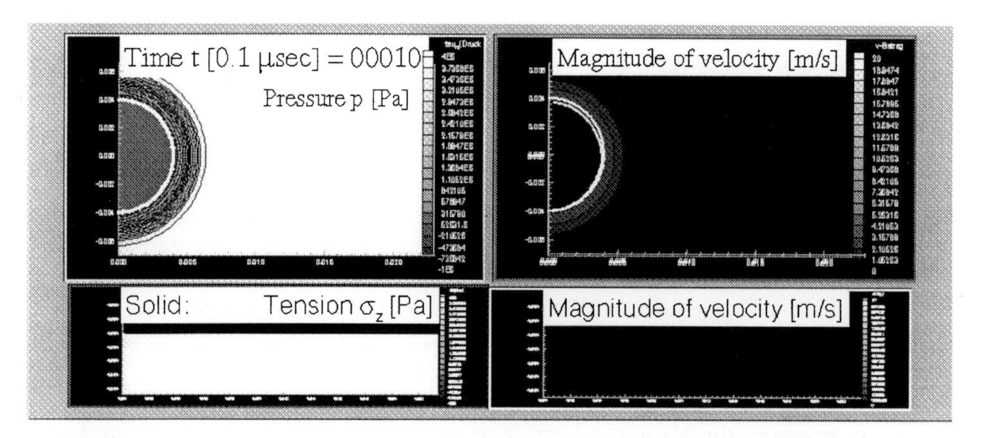

Figure 22.5: The collapsing bubble emits a rarefaction wave inside the liquid.

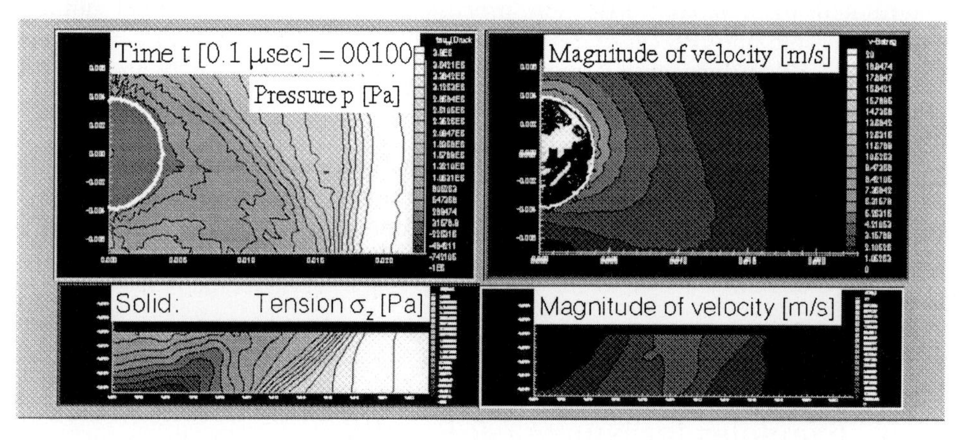

Figure 22.6: The region of underpressure between bubble and solid disappeared; the load of the solids surface changes.

phases were assumed to be at rest with the solid in equilibrium and the following, homogeneous state conditions in the fluid phases:

phase	density ρ	temperature T	pressure p
gas (air)	1.2 kg/m^3	293 K	1 bar
liquid (water)	1000.0 kg/m^3	293 K	39.2 bar.

Because the initial gas pressure inside the bubble is chosen lower than the liquid pressure around the bubble, a rarefaction wave is emitted into the liquid during the first phase of the collapse. Conversely, the gas inside the bubble is accelerated by a weak shock wave starting at the bubble wall and propagating towards the center of the bubble with advancing time. The rarefaction wave reaches the boundary of the solid after about 2 μs and is reflected there causing a tractive loading on the solid surface and stress waves inside the solid. On the liquid side, the underpressure initiates an acceleration of the liquid towards the bubble center with an additional component versus the axis of symmetry. The consequence is the disappearance of the underpressure region and a change of loading on the surface of the solid. So, a wave propagates into the solid starting at its upper boundary. During these processes taking place near the liquid-solid interface, the top side of the bubble is still accelerated without restrictions by effects of the solid. This is the reason for a first flattening of the bubble wall in this region. With advancing time, the curvature of the bubble wall changes its sign. This process can be interpreted as the initiation of a liquid jet towards the solid through the bubble that was also recognized in many experimental studies, for example [14, 15].

The second simulation presented in this section belongs to the liquid flow through a fuel injector channel with a sharp inlet corner where cavitation can occur in local regions of underpressure [1]. This problem is under work in cooperation with an experimentally working research group at the Department of Technical Thermodynamics of the RWTH Aachen. It concerns the generation of cavitation and its dependence on the shape parameters of the injector and the pressure ratio. For this purpose, parts of the experimental configuration have been discretized to allow a transient simulation of the beginning of the injection process.

For the simulation presented in the Figures 22.7 and 22.8, an injection pressure of 50 bar was chosen as the condition at the top boundary of the computational domain, whereas a combustion chamber pressure of 5 bar was assumed at the bottom boundary. Within this simulation, the effects of heat conduction and viscosity were neglected and should be incooperated in a later phase of the simulation. As initial conditions, the liquid was assumed at rest and the injection pressure was prescribed in the whole computational domain. Because the lower combustion chamber pressure is set as a boundary condition at the outflow of the injector, the acceleration of the liquid within a first phase of the simulation is caused via a rarefaction wave propagating towards the top

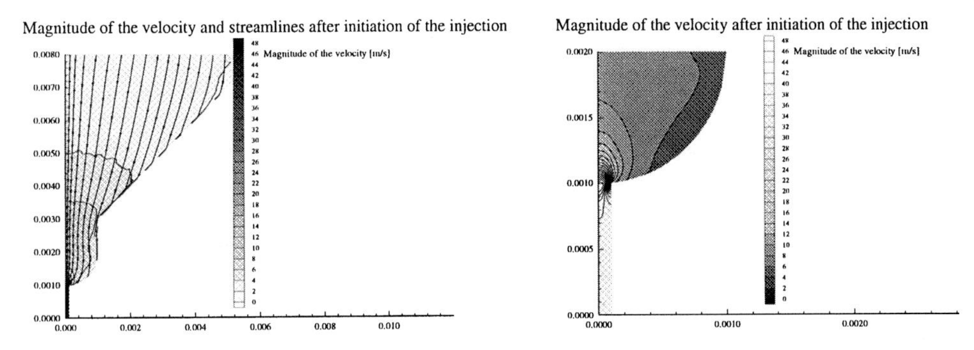

Figure 22.7: Velocity inside the preinjector and the injector channel.

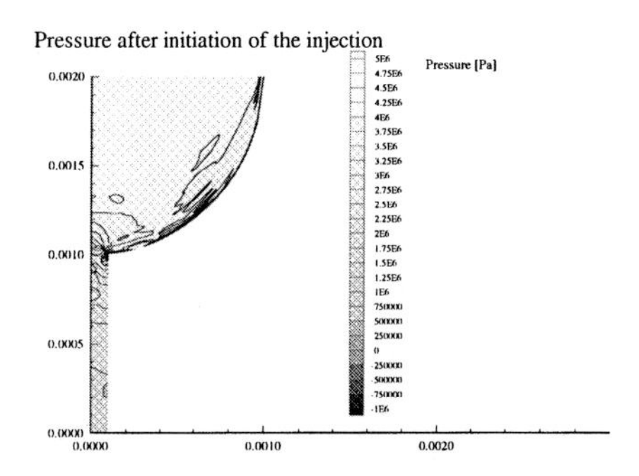

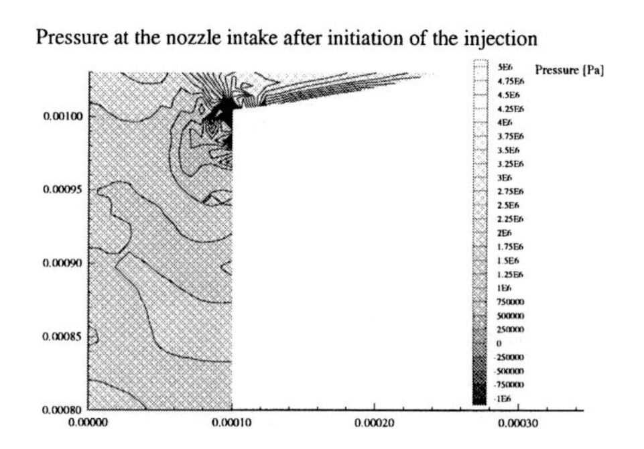

Figure 22.8: Pressure distribution near the intake of the injector channel.

boundary. This builds up the initiation for transient processes. The Figures 22.7 and 22.8 show streamlines, or more precisely, integral curves of the instantaneous velocity field, the magnitude of velocity and the pressure of a flow situation at about 22 μs after beginning. A region of underpressure can be seen behind the sharp inlet corner of the injector. In this region where the liquid pressure undergoes the vapor pressure, the first development of cavitation is expected. So, simulations of single bubbles should be inserted there by separate calculation blocks similar to the calculation presented previously.

22.6 Conclusions

With the goal to investigate the development of cavitation in highly accelerated liquids and its transient behavior, the liquid flow together with single gas- or vapor-filled cavitation bubbles is simulated numerically. The flow problem has been described mathematically by the system of Navier-Stokes equations for compressible fluid flow of both, phases and appropriate transition conditions at the phase interfaces.

For the approximation of the solution of this system of equations, an explicit finite element method has been presented that approximates the solution on an optionally moving, hybrid grid in two space dimensions by linear shape functions. The changing shape of bubble walls is fitted by a moving line of the grid that is regenerated in every time step.

Because of the high importance of numerical stability for the numerical simulation of these sensitive processes, aspects of linear and nonlinear stability of the finite element discretization for the different kinds of the treated one- or two-phase flows have been considered. In this context, the limiter for the control of the spatial order of approximation was improved as demonstrated by test calculations.

Finally, results of first applications of the numerical method to initial phases of a single bubble's collapse near the surface of an elastic solid and of the liquid flow in an injector were presented.

References

[1] P. Roosen: *Entwicklung und Erprobung von Fluoreszensmeßtechniken zur Dieselstrahldiagnostik.* Dissertation, RWTH Aachen, 1989.
[2] J. Bode: *Zum Kavitationseinfluß auf den Zerfall von Flüssigkeitsfreistrahlen.* Report 1/1991, Max-Planck-Institut für Strömungsforschung, Göttingen.
[3] M. Knapp: *Experimentelle Untersuchungen kavitierender Strömungen in Einspritzdüsen und des austretenden Flüssigkeitsstrahls sowie eines turbulenten Gasfreistrahls.* Report 106/1993 des Max-Planck-Instituts für Strömungsforschung, Göttingen.
[4] X. Lin, J. Ballmann: *A numerical scheme for axissymmetric elastic waves in bodies.* Wave Motion **21** (1995) 115–126.
[5] C. Dickopp: *Ein Navier-Stokes-Löser zur Simulation kollabierender Kavitationsblasen in der Nähe von Festkörperoberflächen.* PhD thesis, RWTH Aachen, Shaker Verlag, 1997.
[6] U. Specht, C. Dickopp, J. Ballmann: *A Numerical Scheme for Stress Waves at a Fluid Solid Interface.* NNFM 51, Vieweg Braunschweig, 1995, pp. 175–185.
[7] C. Dickopp, J. Ballmann: *Numerical Simulation of Transient Three-Phase-Systems – Cavitation, Bubble Dynamics.* ZAMM **78** (1998) 347–348.
[8] J.P. Boris, D.L. Book: *Flux-Corrected Transport I., SHASTA: A fluid transport algorithm that works.* J. Comp. Phys. **11** (1973) 38–69.
[9] R. Löhner, K. Morgan, J. Peraire, M. Valdati: *Finite element flux-corrected transport (FEM-FCT) for the Euler- and Navier-Stokes-equations.* Finite Elements in Fluids **7** (1987) 105–121.
[10] W. Rick: *Adaptive Galerkin Finite Elemente Verfahren zur numerischen Strömungssimulation auf unstrukturierten Netzen.* PhD thesis, RWTH Aachen, Shaker Verlag, 1994.
[11] A. Harten: *High resolution schemes for hyperbolic conservation laws.* J. Comput. Phys. **49** (1983) 357–393.
[12] R.J. LeVeque: *Numerical Methods for Conservation Laws.* Lectures in Mathematics, ETH Zürich, Birkhäuser, 1992.
[13] D. Kroener: *Numerical Schemes for Conservation Laws.* John Wiley & Sons and Teubner, 1997.
[14] A. Vogel, W. Lauterborn, R. Timm: *Optical and acoustic investigations of the dynamics of laser-produced cavitation bubbles near a solid boundary.* J. Fluid Mech. **206** (1989) 209–338.
[15] A. Philipp, W. Lauterborn: *Cavitation erosion by single laser-produced bubbles.* J. Fluid Mech. **361** (1998) 75–116.

23 Transient Phenomena in Double Front Detonations: Modelling and Numerical Simulation

Ulrich Uphoff, Markus Rose, Dieter Hänel and Paul Roth *

Abstract

The paper deals with the numerical investigation of detonation waves in gas/particle mixtures, revealing the specific feature of a double front detonation (DFD) complex, which was observed experimentally some years before. The propagation of a DFD is a highly transient phenomenon, driven by chemical reactions and complex processes of heat, momentum, and mass transfer between the phases, thereby incorporating a wide range of characteristic time and length scales. For a reliable numerical simulation, the source terms and interphase fluxes have to be modelled in a proper way and the main scales need to be resolved. Two problems were identified, (1) the elaboration of realistic models describing the physical and chemical processes involved in the DFD phenomenon, (2) the development and implementation of suitable grid adaption mechanisms as well as the implementation and test of high-resolution shock capturing schemes for the discretisation of the governing equations. Simulations over a wide range of parameters reveal the experimentally observed features of double front detonations and show that the distance between the first and the secondary detonation front is time dependent. A stationary structure of a double front detonation can therefore not be postulated a priori.

The numerical methods developed in the project represent advanced tools for studying wave propagation phenomena in reactive gas/particle mixtures.

* Gerhard-Mercator-Universität, Institut für Verbrennung und Gasdynamik, Lotharstr. 1, D-47048 Duisburg, Germany

23.1 Introduction

Double front detonations, which are characterised by a secondary shock following the leading shock of the gas detonation, have been investigated experimentally [1] and numerically [2–4] during the last two decades. Experiments performed in detonation tubes give rise to the assumption that the wave complex is stationary, indicating a constant distance between the first and the secondary shock. Some numerical models are therefore based on quasi-stationary equations describing the propagation of the detonation complex in the gas/particle mixture. A relevant condition to assume a quasi-steady DFD is the existence of two Chapman-Jouguet states at some distance behind the gaseous detonation front in the heterogeneous mixture. Due to the presence of the secondary phase, two additional conditions have to be fulfilled at these points, concerning the effective heat release rate $\mathrm{d}q/\mathrm{d}t$, which is the balance between the heat release rate from chemical reactions $\mathrm{d}q_+/\mathrm{d}t$ and the heat loss rate due to wall effects and interphase fluxes $\mathrm{d}q_-/\mathrm{d}t$:

$$\frac{\mathrm{d}q}{\mathrm{d}t} = \frac{\mathrm{d}q_+}{\mathrm{d}t} - \frac{\mathrm{d}q_-}{\mathrm{d}t} \,. \tag{1}$$

The typical evolution of the heat release rate in a DFD complex is shown in Figure 23.1. The rate is first negative behind the leading shock in the gaseous induction zone. After the beginning of gaseous reactions, $\mathrm{d}q/\mathrm{d}t$ increases, goes through zero and becomes positive. At the end of gaseous reactions, the heat loss rate becomes again predominant, $\mathrm{d}q/\mathrm{d}t$ decreases, goes through zero again, and becomes negative. The ignition of solid particles leads to a further change of sign of $\mathrm{d}q/\mathrm{d}t$. Then $\mathrm{d}q/\mathrm{d}t$ undergoes a new maximum and decreases to negative values again due to the effect of energy loss. Zel'dovich and Kompaneets [5] could show that it is necessary for the steady propagation of the DFD to achieve sonic conditions, $Ma = (D - u)/c = 1$, at two points behind the detonation front, and to have a zero effective heat release rate at these points, $\mathrm{d}q/\mathrm{d}t = 0$, with its second derivative being less than zero $\mathrm{d}^2q/\mathrm{d}t^2 < 0$. It is not

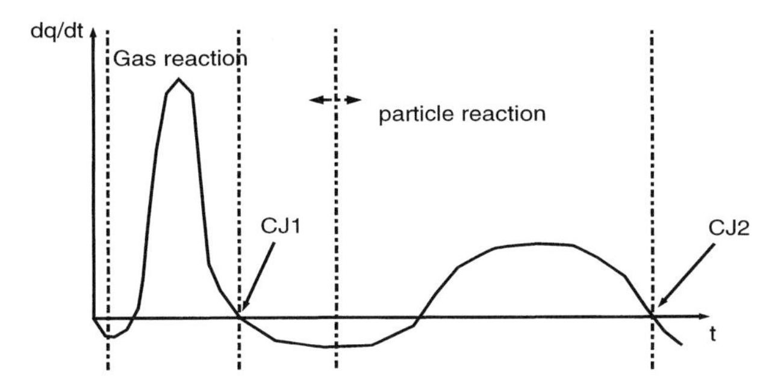

Figure 23.1: Sketch of the heat release rate in a DFD complex.

shown yet, if or under which conditions these requirements are sufficient for the steady propagation of the DFD.

This work is therefore based on the time-dependent reactive Euler equations for the two-phase medium [6]. The dynamics of the physical and chemical processes driving the detonation wave have been investigated and it was tried to answer the question whether the double front complex becomes quasi-stationary or not.

23.2 Modelling

In this approach, the two phases of the reactive gas/particle mixture are considered as two inviscid, interpenetrating continua with low particle loading, high flow velocities, and fast heat release processes in the gas phase. The kinetics of the combustion processes is given by a two-step reaction mechanism, describing the combustion of aluminium particles in a stoichiometric H_2/O_2 mixture:

$$\begin{aligned} &\mathrm{A} \rightarrow \mathrm{B}\,, \quad \omega_{\mathrm{AB}} = 1.5 \times 10^9 \rho_{\mathrm{g}}^{\mathrm{A}} \exp(-40 T_0/T)\,\mathrm{gcm^{-3}s^{-1}}\,, \\ &\mathrm{S} + \mathrm{B} \rightarrow \mathrm{C}\,, \quad \omega_{\mathrm{BC}} = -\frac{3\rho_{\mathrm{p}}}{t_{\mathrm{p}}}\,\mathrm{gcm^{-3}s^{-1}} \quad (T_{\mathrm{p}} \geq 1350\,\mathrm{K})\,. \end{aligned} \tag{2}$$

Mass production rates due to homogeneous and heterogeneous reactions are given by ω_{AB} and ω_{BC}. Heterogeneous reactions at the particle surface are assumed to start if the threshold value of 1350 K for particle temperature is reached. Particle combustion is described by the empirical characteristic burning time taken from [2]:

$$t_{\mathrm{p}} = 400\, d_{\mathrm{p0}}^2/\phi^{0.9}\ \mathrm{s}\,, \tag{3}$$

with d_{p0} and ϕ being the particle diameter and the volume concentration of the oxidizing gas component. Under these assumptions, the dynamics of the system in two-space dimensions is given by the reactive Euler equations:

$$\frac{\partial \boldsymbol{U}_{(\bullet)}}{\partial t} + \frac{\partial \boldsymbol{F}_{(\bullet)}}{\partial x} + \frac{\partial \boldsymbol{G}_{(\bullet)}}{\partial y} = \boldsymbol{Q}_{(\bullet)}\,, \quad (\bullet) = \mathrm{g}, \mathrm{p}\,. \tag{4}$$

The gas phase is assumed to be composed of the three species A, B, and C and of spherical particles of material S. The vectors of conserved variables, fluxes, and source and interphase exchange terms are:

$$\boldsymbol{U}_{\mathrm{g}} = \begin{pmatrix} \rho_{\mathrm{g}}^{(\mathrm{A})} \\ \rho_{\mathrm{g}}^{(\mathrm{B})} \\ \rho_{\mathrm{g}}^{(\mathrm{C})} \\ \rho_{\mathrm{g}} u_{\mathrm{g}} \\ \rho_{\mathrm{g}} v_{\mathrm{g}} \\ \rho_{\mathrm{g}} e_{\mathrm{g}} \end{pmatrix}, \quad \boldsymbol{U}_{\mathrm{p}} = \begin{pmatrix} \rho_{\mathrm{p}} \\ \rho_{\mathrm{p}} u_{\mathrm{p}} \\ \rho_{\mathrm{p}} v_{\mathrm{p}} \\ \rho_{\mathrm{p}} e_{\mathrm{p}} \\ n_{\mathrm{p}} \end{pmatrix},$$

$$\boldsymbol{F}_{\mathrm{g}} = \begin{pmatrix} \rho_{\mathrm{g}}^{(\mathrm{A})} u_{\mathrm{g}} \\ \rho_{\mathrm{g}}^{(\mathrm{B})} u_{\mathrm{g}} \\ \rho_{\mathrm{g}}^{(\mathrm{C})} u_{\mathrm{g}} \\ \rho_{\mathrm{g}} u_{\mathrm{g}}^2 + P_{\mathrm{g}} \\ \rho_{\mathrm{g}} u_{\mathrm{g}} v_{\mathrm{g}} \\ u_{\mathrm{g}} \left(\rho_{\mathrm{g}} e_{\mathrm{g}} + P_{\mathrm{g}}\right) \end{pmatrix}, \quad \boldsymbol{F}_{\mathrm{p}} = \begin{pmatrix} \rho_{\mathrm{p}} u_{\mathrm{p}} \\ \rho_{\mathrm{p}} u_{\mathrm{p}}^2 + P_{\mathrm{p}} \\ \rho_{\mathrm{p}} u_{\mathrm{p}} v_{\mathrm{p}} \\ u_{\mathrm{p}} \left(\rho_{\mathrm{p}} e_{\mathrm{p}} + P_{\mathrm{p}}\right) \\ n_{\mathrm{p}} u_{\mathrm{p}} \end{pmatrix},$$

$$\boldsymbol{G}_{\mathrm{g}} = \begin{pmatrix} \rho_{\mathrm{g}}^{(\mathrm{A})} v_{\mathrm{g}} \\ \rho_{\mathrm{g}}^{(\mathrm{B})} v_{\mathrm{g}} \\ \rho_{\mathrm{g}}^{(\mathrm{C})} v_{\mathrm{g}} \\ \rho_{\mathrm{g}} v_{\mathrm{g}} u_{\mathrm{g}} \\ \rho_{\mathrm{g}} v_{\mathrm{g}}^2 + P_{\mathrm{g}} \\ v_{\mathrm{g}} \left(\rho_{\mathrm{g}} e_{\mathrm{g}} + P_{\mathrm{g}}\right) \end{pmatrix}, \quad \boldsymbol{G}_{\mathrm{p}} = \begin{pmatrix} \rho_{\mathrm{p}} v_{\mathrm{p}} \\ \rho_{\mathrm{p}} v_{\mathrm{p}} u_{\mathrm{p}} \\ \rho_{\mathrm{p}} v_{\mathrm{p}}^2 + P_{\mathrm{p}} \\ u_{\mathrm{p}} (\rho_{\mathrm{p}} e_{\mathrm{p}} + P_{\mathrm{p}}) \\ n_{\mathrm{p}} v_{\mathrm{p}} \end{pmatrix},$$

$$\boldsymbol{Q}_{\mathrm{g}} = \begin{pmatrix} -\omega_{\mathrm{AB}} \\ +\omega_{\mathrm{AB}} - \omega_{\mathrm{BC}} \\ +\omega_{\mathrm{BC}} \\ (F_{\mathrm{D}})_x + \omega_{\mathrm{BC}}\, u_{\mathrm{p}} \\ (F_{\mathrm{D}})_y + \omega_{\mathrm{BC}}\, v_{\mathrm{p}} \\ Q + \boldsymbol{F}_{\mathrm{D}}\, \boldsymbol{v}_{\mathrm{p}} + J\, e_{\mathrm{p}} \end{pmatrix}, \quad \boldsymbol{Q}_{\mathrm{p}} = \begin{pmatrix} -\omega_{\mathrm{BC}} \\ -(F_{\mathrm{D}})_x - \omega_{\mathrm{BC}}\, u_{\mathrm{p}} \\ -(F_{\mathrm{D}})_y - \omega_{\mathrm{BC}}\, v_{\mathrm{p}} \\ -Q - \boldsymbol{F}_{\mathrm{D}}\, \boldsymbol{v}_{\mathrm{p}} - J\, e_{\mathrm{p}} \\ 0 \end{pmatrix}.$$

The indices g and p indicate the gas and particle phase, respectively, and ρ, ρu, ρv and e correspond to mass, momentum in x- and y-direction and total energy of the fluid. $\boldsymbol{F}_{\mathrm{D}}$ and Q represent interphase fluxes of drag forces and energy. An extensive discussion on the range of applicability of this model in view of particle loading and the pressure terms P_{g} and P_{p} can be found in [3].

23.3 Numerical Method

23.3.1 Grid adaption

Detonation waves in gas/particle mixtures involve physical and chemical phenomena, which are characterised by a wide range of length, time and velocity scales. Characteristic lengths range from the size of the detonation tube (≈ 10 m) down to the length of the homogeneous reaction zone (≈ 10 µm). To capture most of the physically important phenomena within a numerical simulation, efficient grid adaption is required. Two basically different methodologies for grid adaption are applied in this work. Mainly for one-dimensional calculations, a computational mesh with constant number of grid cells is used, which are redistributed in the computational domain according to the requirements of the solution. The integration in time of the governing equations is performed by the fully implicit method implemented in the computer code DASSL [7], which is of up to fifth order accuracy. The concept of a dynamically moving grid and its application to detonation and deflagration waves is described in detail in [3, 8]. Therefore, this report concentrates on the second methodology applied to adaptive grid distribution known as adaptive mesh refinement (AMR, [9]).

The basic ideal of the AMR approach is a local superposition of successively refined mesh patches as illustrated in Figure 23.2 for one refinement sequence. A course grid is defined according to the geometrical size of the domain to be resolved. A hierarchical number of refinement levels is also defined, which contain finer and finer grids. The initial and boundary conditions on a refined grid level are interpolated from the grid cells on the courser level below. The temporal adaptation goes hand in hand with a spatial adaption to guarantee consistency and stability of the time integration method. If the time step used for the course grid is Δt_0, the grid on the next finer level with a refinement factor of n has to be integrated in time by a step of $\Delta t_0/n$.

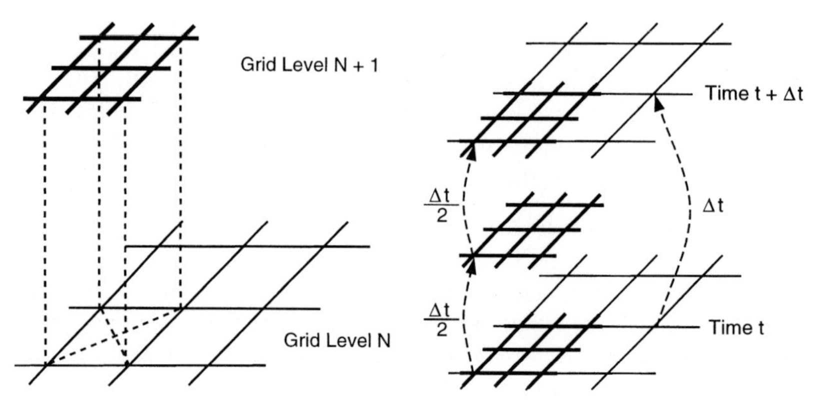

Figure 23.2: Illustration of the adaptive mesh refinement technique.

23.3.2 Spatial discretisation and integration in time

In the AMR solution concept, the mesh patches on each grid level are integrated in time by an operator splitting technique, separating fluxes and hydrodynamic source terms from those originating from chemical reactions:

$$\boldsymbol{U}^{n+1} = \mathcal{L}_{\text{flow}}^{\Delta t/2} \mathcal{L}_{\text{s/i}}^{\Delta t} \mathcal{L}_{\text{flow}}^{\Delta t/2} \boldsymbol{U}^{n} . \tag{5}$$

The operators $\mathcal{L}$ consist of independent integration schemes for the convective part $\mathcal{L}_{\text{flow}}$ and the source terms and interphase-fluxes $\mathcal{L}_{\text{s/i}}$, respectively. The schemes implemented in this work are of second order accuracy in time and space except for the regions containing discontinuities or steep gradients, where a limiter function decreases the order of discretisation to prevent artificial oscillations in the numerical solution.

Different methods were implemented and tested to perform the integration of the set of ordinary differential equations by using the operator $\mathcal{L}_{\text{s/i}}$. The most efficient integration method, which was tested, works in a hybrid way. According to the shortest chemical length scale, this method switches between an explicit and fully implicit integration scheme. The basic idea for this switching procedure was taken from [10]. The hybrid method decreases the computational time by a factor of 12 in comparison to a fully implicit scheme.

The discretisation of the convective parts consists of a Harten-Yee scheme [11] for the gas phase and an approximative Riemann solver according to Harten, Lax, and van Leer [12] combined with a MUSCL [13] extrapolation for the particle continuum. The specific combination of these discretisation schemes with an adaptive mesh refinement guarantees a proper and oscillation-free resolution of discontinuities in the flow field if the methods are applied correctly.

23.4 Results

The main results of this research project are described in detail in [3, 14, 15]. Specific features of these calculations illustrating the AMR technique are summarized here.

23.4.1 Simulations in one dimension

For a typical simulation in one space dimension, the size of the coursest mesh was chosen to be 0.1 cm. Three levels of refinement were defined with refinement factors of 10, 5, and 6, respectively, resulting in a mesh size of about 3 µm on the finest level. The factors of refinement were selected according to careful parameter studies.

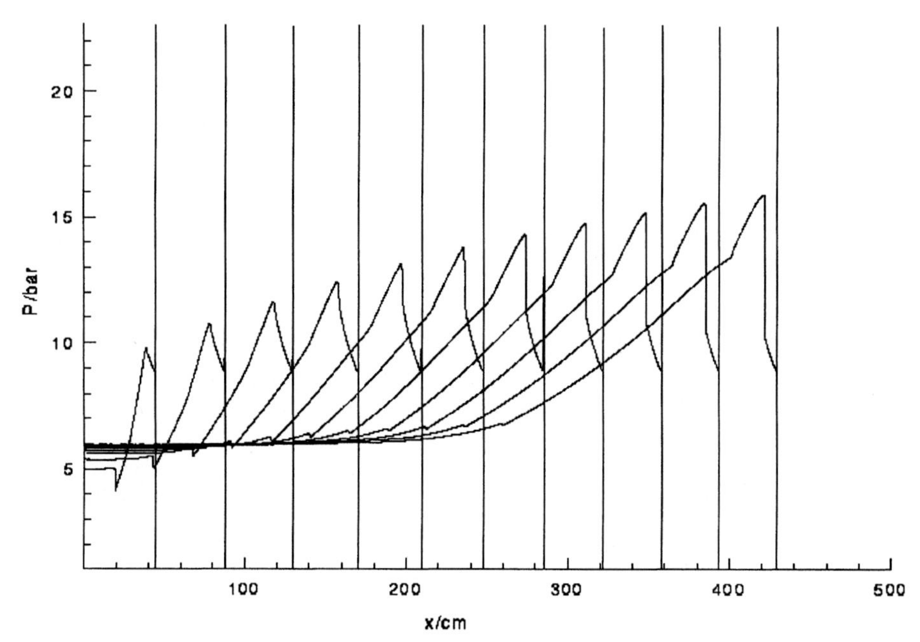

Figure 23.3: Pressure profiles calculated from a H_2/O_2-mixture enriched with Al-particles with loading of $n_p = 1.21 \times 10^4$ cm^{-3} and particle diameter of $d_p = 13$ μm.

Particles dispersed in the gas mixture are aluminium having an initial diameter of $d_{p0} = 13$ μm and a particle loading of $n_p = 1.21 \times 10^4$ cm^{-3}. The initial pressure in the system is 1 bar, and the heat of homogeneous reaction was fixed to $Q_g = 20\ RT_0$. The gas/particle mixture was ignited by a temperature gradient close to the left end of the tube. A high temperature of 1350 K decreases on a length of 2 cm down to 450 K. In combination with the parameters in Equation (2), a detonation wave is established in the mixture, which consists temporally of two detonation fronts. For the given configuration, a stationary character of the double front complex was not observed, but the secondary shock front approaches the first shock as indicated by the pressure plots in Figure 23.3. The increase in maximum pressure of the secondary shock indicates an acceleration of the secondary wave, whereas the first shock front travelles with constant velocity ($\approx$ 2000 m/s). When the secondary shock reaches the first one, the detonation complex continues propagating as a single front detonation wave. The same results were obtained after changing the main parameters of the combustion process, i. e. activation energy, ignition temperature of particles, and particle burning velocity. Details on the corresponding calculations can be taken from [14].

23.4.2 Simulations in two dimensions

The computational costs of the calculations in one space dimension are in the order of days on a single processor work-station. If the same requirements for accuracy and resolution of all relevant time and length scales have to be fulfilled also for simulations in two dimensions, these calculations would last for months. Nevertheless, simulations were performed for a two-dimensional set-up, mainly to test the implementation of the adaption technique, but also to get a first impression of the time-dependent character of the double front detonation in two dimensions. Of main interest are the instabilities in longitudinal and transversal direction as known from a typical gas detonation. The computational domain is of size 40 cm×1 cm, and the parameters describing the homogeneous and heterogeneous reactions are the same as used for the one-dimensional calculations. The initial particle loading is $n_p = 2.42 \times 10^4$ cm^{-3}, with a particle diameter of $d_{p0} = 13$ μm. Ignition of the mixture was initiated by a jump in pressure and temperature, calculated from the von Neumann conditions of the corresponding stationary ZND-detonation.

The existence of the secondary wave is indicated in Figure 23.4. The figure shows pressure distributions in a section of 4 cm length capturing the reaction front after 500 and 1000 computational steps. Contrary to the one-dimensional calculations, the secondary pressure wave seems to be slower than the

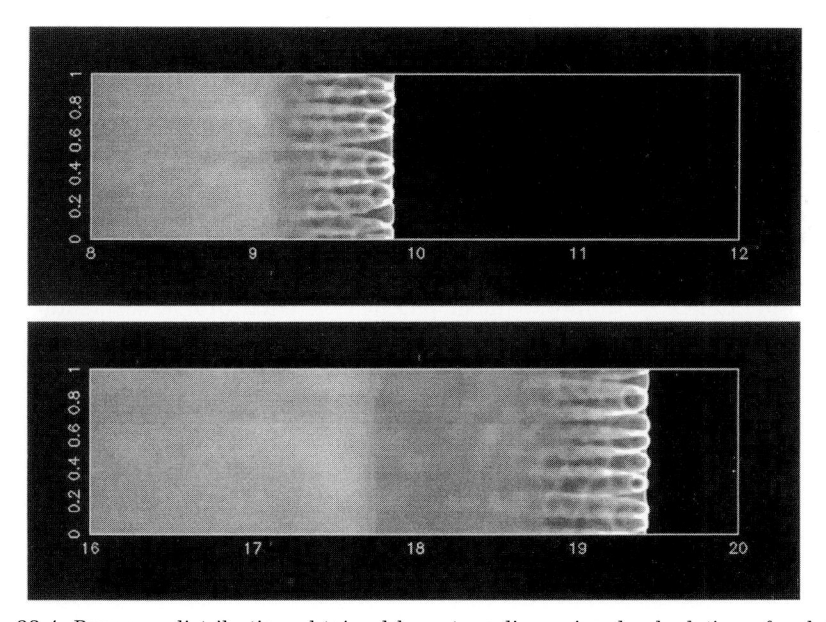

Figure 23.4: Pressure distribution obtained by a two-dimensional calculation of a detonation wave in a stoichiometric H_2/O_2-mixture enriched with Al-particles with loading of $n_p = 2.42 \times 10^4$ cm^{-3} and particle diameter of $d_p = 13$ μm. The upper and lower part show the calculated pressure field after 500 and 1000 time steps, respectively.

primary detonation front. But in the long run of the simulation, the secondary wave is accelerated and turns into a compression wave. The highly instationary, cellular structure of the first detonation front becomes obvious, whereas the secondary discontinuity seems to be much more stable.

23.5 Conclusion

The experimentally observed evolution of a secondary compression wave following the gas detonation front in a stoichiometric H_2/O_2 mixture with dispersed aluminium particles was investigated numerically. Hydrodynamic discontinuities and regions, which are influenced by homogeneous and heterogeneous reactions, were calculated with high accuracy by the combination of high resolution shock capturing schemes with an adaptive mesh refinement technique. The conclusion of a stationary double front detonation complex obtained by experiments by Veyssiere and Manson [1] does not coincide with our results. Our parametric studies always showed a secondary detonation front, which travelles with a slightly higher velocity than the first detonation front. Finally, this difference in the velocities led to a unification of both detonation waves. Calculations in two dimensions also revealed the temporal existence of the secondary discontinuity, which seemed to be much more stable against disturbances than the first one. Due to the lower heat release rate by heterogeneous reactions, the transversal and longitudinal instabilities did not become significant for the secondary wave. These instabilities led to the cellular structure of the gas detonation front.

The sophisticated computational code, which was developed within this research project, represents an advanced tool for performing reliable calculations of multiphase combustion processes in one and two dimensions. To improve the efficiency of the current code, advantage can be taken of the inherent independency of different mesh patches on each grid level, resulting in a parallel execution of integrating the mesh patches. The complex task of an efficient and proper data transfer between parallel processes is best be achieved by specialized message passing systems like MPI.

The developed computer code was also used for investigating the ignition of combustible gases by focussed shock waves [16]. This topic is of great importance not only from a theoretical point of view but also due to safety aspects.

References

[1] B. Veyssiere, N. Manson: *Sur l'Existence d'un Second Front de Détonation des Mélanges Biphasiques Hydrogène-Oxygène-Azote-Particules d'Aluminium.* Comptes rendus Acad. Sci. **295** (1982) 335–338.

[2] B. Veyssiere, B.A. Khasainov: *Structure and multiplicity of detonation regimes in heterogeneous hybrid mixtures.* Shock Waves **4** (1995) 171–186.

[3] U. Uphoff, D. Hänel, P. Roth: *Influence of reactive particles on the formation of a one-dimensional detonation wave.* Combust. Sci. and Tech. **110–111** (1995) 419–441.

[4] B.A. Khasainov, B. Veyssiere: *Initiation of detonation regimes in hybrid two-phase mixtures.* Shock Waves **6** (1996) 9–15.

[5] Ya.B. Zel'dovich, A.S. Kompaneets: *Theory of Detonations.* Academic Press, New York, 1960.

[6] B. Reichelt, P. Roth, L. Wang: *Stoßwellen in Gas-Partikel-Gemischen mit Massenaustausch zwischen den Phasen.* Wärme und Stoffübertragung **19** (1985) 101–111.

[7] L.R. Petzold: *A Description of DASSL: A Differential/Algebraic System Solver.* Amsterdam, North-Holland, 1983.

[8] T. Ludwig, P. Roth: *Modelling of Combustion Wave Propagation in Reactive Laminar Gas/Particle Mixtures.* Int. J. Multiphase Flow **23** (1997) 93–111.

[9] J.J. Quirk: *An Adaptive Grid Algorithm for Computational Shock Hydrodynamics.* Ph. D. thesis, Cranfield Institute of Technology, 1991.

[10] T.R. Young: *CHEMEQ – A subroutine for solving stiff ordinary differential equations.* NRL Memorandum Report 4091, 1987.

[11] H.C. Yee: *Upwind and Symmetric Shock-Capturing Schemes.* NASA TM 89464, 1987.

[12] A. Harten, P.D. Lax, B. van Leer: *On Upstream Differencing and Godunov-Type Schemes for Hyperbolic Conservation Laws.* SIAM Review **25** (1983) 35–61.

[13] B. van Leer: *Towards the Ultimate Conservative Difference Scheme. V: A Second Order Sequel to Godunov's Method.* J. Comp. Phys. **32** (1979) 101–136.

[14] U. Uphoff, D. Hänel, P. Roth: *Numerical Modelling of Detonation Structure in Two-Phase Flows.* Shock Waves **6** (1996) 17–20.

[15] U. Uphoff: *Numerische Simulation von Verbrennungswellen in Gas/Partikel-Gemischen.* Dissertation, Institut für Verbrennung und Gasdynamik, Gerhard-Mercator-Universität Duisburg, 1997.

[16] M. Rose, U. Uphoff, P. Roth: *Ignition of a Reactive Gas by Focussing of a Shock Wave.* Proc. of the 16th Int. Coll. on the Dynamics on Explosions and Reactive Systems, Cracow, 1997, pp. 554–556.